M4 Carbine
M16A2 W/E

Armorer's Guide

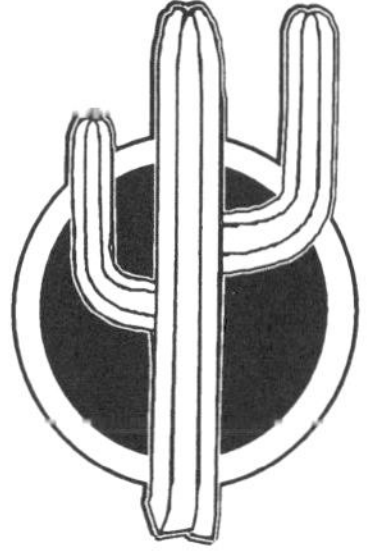

Desert Publications
El Dorado, AR 71731-1751 U. S. A.

M4 Carbine
M16A2 W/E
Armorer's
Guide

Published by Desert Publications
505 East Fifth Street
El Dorado,AR 71730 U.S.A.
870-862-3811
www.deltapress.com

ISBN 978-0-87947-198-0
2 4 6 8 9 7 5 3
Printed in U.S.A.

The Publisher Desert Publications is a division of
The DELTA GROUP, Ltd.
Direct all inquiries & orders to the above address.

WARNING

ALL WARNINGS in this technical manual pertain to both the rifle and the carbine unless otherwise specified.

Before starting an inspection, be sure to clear the rifle. Do not pull the trigger until the rifle has been cleared. Inspect the chamber to ensure that it is empty and no ammunition is in position to be chambered.

Do not keep live ammunition near work area.

To avoid injury to your eyes, use care when removing and installing spring-loaded parts.

All M16A2 rifles must be inspected and gaged at least once annually for safety and serviceability. Air Force users refer to inspection requirements in Air Force Regulation (AFR) 50-36, Volume 1.

All Army Reserve and Army National Guard M16A2 rifles must be inspected and gaged at least once every 2 years, after the initial inspection/gaging procedures have been accomplished. This 2 year interval may be maintained unless preventive maintenance checks and services (PMCS) or other physical evidence indicates that an individual unit's M16A2 rifles require inspection/ gaging at a more frequent interval. If it is determined that a yearly inspection is necessary for an individual unit, only that unit will be affected. This will not affect other units in regard to the interval of inspection.

It is recommended that training units inspect/gage all rifles at the end of each training cycle. Training units will inspect/gage all rifles at least once annually.

Below direct support maintenance, **DO NOT** interchange bolt assemblies from one rifle to another. Doing so may result in injury to, or death of, personnel.

Bolt cam pin must be installed or rifle will below up while firing the first round. If the bolt cam pin is not installed, injury to or death of, personnel may result.

Dry cleaning solvent is flammable and toxic and should be used in a well ventilated area. The use of rubber gloves is necessary to protect the skin when washing rifle parts.

When using solid film lubricant or dichloromethane, be sure the area is well ventilated.

When using carbon removing compound (item 8, app D), avoid skin contact. If carbon removing compound comes in contact with the skin, wash thoroughly with running water. using a good lanolin base cream after exposure to the compound is helpful. Using gloves and protective equipment is required.

The lock plate prevents the selector lever from being placed in BURST and will be installed at the discretion of the unit commander. It is mandatory for use in civil disturbance (riot control).

Only blank cartridge M200 is to be used when the blank firing attachment is attached to the rifle.

WARNING (CONT)

Do not fire blank ammunition at a representative enemy at distances of less than 20 feet (6.10 m). The unburned propellant grains can cause injury within this distance.

For further information on safety, care, and handling of ammunition: Army and Air Force users refer to M16A2 Rifle Operator's Manual.

For additional first aid data, see Field Manual (FM) 21-11.

CHANGE

No. 3

UNIT AND DIRECT SUPPORT MAINTENANCE MANUAL
(INCLUDING REPAIR PARTS AND SPECIAL TOOLS LIST)
RIFLE, 5.56MM, M16A2 W/E
(1005-01-128-9936) (EIC:4GM)

TM 9-1005-319-23&P, May 1991, is changed as follows:

1. Remove old pages and insert new pages as indicated below.

2. New or changed material is indicated by a vertical bar adjacent to the material.

3. New or changed illustrations are indicated by a miniature pointing hand highlighting the change.

Remove Pages	Insert Pages
a thru iv/v(blank)	a thru iv/v(blank)
	1-0.1/(1-0.2 blank)
1-3 thru 1-6	1-3 thru 1-6
2-3 and 2-4	2-3 and 2-4
2-13 thru 2-18	2-13 thru 2-18.2
2-21 thru 2-24	2-21 thru 2-24
2-27 and 2-28	2-27 and 2-28
2-47 and 2-48	2-47 thru 2-48
2-53 and 2-54	2-53 and 2-54
2-57 thru 2-60	2-57 thru 2-60
2-63 thru 2-71/(2-72 blank)	2-63 thru 2-72
3-39 and 3-40	3-39 and 3-40
3-57 and 3-58	3-57 and 3-58
3-77 thru 3-82	3-77 thru 3-82
4-1 and 4-2	4-1 and 4-2
4-5 thru 4-10	4-5 thru 4-10
B-5 thru B-8	B-5 thru B-8
C-7 and C-8/(C-9 blank)	C-7 and C-8/(C-9 blank)
Fig C-1 thru Fig C-2	Fig C-1 thru Fig C-2
C-5-1 thru Fig C-9	C-5-1 thru Fig C-9
C-10-1 thru Fig C-14	C-10-1 thru Fig C-14
C-15-1 thru C-17-1	C-15-1 thru C-17-1
I-1 thru I-10	I-1 thru I-10
Cover	Cover

4. File this change sheet in the front of the publication for reference purposes.

CHANGE

No. 2

UNIT AND DIRECT SUPPORT MAINTENANCE MANUAL
(INCLUDING REPAIR PARTS AND SPECIAL TOOLS LIST)
RIFLE, 5.56MM, M16A2 W/E
(1005-01-128-9936)(EIC:4GM)

TM 9-1005-319-23&P, May 1991, is changed as follows:

1. Remove old pages and insert new pages as indicated below.

2. New or changed material is indicated by a vertical bar adjacent to the material.

3. New or changed illustrations are indicated by a miniature pointing hand highlighting the change.

Remove Pages	**Insert Pages**
I-3 and I-4	I-3 and I-4

4. File the change sheet in the front of the publication for reference purposes.

CHANGE

NO. 1

HEADQUARTERS
DEPARTMENTS OF THE ARMY
AND AIR FORCE

Washington, DC   2 April 1992

UNIT AND DIRECT SUPPORT MAINTENANCE MANUAL
(INCLUDING REPAIR PARTS AND SPECIAL TOOLS LIST)
RIFLE, 5.56MM, M16A2 W/E
(1005-01-128-9936) (EIC:4GM)

TM 9-1005-319-23&P, May 1991, is changed as follows:

1. Remove old pages and insert new pages as indicated below.

2. New or changed material is indicated by a vertical bar adjacent to the material.

3. New or changed illustrations are indicated by a miniature pointing hand highlighting the change.

Remove Pages	Insert Pages
2-7 and 2-8	2-7 and 2-8
2-57 thru 2-60	2-57 thru 2-60
3-35 and 3-36	3-35 and 3-36
3-65 and 3-66	3-65 and 3-66
3-69 and 3-70	3-69 and 3-70
3-87 and 3-88	3-87 and 3-88
3-93 thru 3-96	3-93 thru 3-96
C-2-1 and Figure C-3	C-2-1 and Figure C-3
C-6-1 thru Figure C-8	C-6-1 thru Figure C-8
C-10-1 and Figure C-11	C-10-1 and Figure C-11
C-17-1	C-17-1
I-1 thru I-4	I-1 thru I-4
I-7 and I-8	I-7 and I-8

4. File this change sheet in the front of the publication for reference purposes.

By Order of the Secretary of the Army:

GORDON R. SULLIVAN
General, United States Army
Chief of Staff

Official:

MILTON H. HAMILTON
Administrative Assistant to the
Secretary of the Army

00851

DISTRIBUTION:

To be distributed in accordance with DA Form 12-40-E, (Block 0020),
Unit, Direct and General Support Maintenance requirements for TM 9-1005-319-23&P.

**DEPARTMENTS OF THE ARMY
AND AIR FORCE**

Washington, DC,   1 May 1991

**Unit and Direct Support Maintenance Manual
(Including Repair Parts and Special Tools List)
For
RIFLE, 5.56MM, M16A2, W/E
(1005-01-128-9936)
AND
CARBINE, 5.56MM, M4
(1005-01-231-0973)**

Current as of September 1993

DISTRIBUTION STATEMENT. APPROVED FOR PUBLIC RELEASE; DISTRIBUTION IS UNLIMITED.

REPORTING ERRORS AND RECOMMENDING IMPROVEMENTS

You can help improve this manual. If you find any mistakes or if you know of a way to improve the procedures please let us know.

Army users mail your letter, DA Form 2028 (recommended Changes to Publications and Blank Forms), or DA Form 2028-2 located in the back of this manual direct to: Commander, US Army Armament, Munitions and Chemical Command, ATTN: AMSMC-MAS, Rock Island, IL 61299-6000.

Air Force users submit AFTO Form 22, Technical Order System Publication Improvement Report and Reply, to: WR-ALC/MMDET, Robins AFB, GA 31098-5609.

A reply will be furnished to you.

---

*This manual supersedes ARMY TM 9-1005-319-23&P dated 28 August 1987, including all changes.

Page   Illus
     Figure

HOW TO USE THIS MANUAL

Read this manual carefully before performing required maintenance. This manual will be referred to for Inspection/Maintenance and Repair procedures.

GENERAL

There are several things you need to know to use this manual efficiently.

1. All references in the manual are to pages only. Reference to maintenance procedures is to the page where the respective initial setup appears.

2. Illustrations for the maintenance procedures show only those parts affected by the operation being performed.

3. Whenever the male gender is mentioned in the manual (i.e., crewman, repairman), it also pertains to females.

4. When the term "evacuate to support maintenance" is used, the entire rifle must be evacuated.

5. When a procedure is common to M16A2 rifle and M4 carbine, **ONLY** the M16A2 configuration will be depicted. If a procedure is not common to both weapons, the procedure will be incorporated.

6. When the word rifle is referenced in text, it will reference the rifle and the carbine.

INDEXES

This manual is organized to help you find the information you need quickly. There are several useful indexes.

1. **Table of Contents.** Lists in order all chapters, sections, and appendixes. Gives page references.

2. **Nomenclature Cross-References List.**

3. **Chapter Overviews.** Summarize material covered in the chapter. Are located at the beginning of each chapter.

4. **Symptom Index.** Located just before the troubleshooting table in each maintenance chapter. Lists, in alphabetical order, parts of the rifle with possible malfunctions. References pages of the troubleshooting table.

5. **Alphabetical Index.** Located at the end of the manual. An extensive subject index for everything in the manual. Gives page references.

MAINTENANCE PROCEDURES

There are two maintenance chapters:

Army personnel use chapter two for unit maintenance procedures and chapter three for direct support maintenance procedures.

Air Force personnel: Only Air Force Specialty Code 753XX Combat Arms Training and Maintenance (CATM) specialists, technicians, and gunsmiths are authorized to perform maintenance procedures contained in this manual.

Each maintenance task has an initial setup containing a list of the following things you will need in order to do your maintenance task:

1. Tools and Special Tools. For standard and special tools, see appendixes B and C. Army users are to use the Tool Set, Gage Set, and/or Shop Set listed in the initial setup.

2. Materials/Parts. Lists expendable materials and 100 percent replaceable parts. Each material or part is followed by a part number or appendix reference.

3. References. Lists other publications containing necessary information.

4. Equipment Condition. Lists conditions to be met before starting the procedure. The reference on the left of the condition is a page reference to instructions for setting up the condition.

5. General Safety Instructions. Lists safety instructions to follow before performing maintenance procedures.

EXTERNAL VIEW OF 5.56MM CARBINE, M4

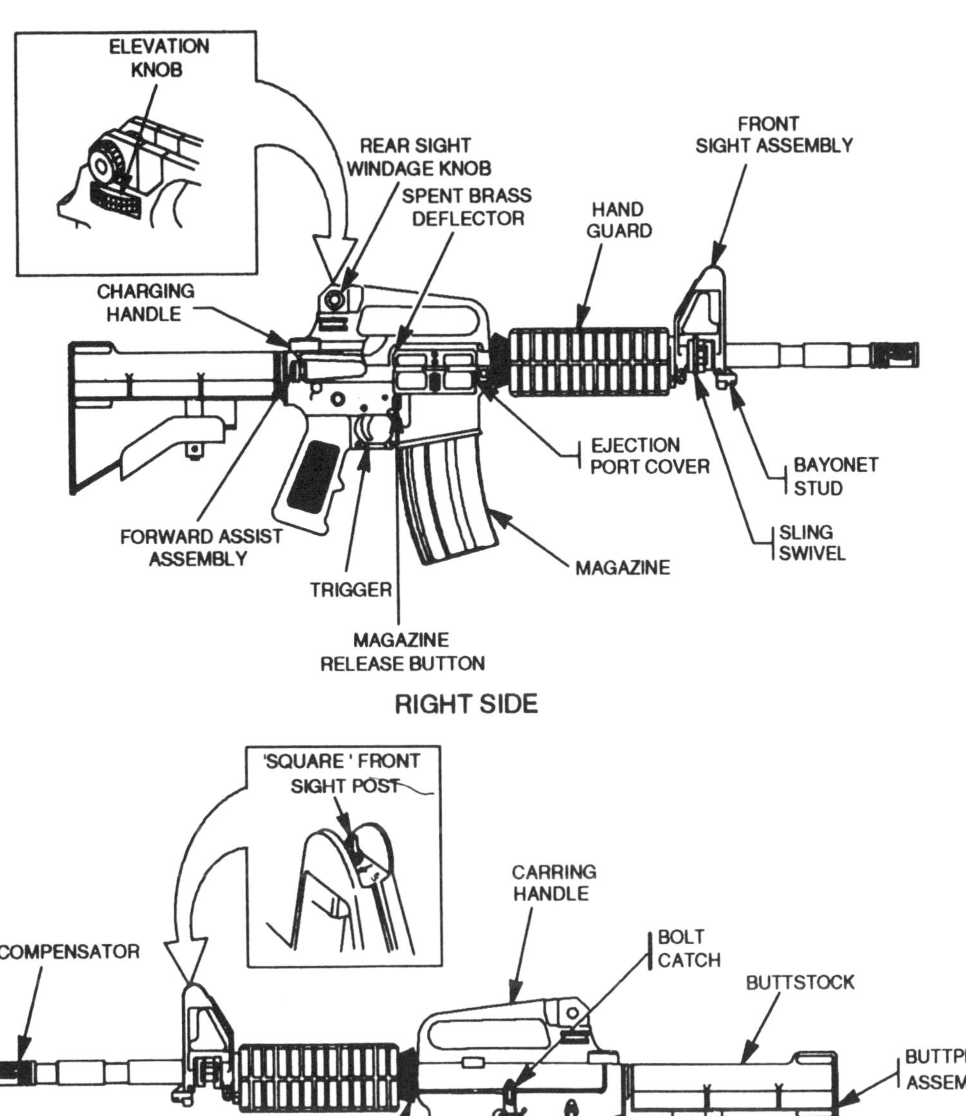

CHAPTER 1
INTRODUCTION

CHAPTER OVERVIEW

This chapter contains general information, equipment description and data, and principles of operation for the M16A2 rifle.

Section I. GENERAL INFORMATION

1-1. SCOPE.

  a. Type of Manual: Unit and Direct Support Maintenance.

  b. Model Number and Equipment Name: 5.56mm Rifle M16A2.

  c. Purpose of Equipment. Provides personnel an offensive/defensive capability to engage targets with small arms fire.

1-2. MAINTENANCE FORMS, RECORDS, AND REPORTS. Department of the Army forms and procedures used for equipment maintenance will be those prescribed by DA PAM 738-750, The Army Maintenance Management System.

Air Force users refer to TO 11W-1-10 for applicable forms and records.

1-3. DESTRUCTION OF ARMY MATERIEL TO PREVENT ENEMY USE. See TM 750-244-7.

1-4. PREPARATION FOR STORAGE OR SHIPMENT. Refer to page 2-70.

Air Force users refer to Special Package Instruction (SPI) 00-856-6885.

1-5. OFFICIAL NOMENCLATURE, NAMES AND DESIGNATIONS.

NOMENCLATURE CROSS-REFERENCE LIST

Common Name	Official Nomenclature
Action Spring	Compression Helical Spring
Ball Bearing	Bearing Ball
Bolt Catch Spring	Compression Helical Spring
Burst Disconnector	Lock-Release Lever
Cam Clutch Spring	Helical Spring
Charging Handle Assembly	Handle Assembly

1-5. OFFICIAL NOMENCLATURE, NAMES AND DESIGNATIONS (CONT).

NOMENCLATURE CROSS-REFERENCE LIST

Common Name	Official Nomenclature
Disconnector Springs	Compression Helical Spring
Ejector Spring	Helical Spring
Extractor Spring Assembly	Spring Assembly
Hammer Spring	Torsion Helical Spring
Lower Receiver Extension	Spring Receiver Holder
Magazine	Cartridge Magazine
Magazine Catch Spring	Compression Helical Spring
Peel Washer	Shim
Pistol Grip	Rifle Grip
Pivot Pin Detent	Takedown Pin Detent
Rifle	Rifle, 5.56mm, M16A2
Rifle Barrel Assembly	Barrel Assembly
Selector Lever	Fire Control Selector
Semiautomatic Disconnector	Lock-Release Lever
Sling	Small Arms Sling
Trigger Spring	Torsion Helical Spring
Upper Receiver	Upper Cartridge Receiver

1-6. REPORTING EQUIPMENT IMPROVEMENT RECOMMENDATIONS (EIR).

If your M16A2 rifle needs improvement, let us know. Send us an EIR. You, the user, are the only one who can tell us what you don't like about your equipment. Let us know why you don't like the design.

Army users submit SF 368 (Product Quality Deficiency Report) to: Commander, US Army Armament, Munitions and Chemical Command, ATTN: AMSMC-QAD-I, Rock Island, IL 61299-6000.

Air Force users submit Materiel Deficiency Report (MDR) to: DIR MAT MGT ROBINS AFB GA//MMIBTC// and Product Quality Deficiency Report to: DIR MAT MGT ROBINS AFB GA//MMQA//IAW Technical Order OO-35D-54.

A reply will be sent to you.

1-7. CORROSION PREVENTION AND CONTROL (CPC).

CPC of Army materiel is a continuing concern. It is important that any corrosion problems with this item be reported so that the problem can be corrected and improvements can be made to prevent the problem in future items.

While corrosion is typically associated with rusting of metals, it can also include deterioration of other materials such as rubber and plastic. Unusual cracking, softening, swelling, or breaking of these materials may be a corrosion problem.

If a corrosion problem is identified, it can be reported using Standard Form 368, Product Quality Deficiency Report. Use of key words such as ''corrosion'', ''rust'', ''deterioration'', or ''cracking'' will assure that the information is identified as a CPC problem.

Army users submit Product Quality Deficiency Report (SF 368) to:

Commander
US Army Armament, Munitions and Chemical Command
ATTN: AMSMC-QAD/Customer Feedback Center
Rock Island, Illinois 61299-6000

Air Force users submit Materiel Deficiency Report (MDR) to:

DIR MAT MGT
ATTN: MMIBTC
Robins AFB, GA

and Product Quality Deficiency Report to:

DIR MAT MGT
ATTN: MMQA
Robins AFB, GA

Section II.   EQUIPMENT DESCRIPTION AND DATA

1-8.   EQUIPMENT CHARACTERISTICS, CAPABILITIES, AND FEATURES.

a.   Characteristics.

(1) Lightweight	**(4)** Magazine-fed
(2) Air-cooled	**(5)** Semiautomatic or burst fire
(3) Gas-operated	

b.   Capabilities. Provides personnel an offensive/defensive capability to engage targets with direct small arms fire.

c.   Features.

(1)   Receivers are made of light-weight aluminum alloys; however, the safety, durability, and function of the rifles are in no way reduced. The portability and logistical values are greatly increased, particularly when air transport is used.

(2)   The bolt locking action is one of the mechanical features of the rifle. The bolt assembly and barrel extension contain locking lugs which engage and lock the bolt assembly firmly in the barrel extension. The initial force of the explosion of the cartridge is absorbed by the barrel, barrel extension, and bolt assembly.

1-8. EQUIPMENT CHARACTERISTICS, CAPABILITIES, AND FEATURES (CONT).

(3) The trigger guard is easily adaptable to winter operations. A spring-loaded retaining pin is depressed to allow ready access to the trigger when wearing arctic mittens.

(4) The ejection port cover prevents dirt or sand from getting into the ejection port. The ejection port cover must be closed during periods when firing is not anticipated. It opens automatically by the forward or rearward movement of the bolt carrier assembly.

1-9. LOCATION AND DESCRIPTION OF MAJOR COMPONENTS.

(A) **MAGAZINE.** 30 cartridge capacity.

(B) **SLING.** The sling is adjustable and provides a means to carry the rifle.

(C) **BOLT CARRIER ASSEMBLY.** Carries bolt assembly to chamber and fires the rifle. Contains the firing pin, cartridge extractor, bolt assembly, cartridge ejector, and bolt cam pin.

(D) **CHARGING HANDLE ASSEMBLY.** Provides a means of charging the rifle.

(E) **UPPER RECEIVER AND BARREL ASSEMBLY.** Upper receiver contains rear sight assembly, ejection port, ejection port cover, and a housing for the key and bolt carrier assembly and bolt assembly. Rifle barrel assembly is air-cooled, contains compensator and front sight assembly, and holds the two handguard assemblies and the sling swivel.

(F) **LOWER RECEIVER AND BUTTSTOCK ASSEMBLY.** Lower receiver contains the trigger assembly, sear, hammer assembly, selector lever, rifle grip, bolt catch, and buttstock assembly. The buttstock assembly houses the action spring, buffer assembly, and extension assembly.

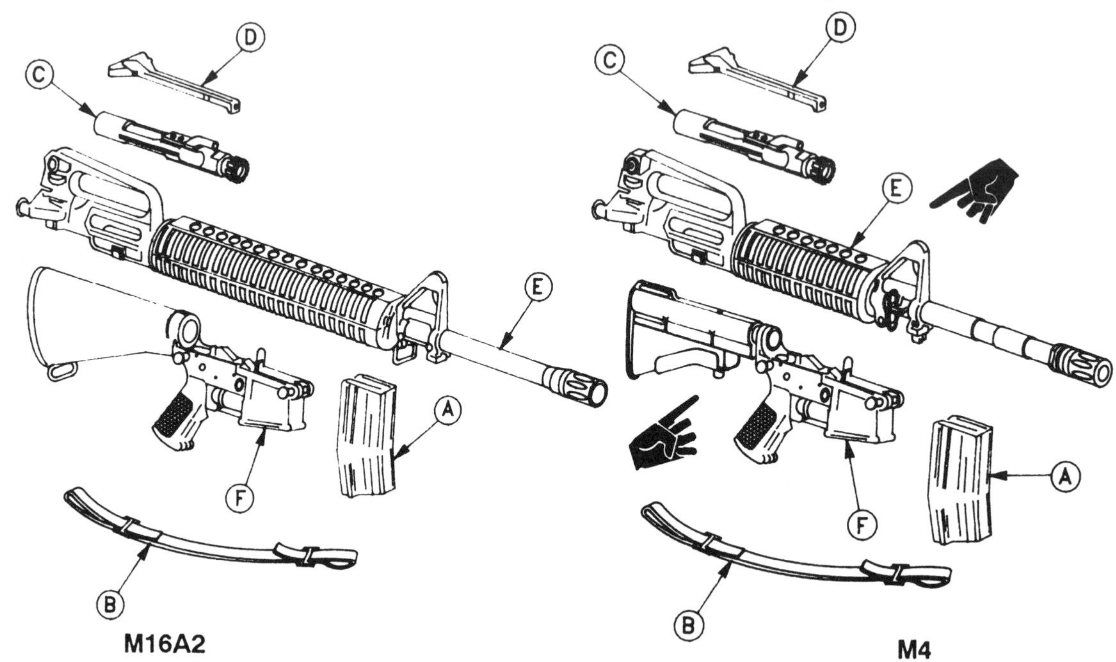

1-10. EQUIPMENT DATA.

	US CUSTOMARY	METRIC
Weight:		
Carbine, M4 without magazine and sling	6 lb	2.72 kg
Rifle, M16A2 without magazine and sling	7 lb 8 oz	3.40 kg
Sling, adjustable	4 oz	0.11 kg
Empty magazine	4 oz	0.11 kg
Loaded magazine	1 lb 1 oz	0.48 kg
Carbine, M4 w/sling and loaded magazine	7 lb 5 oz	3.32 kg
Rifle M16A2 w/sling and loaded magazine	8 lb 13 oz	4.00 kg
Bayonet-Knife M7	10.5 oz	0.30 kg
Scabbard M10	5 oz	0.14 kg
Length:		
Carbine with compensator, buttstock extended	33.0 in.	83.82 cm
Carbine with compensator, buttstock closed	29.75 in.	75.57 cm
Rifle with compensator	39.63 in.	100.66 cm
Barrel (Carbine)	14.5 in.	36.83 cm
Barrel (Rifle)	20 in.	50.8 cm
Barrel with compensator (Carbine)	15.5 in.	39.37 cm
Barrel with compensator (Rifle)	21 in.	53.34 cm
Mechanical features:		
Rifling	right-hand twist 6 grooves, 1 turn in 7 inches (17.78 cm)	
Method of operation	direct gas	
Type of breech mechanism	rotating bolt	
Method of feeding	magazine	
Cooling	air	
Trigger pull	5.5 to 9.5 lb	2.49 to 4.31kg
Ammunition:		
Caliber	223	5.56mm
Type	ball, blank, dummy, and tracer	
Firing characteristics:		
Muzzle velocity (Carbine) (approximate)	2,970 fps	905.85 mps
Muzzle velocity (Rifle) (approximate)	3,100 fps	945.5 mps
Chamber pressure	52,000 psi	358,540 kPa
Cyclic rate of fire (Carbine) (approximate)	700-970 rds/m	
Cyclic rate of fire (Rifle) (approximate)	700-900 rds/m	

1-10. EQUIPMENT DATA -- (Cont).

	US CUSTOMARY	**METRIC**
Maximum rate of fire:		
Semiautomatic	45 rds/m	
Burst	90 rds/m	
Sustained rate of fire	12/15 rds/m	
Maximum range	3,938 yards	Approximately 3,600 meters
Maximum effective range:		
Individual/point targets (Carbine)	547 yards	500 meters
Individual/point targets (Rifle)	602 yards	550 meters
Area targets (Carbine)	875 yards	800 meters
Area targets (Rifle)	875 yards	800 meters

Section III. PRINCIPLES OF OPERATION

1-11. GENERAL. The 5.56mm M16A2 rifle and M4 carbine:

a. Is gas-operated. It fires in either the semiautomatic or burst mode.

b. Has positive locking of the bolt. Firing pin is part of the bolt carrier assembly and cannot strike the primer until the bolt assembly is fully locked.

1-12. PRINCIPLES OF OPERATION.

(A) **MAGAZINE.** Holds cartridges ready for feeding and provides a guide for positioning cartridges for stripping. Provides quick reload capabilities for sustained firing.

(B) **SLING.** Provides the means for carrying the rifle.

(C) **BOLT CARRIER ASSEMBLY.** Provides stripping, chambering, locking, firing, extraction, and ejection of cartridges using the drive springs and projectile propelling gases for power.

(D) **CHARGING HANDLE ASSEMBLY.** Provides initial charging of the rifle. The handle latch locks the charging handle assembly in the forward position during sustained fire to prevent injury to the operator.

(E) **UPPER RECEIVER AND BARREL ASSEMBLY.** Provides support for the bolt carrier assembly. The barrel chambers the cartridge for firing and directs the projectile.

(F) **LOWER RECEIVER AND BUTTSTOCK ASSEMBLY.** Provides firing control for the rifle and carbine. **M16A2 ONLY** provides storage for basic cleaning materials.

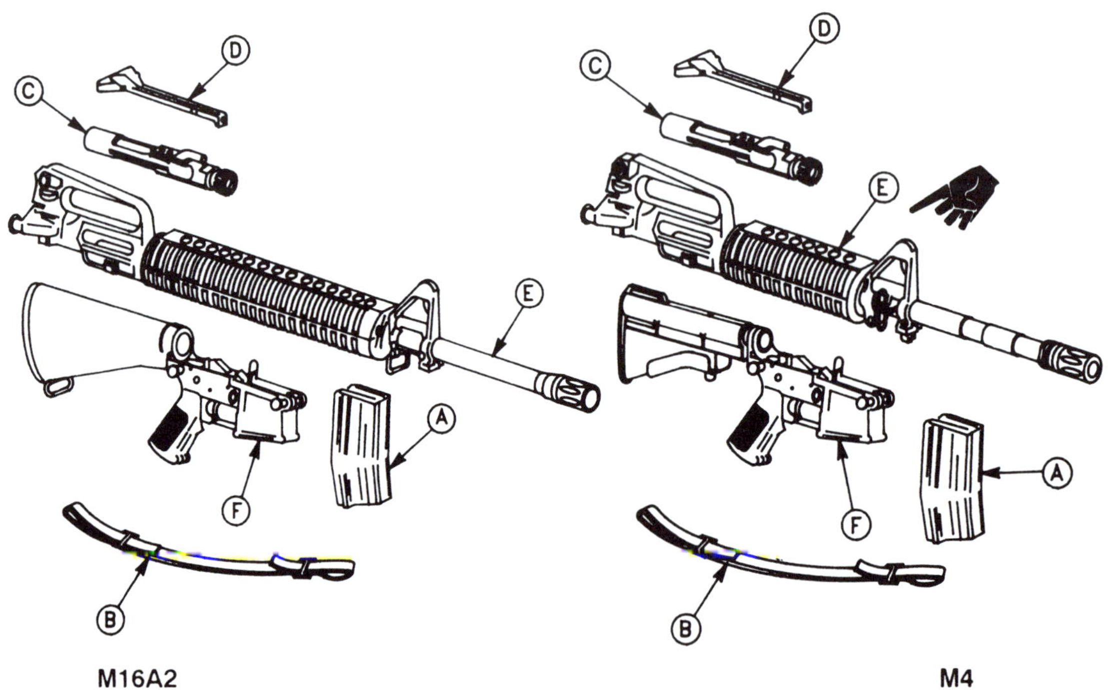

CHAPTER 2
UNIT MAINTENANCE INSTRUCTIONS

CHAPTER OVERVIEW

This chapter provides information and instructions to help keep the rifle in good repair and contains the following sections:

a. Repair Parts, Special Tools, TMDE, and Support Equipment
b. Service Upon Receipt
c. Preventive Maintenance Checks and Services (PMCS)
d. Troubleshooting
e. Maintenance Procedures

Section I. REPAIR PARTS, SPECIAL TOOLS, TMDE, AND SUPPORT EQUIPMENT

2-1. COMMON TOOLS AND EQUIPMENT. For authorized common tools and equipment, refer to the Modified Table of Organization and Equipment (MTOE) applicable to your unit.

Air Force users must maintain the following common tools:

3-ounce soft-brass hammer	Tweezers/round nose pliers
Vise	Hammer
Flat tip screwdriver	Needle nose pliers
Punch	

2-2. SPECIAL TOOLS, TMDE, AND SUPPORT EQUIPMENT. Special tools required for unit support are listed in appendixes B and C. Fabricated tools are listed and illustrated in appendix E.

2-3. REPAIR PARTS. Repair parts are listed and illustrated in appendix C of this manual.

Section II. SERVICE UPON RECEIPT

2-4. GENERAL.

a. Inspect the rifle for damage incurred during shipment. If rifle has been damaged, report the damage on SF 364, Report of Discrepancy (ROD).

b. Check the rifle against the packing slip to see if shipment is complete. Army users report all discrepancies in accordance with DA PAM 738-750.

Air Force users submit Materiel Deficiency Report (MDR) to: DIR MAT MGT ROBINS AFB GA//MMIBTC// and Product Quality Deficiency Report to: DIR MAT MGT ROBINS AFB GA//MMQA//. IAW Technical Order OO-35D-54.

c. Check to see whether the equipment has been modified.

2-5. SERVICE UPON RECEIPT OF MATERIEL.

WARNING

Before starting an inspection, be sure to clear the rifle. Do not actuate the trigger before clearing the rifle. Inspect the chamber to make sure it is empty and free of obstructions. Check to see there are no obstructions in the barrel and no ammunition is in position to be chambered.

SERVICE UPON RECEIPT

LOCATION	ITEM	ACTION	REMARKS
1. Container	a. M16A2 rifle	a. Remove rifle from containers.	
		b. Inspect the equipment for damage incurred during shipment.	If the equipment has been damaged, report the damage on SF Form 364, Report of Discrepancy (ROD).
		c. Check the equipment against the packing list to see if the shipment is complete.	Report all discrepancies in accordance with the instructions of DA PAM 738-750.
	b. Basic issue items	Check for missing items.	Refer to TM 9-1005-319-10 (operator's manual)
2. M16A2 rifle	a. Barrel assembly	If volatile corrosion inhibitor (VCI) is in barrel, remove and discard.	
	b. All parts	a. Field-strip rifle and inspect for missing, damaged, and rusted or corroded parts.	Refer to operator's manual.
		b. Clean and lubricate.	Refer to operator's manual.
		c. Reassemble.	Refer to operator's manual.
		d. Function check.	Refer to page 2-69.

SERVICE UPON RECEIPT (CONT)

LOCATION	ITEM	ACTION	REMARKS
		e. Check to see whether the equipment has been modified.	Refer to DA PAM 25-30.
	c. Magazine	Check for positive retention and functioning of bolt catch.	Refer to operator's manual.

Section III. PREVENTIVE MAINTENANCE CHECKS AND SERVICES (PMCS)

2-6. GENERAL. This section contains the procedures and instructions necessary to perform unit preventive maintenance checks and services. These services are to be performed by unit maintenance personnel with the assistance of the operator where practical.

2-7. PREVENTIVE MAINTENANCE CHECKS AND SERVICES

WARNING
Before starting an inspection, be sure to clear the rifle. Do not keep live ammunition near the work area.

a. General. The PMCS procedures are contained in the table following. They are arranged in logical sequence requiring a minimum amount of time and motion on the part of the persons performing them and are arranged so that there will be minimum interference between persons performing checks simultaneously on the same end item.

b. Item No. Column. Checks and services are numbered in disassembly sequence. This column shall be used as a source of item numbers for the "TM Number" column on DA Form 2404, Equipment Inspection and Maintenance Worksheet, in recording results of PMCS.

c. Interval Column. This column gives the designated interval when each check is to be performed.

d. Item To Be Checked Or Serviced Column. This column lists the items to be checked or serviced.

e. Procedure Column. This column contains a brief description of the procedure by which the check is to be performed. It contains all the information required to accomplish the checks and services. Information marked SH indicates a specific equipment shortcoming and the procedure needed to correct the shortcoming.

NOTE
For the purpose of this technical manual, the following definition is supplied. This definition is not intended to apply to any other document

Shortcoming (SH): A fault that requires maintenance or supply action on a piece of equipment, but does not render equipment Not Mission Capable.

f. Not Fully Mission Capable If: Column. This column contains a brief statement of the condition (e.g., malfunction, shortage) that would cause the covered equipment to be less than fully ready to perform its assigned mission.

2-7. PREVENTIVE MAINTENANCE CHECKS AND SERVICES (CONT).

PREVENTIVE MAINTENANCE CHECKS AND SERVICES FOR M16A2 RIFLE (CONT)

Item No.	Interval	Item To Be Checked Or Serviced	Procedure	Not Fully Mission Capable If:

WARNING

Before starting an inspection, be sure to clear the rifle. Do not pull the trigger until the rifle has been cleared. Inspect the chamber to ensure that it is empty and no ammunition is in position to be chambered. Do not keep live ammunition near work area.

NOTE

An inactive rifle is a rifle which has been stored in an arms room for a period of 90 days without use. The rifle may or may not have been assigned to an individual.

Inactive rifles shall receive quarterly PMCS unless inspection reveals more frequent servicing is necessary.

Normal cleaning (PMCS) of an inactive rifle will be performed every 90 days. Should the unit armorer detect corrosion on a rifle prior to the end of the 90-day period, the PMCS should be performed immediately.

Solid Film Lubricant (SFL) is the authorized touch up for the M16A2 Rifle and may be used on up to one third of the exterior finish of the rifle.

FOR ARMY CONUS USE ONLY AND AIR FORCE TRAINING RIFLES ONLY:
Solid Film Lubricant may be used as a touch up without limitation on the upper receiver and barrel assembly. This is to say that units which **DO NOT** fall under the category of Divisional Combat Units or rapid deployment type units may have up to 100 percent of the exterior surface of the upper receiver and barrel assembly protected with SFL. Prior to application of SFL, the surface must be thoroughly cleaned and inspected for corrosion and/or damage. If corroded or damaged, the part must be repaired or replaced prior to application of SFL. Continued use under combat conditions would result in an unprotected surface when the SFL wears off. This would result in a large light reflecting surface and accelerated deterioration of the unprotected surface. Therefore, Divisional Combat Units and units which fall under the definition of Rapid Deployment type must adhere to the limitation of **NOT** over ⅓ of their exterior surface covered by SFL.

When determining mission capability, deadline if it is a deficiency.

PREVENTIVE MAINTENANCE CHECKS AND SERVICES FOR M16A2 RIFLE (CONT)

Item No.	Interval	Item To Be Checked Or Serviced	Procedure	Not Fully Mission Capable If:
1	Quarterly	Magazine (serviceability check)	Disassemble as in TM 9-1005-319-10 (operator's manual). Inspect tube (1) for bulges, dents, or damaged feeder lips (2). Inspect spring (3) and follower (4) for kinks or damage. SH—Replace the magazine if any of these conditions exist. Reassemble magazine and check for binding during operation of follower (4). SH—Replace the magazine if the follower binds.	A magazine is not available for use with the rifle.
2	Quarterly	Charging handle assembly and selector lever	**WARNING** If the rifle fails any of the following selector lever tests, evacuate it to support maintenance. Continued use of the rifle could result in injury to, or death of, personnel.	

2-7. PREVENTIVE MAINTENANCE CHECKS AND SERVICES (CONT).

PREVENTIVE MAINTENANCE CHECKS AND SERVICES FOR M16A2 RIFLE (CONT)

Item No.	Interval	Item To Be Checked Or Serviced	Procedure	Not Fully Mission Capable If:
2	Quarterly (cont)	Charging handle assembly and selector lever (cont)	Pull charging handle (1) to rear. Check that chamber is clear. Let bolt carrier assembly (2) close. Leave hammer in cocked position. Do not pull trigger.	Charging handle does not lock in place when in the forward position.
			Place selector lever (3) in SAFE position. Pull trigger.	Hammer falls.
			NOTE For the purpose of the following test, ''SLOW'' is defined as 1/4 to 1/2 the normal rate of trigger release.	
			Place selector lever (3) in SEMI position. Pull trigger.	Hammer does not fall.
			Hold trigger to the rear, charge rifle, and release the trigger with a slow, smooth motion, without hesitations or stops, until the trigger is fully forward (an audible click should be heard).	Hammer falls.
			Repeat the above SEMI position test five times.	The rifle malfunctions during any of these five tests.

PREVENTIVE MAINTENANCE CHECKS AND SERVICES FOR M16A2 RIFLE (CONT)

Item No.	Interval	Item To Be Checked Or Serviced	Procedure	Not Fully Mission Capable If:
2	Quarterly (cont)	Charging handle assembly and selector lever (cont)	Place selector lever (3) in BURST position. Charge rifle and squeeze trigger.	Hammer does not fall.
			Hold trigger to the rear, charge rifle, and release trigger. Squeeze trigger.	Hammer falls.
			NOTE Automatic sear should have released hammer while holding trigger in the squeezed position before releasing and resqueezing the trigger.	
			With hammer in forward position, using moderate finger/thumb pressure attempt to place the selector lever (3) in SAFE position.	Moderate finger/thumb pressure moves selector lever to SAFE position.

2-7. PREVENTIVE MAINTENANCE CHECKS AND SERVICES (CONT).

PREVENTIVE MAINTENANCE CHECKS AND SERVICES FOR M16A2 RIFLE (CONT)

Item No.	Interval	Item To Be Checked Or Serviced	Procedure	Not Fully Mission Capable If:
3	Quarterly	Upper receiver and barrel assembly (handguard assemblies) 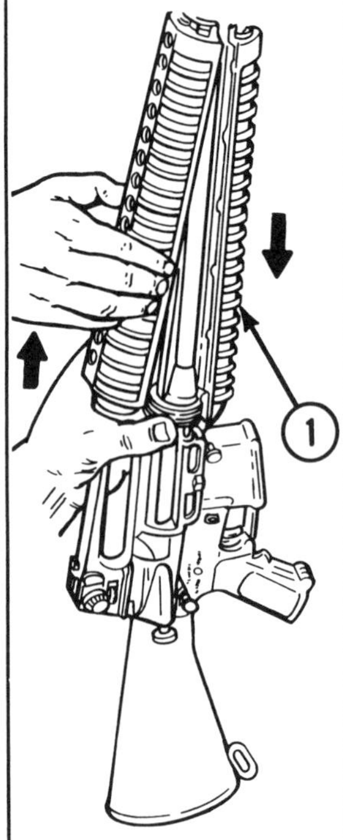	**CAUTION** Do not use screwdriver or any other tool when removing the handguard assemblies, doing so may damage the handguard assemblies and/or slip. **NOTE** Refer to operator's manual for ''buddy system'' procedure on removing handguard assemblies. Remove and inspect handguard assemblies (1) internally and externally for cracks and/or damage. Cracks are acceptable providing they do not extend into the handguard retaining flange, or adversely affect rifle operation or operator safety or proper retention of handguard assembly. Discard and replace the handguard assembly (1) if the heatshield is loose enough to rattle when installed on rifle.	Handguard missing or unserviceable.

PREVENTIVE MAINTENANCE CHECKS AND SERVICES FOR M16A2 RIFLE (CONT)

Item No.	Interval	Item To Be Checked Or Serviced	Procedure	Not Fully Mission Capable If:
4	Quarterly	Upper receiver and barrel assembly (serviceability check)	**WARNING** Dry cleaning solvent is flammable and toxic and should be used in a well ventilated area. The use of rubber gloves is necessary to protect the skin when washing rifle parts. **CAUTION** Damage may occur if excessive force is used to release takedown pin or pivot pin. Use hand pressure ONLY. Release takedown pins and open and separate receivers. Hand check compensator (1) for looseness on barrel (2), then hand check barrel (2) for looseness on upper receiver (3). Check center slot of compensator for alignment (p 2-50). If compensator or barrel is loose, evacuate to support maintenance. Check gas tube (4), forward assist assembly (5), and rear sight assembly (6) for damage. The rear sight spring should retain the rear sight assembly (6) in either position with firmness. SH—If damaged, evacuate to support maintenance.	Compensator or barrel is loose.

2-7. PREVENTIVE MAINTENANCE CHECKS AND SERVICES (CONT).

PREVENTIVE MAINTENANCE CHECKS AND SERVICES FOR M16A2 RIFLE (CONT)

Item No.	Interval	Item To Be Checked Or Serviced	Procedure	Not Fully Mission Capable If:
4	Quarterly (cont)	Upper receiver and barrel assembly (service-ability check) (cont)	**NOTE** If front or rear sight is moved, return to original position. Check front sight post, detent, and spring (7) for damage and corrosion. Clean and lubricate. Check charging handle (8) and ejection port cover (9) for defects and proper function. Check sling swivel (10) and rivet (11) for damage and proper function. SH—Other components are defective, replace as necessary.	Charging handle (8) is defective.

CAUTION

Do not use a wire brush to roughen surfaces. Use a well-ventilated area during cleaning and application of solid film lubricant. If solid film lubricant comes in contact with moving parts or functioning surfaces of the rifle, remove lubricant immediately by washing with dry cleaning solvent.

NOTE

Shiny metal exterior surfaces of the rifle should be recoated with solid film lubricant (item 21, app D). Clean surface with dry cleaning solvent (item 16, app D); dry, roughen with abrasive cloth (item 13, app D) and apply solid film lubricant.

PREVENTIVE MAINTENANCE CHECKS AND SERVICES FOR M16A2 RIFLE (CONT)

Item No.	Interval	Item To Be Checked Or Serviced	Procedure	Not Fully Mission Capable If:
4	Quarterly (cont)	Upper receiver and barrel assembly (serviceability check) (cont)	Inspect upper receiver (3) finish for scratches or worn shiny spots. If scratched or worn shiny in spots, disassemble and remove all lubricant from surface with dry cleaning solvent (item 16, app D). Wear rubber gloves (item 18, app D) and use a wash pan (item 24, app D) to apply solvent. Let parts dry thoroughly. Roughen the surface using abrasive cloth (item 13, app D) and apply solid film lubricant (item 16, app D). Allow 16 to 24 hours to dry before handling. Hold barrel (2) at 40-degree angle (muzzle down). Pull charging handle (8) to rear. Hold bolt carrier assembly (12) to rear and push charging handle forward. Release bolt carrier assembly (12). The bolt carrier assembly should close and lock under its own weight. If it does not, remove the bolt assembly (13) from the key and bolt carrier assembly (14) and slide the key and bolt carrier assembly (without bolt) back and forth in the upper receiver and barrel assembly. If the gas tube (4) hits the carrier key (15), or if the gas tube binds in the carrier key, try to correct the malfunction by adjusting (slightly bending) the gas tube in the area of the handguard assemblies. If unable to adjust, evacuate to support maintenance.	Adjustment does not correct the malfunction.

2-7. PREVENTIVE MAINTENANCE CHECKS AND SERVICES (CONT).

PREVENTIVE MAINTENANCE CHECKS AND SERVICES FOR M16A2 RIFLE (CONT)

Item No.	Interval	Item To Be Checked Or Serviced	Procedure	Not Fully Mission Capable If:
			WARNING Below direct support maintenance, do not interchange bolt assemblies from one rifle to another. Doing so may result in injury to, or death of, personnel.	
5	Quarterly	Key and bolt carrier assembly and bolt assembly (serviceability check)	Remove and disassemble. Visually inspect bolt assembly (1) for cracks, especially in the area of the bolt cam pin hole (2). Check for cracks on locking lugs (3), for a cluster of pits or chipped bolt face (4), and for an elongated firing pin hole (5). If cracked or broken, evacuate to support maintenance for repair.	Defects are found.
			Check for worn or missing bolt rings (6). Check for proper staggering of bolt rings. Insert the bolt assembly (1) into the key and bolt carrier assembly (7). Turn key and bolt carrier assembly (7) so the bolt assembly (1) points down. The bolt assembly must not drop out. Remove bolt assembly (p 2-35). Check for broken or missing firing pin retaining pin (8) and bolt cam pin (9); evacuate to support maintenance for replacement.	The bolt assembly drops out of the key and bolt carrier assembly due to its own weight. Missing or broken firing pin retaining pin or bolt cam pin.

PREVENTIVE MAINTENANCE CHECKS AND SERVICES FOR M16A2 RIFLE (CONT)

Item No.	Interval	Item To Be Checked Or Serviced	Procedure	Not Fully Mission Capable If:
5	Quarterly (cont)	Key and bolt carrier assembly and bolt assembly (serviceability check) (cont)	Check cartridge extractor (10), extractor spring assembly (11), cartridge ejector (12), and ejector spring (13) for dirt and serviceability. If dirty, clean, lubricate, and assemble. If unserviceable, replace as necessary.	Parts are missing or unserviceable.
			Check key and bolt carrier assembly (7) and carrier key (14) for damage and looseness. If damaged or loose, evacuate to support maintenance.	Key and bolt carrier assembly or carrier key is damaged, or carrier key is loose.
			NOTE If carrier key is dented, evacuate to support maintenance.	
			Check firing pin (15) for chips or breaks. If damaged, evacuate to support maintenance.	Firing pin is damaged.
			Pits or wear in area illustrated (16) is permissable.	

2-7. PREVENTIVE MAINTENANCE CHECKS AND SERVICES (CONT).

PREVENTIVE MAINTENANCE CHECKS AND SERVICES FOR M16A2 RIFLE (CONT)

Item No.	Interval	Item To Be Checked Or Serviced	Procedure	Not Fully Mission Capable If:
6	Quarterly	Lower receiver and buttstock assembly (serviceability check)	Remove buffer assembly (1) and action spring (2). Check buffer assembly for cracks. SH - Buffer assembly is cracked. Check action spring (2) for kinks and free length. Free length should be **RIFLE:** 11 3/4 inches (29.85 cm) minimum to 13 1/2 inches (34.29 cm) maximum. SH - If action spring is kinked or does not meet free length requirements. **CARBINE:** 10 1/16 inches (25.56 cm) minimum to 11 1/4 inches (28.58 cm) maximum. Do not attempt to adjust spring length. SH - If action spring is kinked or does not meet free length requirements.	

PREVENTIVE MAINTENANCE CHECKS AND SERVICES FOR M16A2 RIFLE (CONT)

Item No.	Interval	Item To Be Checked Or Serviced	Procedure	Not Fully Mission Capable If:
6	Quarterly (cont)	Lower receiver and buttstock assembly (serviceability check) (cont)	Remove pistol grip screw (3), lockwasher (4), pistol grip (5), helical spring (6), safety detent (7), pivot pin (8), pivot pin detent (9), and helical spring (10). Clean and lubricate metal components. Also clean and generously lubricate pivot pin holes and spring/detent holes. Replace defective/ damaged components as necessary.	Replace defective/damaged components as necessary.
			Disengage takedown pin (11) and pull out, push back in to re-engage takedown pin (an audible click should be heard). If an audible click is not heard, see page 2-57 for repair.	Replace defective/damaged components as necessary.

2-7. PREVENTIVE MAINTENANCE CHECKS AND SERVICES (CONT).

PREVENTIVE MAINTENANCE CHECKS AND SERVICES FOR M16A2 RIFLE (CONT)

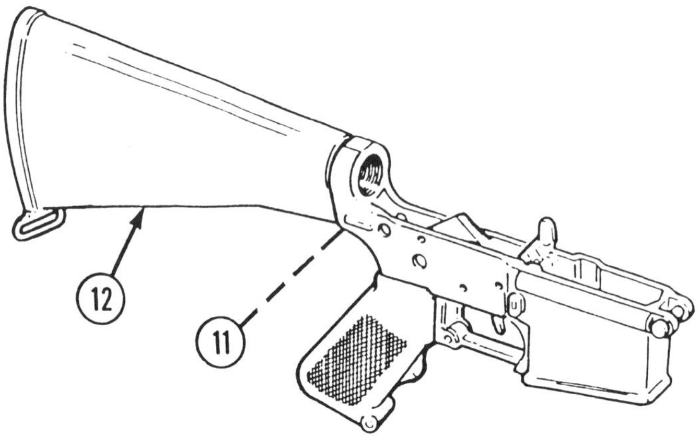

Item No.	Interval	Item To Be Checked Or Serviced	Procedure	Not Fully Mission Capable If:
6	Quarterly (cont)	Lower receiver and buttstock assembly (serviceability check) (cont)	Lubricate helical compression spring and takedown pin detnt (11) by placing one drop of lubricant on take-down pin detent and lowering the buttstock assembly (12) to vertical position. Allow the lubricant to work its way around the helical compres-sion spring and takedown pin detent (11). Check buttstock assembly (12) com-ponents for damage.	

RIFLE ONLY
Under the following conditions, hair-line cracks (no chipped away material allowed) originating from the buttplate end of the buttstock are acceptable.

Replace damaged components as necessary.

PREVENTIVE MAINTENANCE CHECKS AND SERVICES FOR M16A2 RIFLE (CONT)

Item No.	Interval	Item To Be Checked Or Serviced	Procedure	Not Fully Mission Capable If:
6	Quarterly (cont)	Lower receiver and buttstock assembly (serviceability check) (cont)	**RIFLE ONLY** a. One hairline crack, not to exceed 1 in. (2.54 cm) in length, per side of buttstock. b. Two additional hairline cracks up to 0.25 in. (0.64 cm) in length, per side of buttstock. c. A total of three cracks per side of the buttstock, originating from the buttplate end, are allowable. Cracks in the critical area at the front end of the buttstock are not acceptable. Check buttstock assembly (12) for forward to rear movement and/or a 1/32 in. (0.079 cm) gap between the buttstock assembly (12) and the lower receiver (13). If forward to rear movement and/or a 1/32 in. (0.079 cm) gap appears, replace self-locking screw. If still not tight, remove buttstock assembly and check for loose lower receiver extension. If loose, evacuate to support maintenance. If not loose, replace buttplate (14).	The buttstock is cracked in the critical area or does not meet the crack criteria. Lower receiver extension cannot be tightened.

2-7. PREVENTIVE MAINTENANCE CHECKS AND SERVICES (CONT).

PREVENTIVE MAINTENANCE CHECKS AND SERVICES FOR M16A2 RIFLE (CONT)

Item No.	Interval	Item To Be Checked Or Serviced	Procedure	Not Fully Mission Capable If:
6	Quarterly (cont)	Lower receiver and buttstock assembly (serviceability check) (cont)	**RIFLE ONLY** Small amounts of side-to-side, up-and-down or rotational movement of the buttstock assembly is acceptable. (1) Cracks visible around the butt-plate mounting holes while screws are mounted. SH - Cracks are visible around mounting holes when installed on rifle. (2) Cracks or separations around the door assembly are visible when the door assembly is closed. SH - Cracks are visible when door assembly is closed. (3) If buttplate is cracked in excess of 0.25 in. (0.64 cm) in length and extends through the buttplate (14), see pg. 2-64 for repair. SH - If cracked in excess of 0.25 in. extends thru buttplate. (4) The buttplate (14) should not be removed other than for repair or replacement of parts at which time a new self-locking screw, NSN 5305-01-147-8585, must be used.	

2-7. PREVENTIVE MAINTENANCE CHECKS AND SERVICES (CONT).

PREVENTIVE MAINTENANCE CHECKS AND SERVICES FOR M16A2 RIFLE (CONT)

Item No.	Interval	Item To Be Checked Or Serviced	Procedure	Not Fully Mission Capable If:
6	Quarterly (cont)	Lower receiver and buttstock assembly (serviceability check) (cont)	**CARBINE ONLY** Extend buttstock assembly (14.1). Grasp the lock release lever (14.2) in the area of the retaining nut (14.3) and pull downward and slide the buttstock assembly to the rear to separate the buttstock assembly from the lower receiver extension (14.4). Clean and lubricate the takedown pin (14.5). Lubricate the takedown pin detent (*) and spring (*) by placing lubricant on the detent and exercising it with a small screw driver. Clean buttstock assembly inside and out. Check lock release lever for free movement. Check for cracks, dents, and damage to buttstock assembly. Notify support maintenance if buttstock assembly is damaged or lock release lever does not move freely.	If lock release lever is cracked, does not move free, or is dented, or damaged badly enough to interfere with functioning.

***NOT REMOVED. NOT DEPICTED**

2-7. PREVENTIVE MAINTENANCE CHECKS AND SERVICES (CONT).

PREVENTIVE MAINTENANCE CHECKS AND SERVICES FOR M16A2 RIFLE (CONT)

Item No.	Interval	Item To Be Checked Or Serviced	Procedure	Not Fully Mission Capable If:
6	Quarterly (cont)	Lower receiver and buttstock assembly (serviceability check) (cont)	**CARBINE ONLY** Buttstock assembly can be repaired at unit maintenance. Hand check lower receiver extension for looseness and corrosion. If loose, evacuate to support maintenance. Clean and lubricate the lower receiver extension. Grasp the release lever in the area of the retaining nut and pull to reinstall the buttstock assembly onto the lower receiver extension. **BOTH WEAPONS** Function check the magazine catch (15) and bolt catch (16). If defective, evacuate to support maintenance. Check lower receiver (13) finish for scratches and worn shiny spots.	Lower Receiver extension is loose. Magazine catch or bolt catch is defective.

PREVENTIVE MAINTENANCE CHECKS AND SERVICES FOR M16A2 RIFLE (CONT)

Item No.	Interval	Item To Be Checked Or Serviced	Procedure	Not Fully Mission Capable If:
6	Quarterly (cont)	Lower receiver and buttstock assembly (serviceability check) (cont)	**NOTE** If a M16A2 Rifle lower receiver is missing one third or more of its exterior protective finish, resulting in an unprotected/light reflecting surface, it is candidate for overhaul. This missing finish will be considered a shortcoming. This shortcoming requires action to obtain a replacement rifle. Once a replacement has been received, evacuate the original rifle to depot for overhaul. If scratched or worn shiny in spots, repair in the same manner as outlined for upper receiver (see item 4 above).	
7	Quarterly	M16A2 Rifle	Assemble as in TM 9-1005-319-10 (operator's manual). Check sling for damage. If damaged, replace. Check for improperly assembled, broken, missing, or damaged parts. Check overall general appearance. Replace parts as required and authorize evacuation to support maintenance for repair.	

2-7. PREVENTIVE MAINTENANCE CHECKS AND SERVICES (CONT).

PREVENTIVE MAINTENANCE CHECKS AND SERVICES FOR M16A2 RIFLE (CONT)

Item No.	Interval	Item To Be Checked Or Serviced	Procedure	Not Fully Mission Capable If:
8	Quarterly	Annual DS safety and service-ability inspection and gaging	Check the DD Form 314 to ensure annual DS safety and serviceability inspection and gaging has been done and that the next gaging and inspection is scheduled. If annual gaging has not been performed within the last year, notify support maintenance.	Annual gaging has not been performed.

Section IV. TROUBLESHOOTING

2-8. GENERAL.

a. This section contains unit level troubleshooting information for locating and correcting most of the operating troubles which may develop in the M16A2 rifle. Each malfunction for the individual part or assembly is followed by a list of tests or inspections which will help you to determine the corrective actions to take. You should perform the tests/inspections and corrective actions in the order listed.

b. This manual cannot list all malfunctions that may occur, nor all tests or inspections and corrective actions. If a malfunction is not listed or is not corrected by listed corrective actions, see individual repair sections in the maintenance procedures on each major assembly.

2-9. TROUBLESHOOTING PROCEDURES. Refer to troubleshooting table for malfunctions, tests, and corrective actions. The symptom index is provided for a quick reference of the malfunctions covered in the table.

SYMPTOM INDEX

Troubleshooting
Procedures
Page

TROUBLESHOOTING

MALFUNCTION
    TEST OR INSPECTION
        CORRECTIVE ACTION

---

1. FAILURE OF MAGAZINE TO LOCK IN RIFLE.

        Step 1. Dirty or corroded magazine catch (1).

                Disassemble and clean.

        Step 2. Defective magazine catch spring (2).

                Evacuate to support maintenance.

        Step 3. Worn or broken magazine catch (1).

                Evacuate to support maintenance.

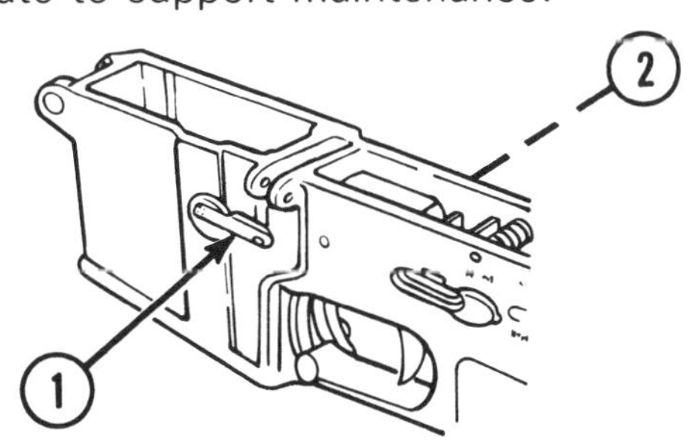

2-9. TROUBLESHOOTING PROCEDURES (CONT).

TROUBLESHOOTING (CONT)

MALFUNCTION
    TEST OR INSPECTION
        CORRECTIVE ACTION

---

2. FAILURE TO FEED.

    Step 1. Magazine catch spring weak or broken.

        Evacuate to support maintenance.

    Step 2. Magazine catch (1) defective.

        Evacuate to support maintenance.

    Step 3. Magazine catch (1) out of adjustment (will not retain magazine).

        Refer to operator's manual.

    Step 4. Short recoil.

        Refer to page 2-28.

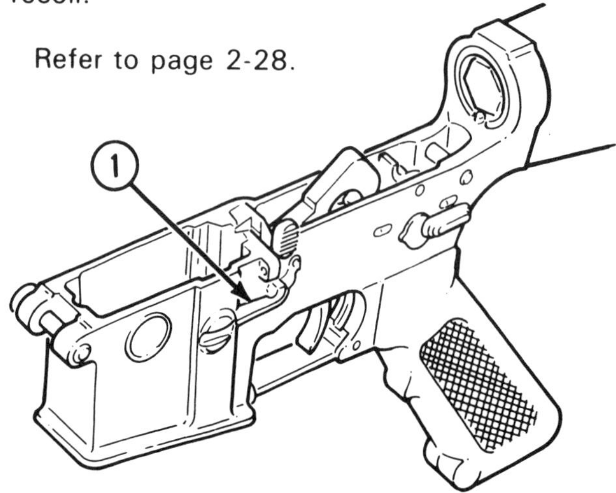

3. FAILURE TO CHAMBER.

    Step 1. Weak or broken action spring (1); RIFLE ONLY: (free length 11 3/4 inches (29.85 cm) minimum to 13 1/2 inches (34.29 cm) maximum). CARBINE ONLY: (10 1/16 inches (25.56 cm) minimum to 11 1/4 inches (28.58 cm) maximum.)

        Replace action spring (p 2-57).

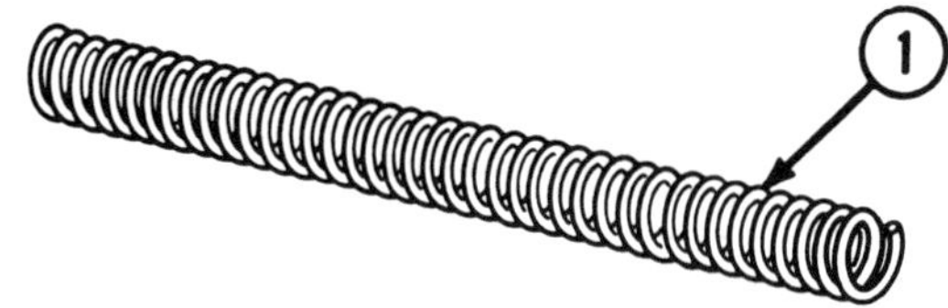

TROUBLESHOOTING (CONT)

MALFUNCTION
    TEST OR INSPECTION
        CORRECTIVE ACTION

    Step 2.   Short recoil.

        Refer to page 2-28.

4.   FAILURE TO LOCK.

    Step 1.   Bolt cam pin (1) missing.

        Replace (p 2-35).

    Step 2.   Loose or damaged bolt carrier key (2).

        a.   Evacuate to support maintenance.

        b.   Dented bolt carrier key may be repaired (p 2-35).

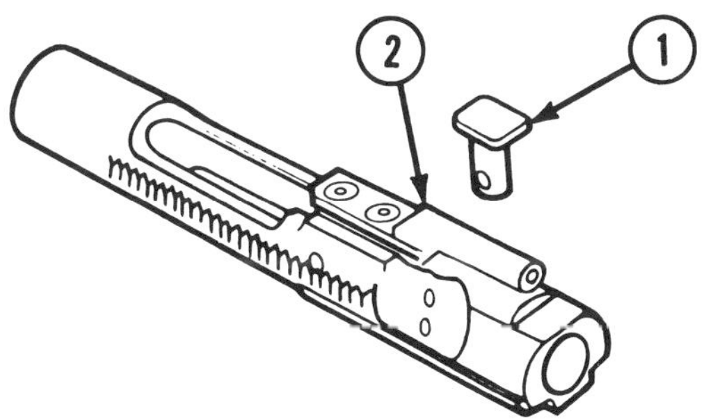

    Step 3.   Improperly assembled extractor spring assembly (3).

        Assemble correctly (p 2-38).

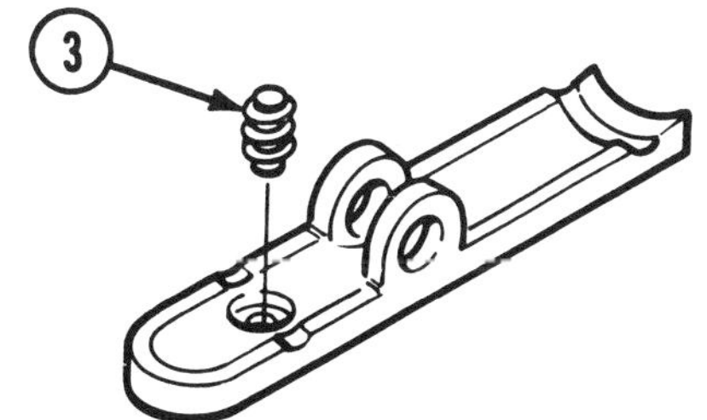

2-9. TROUBLESHOOTING PROCEDURES (CONT).

TROUBLESHOOTING (CONT)

MALFUNCTION
    TEST OR INSPECTION
        CORRECTIVE ACTION

---

4. FAILURE TO LOCK. (CONT)

    Step 4. Bent gas tube (4).

        a. Adjust to its original configuration by bending in area of hand-guard assembly.

        b. If the gas tube cannot be returned to its original configuration, evacuate the rifle to support maintenance.

    Step 5. Weak or broken action spring (5); RIFLE ONLY: (free length 11 3/4 inches (29.85 cm) minimum to 13 1/2 inches (34.29 cm) maximum). CARBINE ONLY: (10 1/16 inches (25.56 cm) minimum to 11 1/4 inches (28.58 cm) maximum.).

    Replace action spring (p 2-57).

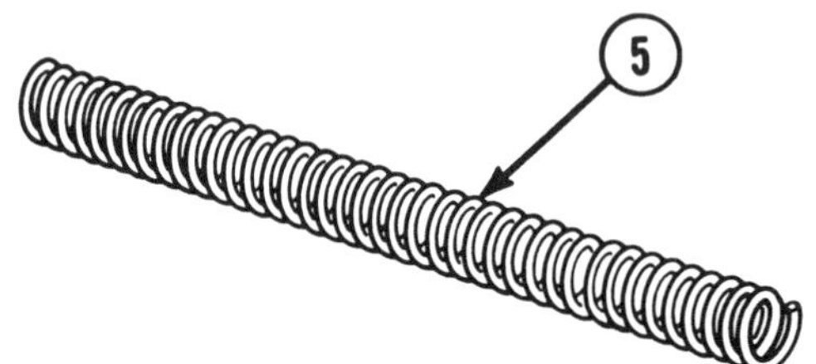

5. FAILURE TO FIRE.

    Step 1. Broken or chipped firing pin (1).

    Evacuate to support maintenance.

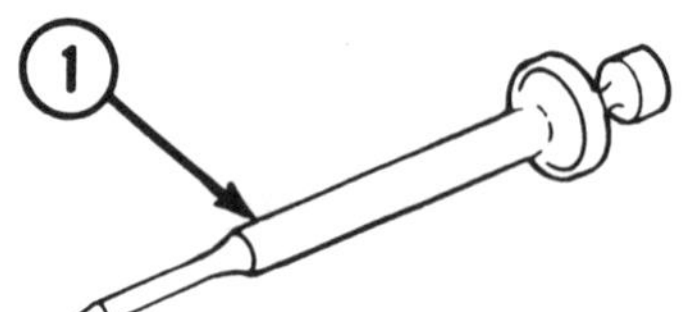

TROUBLESHOOTING (CONT)

MALFUNCTION
    TEST OR INSPECTION
        CORRECTIVE ACTION

Step 2.   Carbon buildup in firing pin recess inside bolt assembly.

        Remove cartridge extractor and clean recess with pipe cleaner (item 11, app D); refer to operator's manual.

Step 3.   Firing mechanism (2) and/or lower receiver assembly (3) improperly assembled or has worn, broken, or missing parts.

        Evacuate to support maintenance.

Step 4.   Broken, defective, or missing firing pin retaining pin (4).

        Replace (p 2-35).

Step 5.   Selector lever (5) frozen on SAFE position.

        Evacuate to support maintenance.

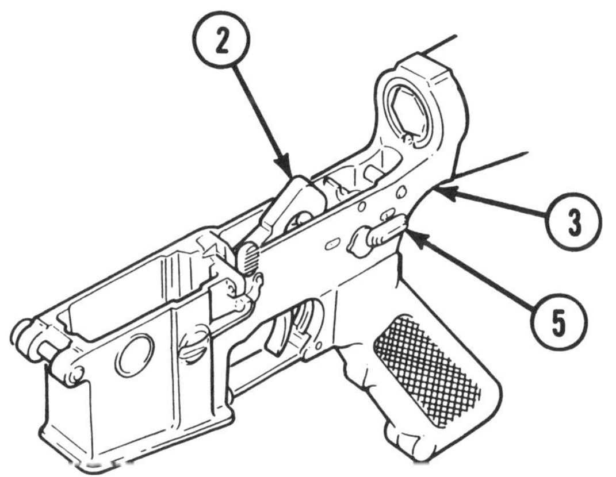

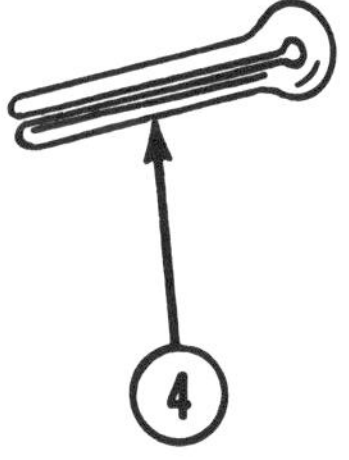

2-9. TROUBLESHOOTING PROCEDURES (CONT).

TROUBLESHOOTING (CONT)

MALFUNCTION
    TEST OR INSPECTION
        CORRECTIVE ACTION

---

6. FAILURE TO UNLOCK.

    Step 1. Burred locking lugs (1) on bolt assembly.

        Remove burrs.

    Step 2. Burred lugs (2) on barrel extension.

        Remove burrs.

    Step 3. Short recoil.

        Refer to page 2-28.

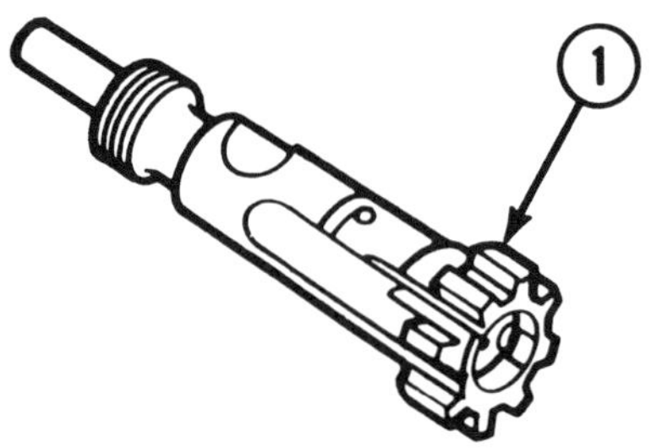
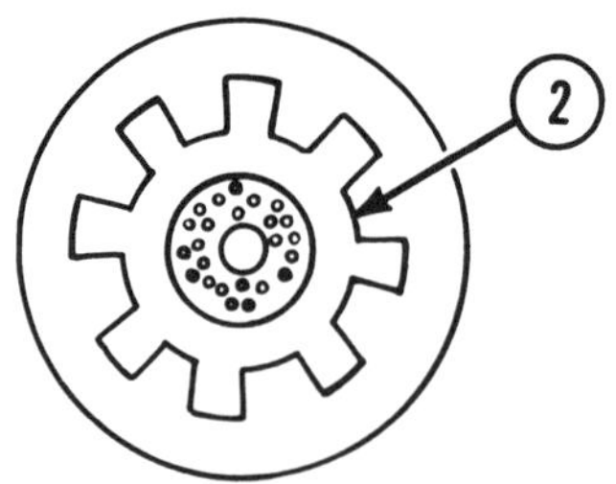

7. FAILURE TO EXTRACT.

    Step 1. Defective extractor pin (1), cartridge extractor (2), and/or extractor spring assembly (3).

        Replace extractor pin (1), cartridge extractor (2), and/or extractor spring assembly (3) (p 2-38).

    Step 2. Short recoil.

        Refer to page 2-28.

NOTE

Rubber insert and spring are an assembly. Illustration shows insert out of assembly for clarification only. Do not remove the rubber insert from the extractor spring assembly.

TROUBLESHOOTING (CONT)

MALFUNCTION
    TEST OR INSPECTION
        CORRECTIVE ACTION

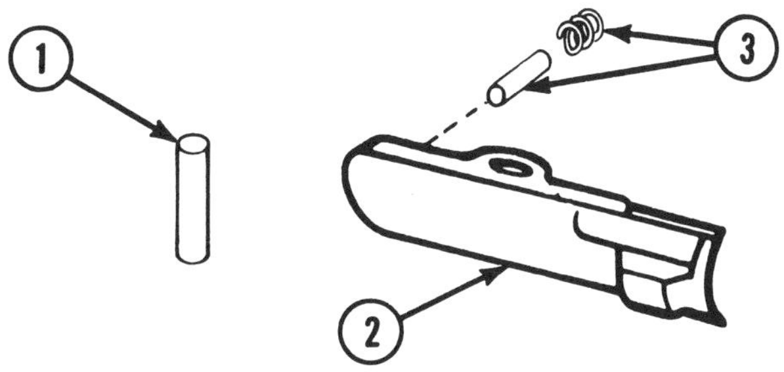

8.   FAILURE TO EJECT.

       Step 1.   Broken cartridge ejector (1).

              Replace (p 2-38).

       Step 2.   Cartridge ejector (1) stuck in bolt body (2).

              Disassemble and clean (p 2-38).

       Step 3.   Weak or broken ejector spring (3).

              Replace (p 2-38).

       Step 4.   Short recoil.

              Refer to page 2-28.

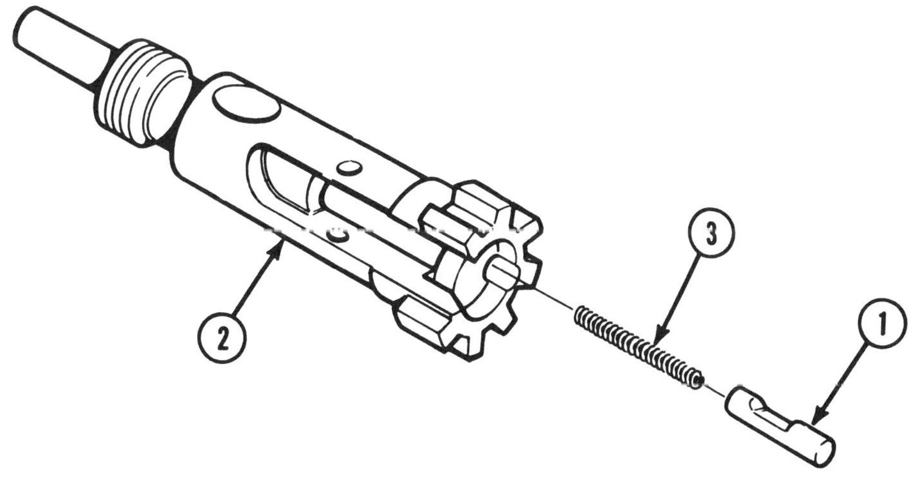

2-9. TROUBLESHOOTING PROCEDURES (CONT).

TROUBLESHOOTING (CONT)

MALFUNCTION
    TEST OR INSPECTION
        CORRECTIVE ACTION

---

9. FAILURE TO COCK.

    Step 1. Worn, broken, or missing parts of firing mechanism.

        Evacuate to support maintenance.

    Step 2. Short recoil.

        Refer to below.

10. SHORT RECOIL.

    Step 1. Broken or damaged action spring (1).

        Replace action spring (p 2-57).

    Step 2. Unlubricated or dirty action spring and receiver extension.

        Clean and lubricate.

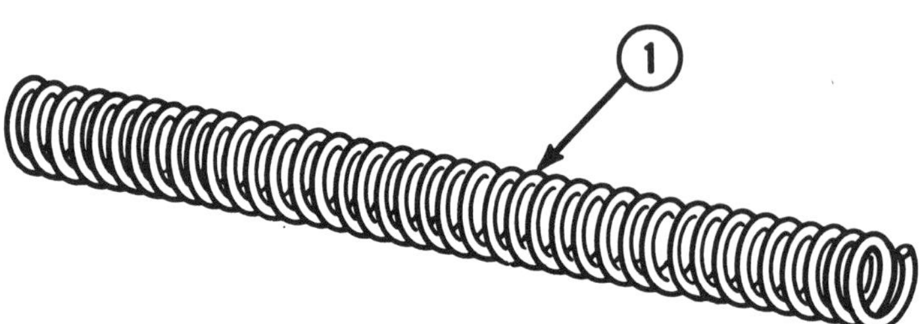

    Step 3. Improper gap space or worn, missing, or broken bolt rings (2).

        a. Stagger bolt ring gaps (approximately ⅓ turn apart).

        b. Evacuate to support maintenance if bolt rings are worn, broken, or missing.

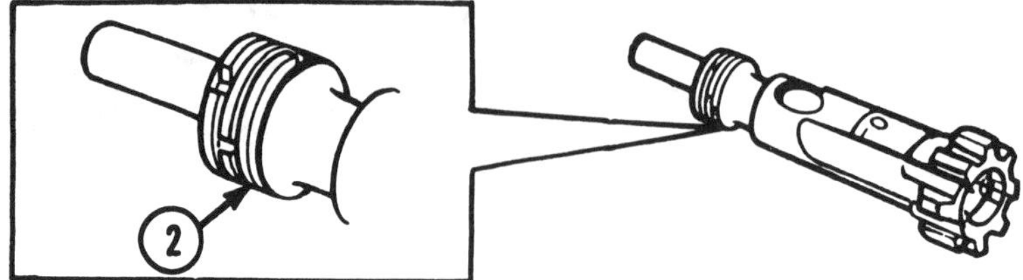

TROUBLESHOOTING (CONT)

MALFUNCTION
    TEST OR INSPECTION
        CORRECTIVE ACTION

Step 4. Carbon build-up or foreign matter in the narrow passage of the bolt carrier key (3).

    Clean with CLP (item 9, app D) and a pipe cleaner (item 11, app D).

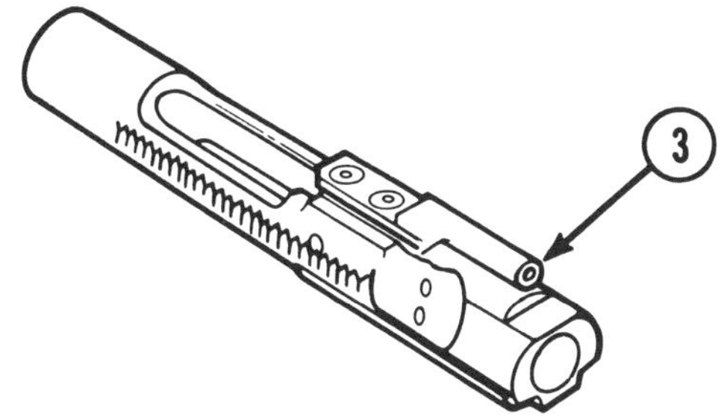

Step 5. Gas leakage caused by broken or loose gas tube (4) around front sight base.

    Evacuate to support maintenance.

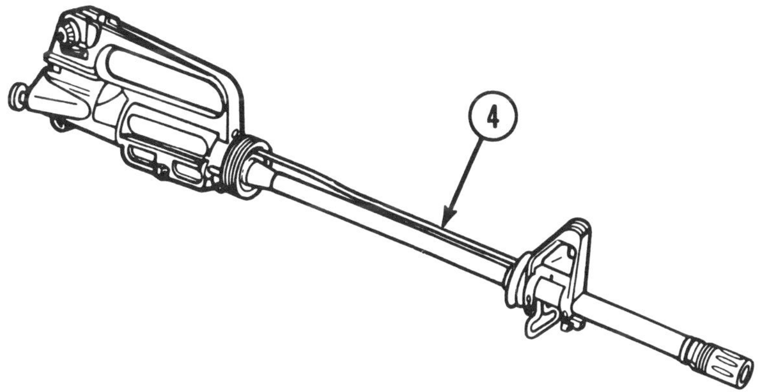

Step 6. Improper alignment of gas tube and carrier key.

    a. Adjust gas tube alignment by bending in area of handguard assembly to its original configuration.

    b. If gas tube cannot be returned to its original configuration, evacuate the rifle to support maintenance.

2-9. TROUBLESHOOTING PROCEDURES (CONT).

TROUBLESHOOTING (CONT)

MALFUNCTION
    TEST OR INSPECTION
        CORRECTIVE ACTION

---

11.   RIFLE CANNOT BE ZEROED.

      Step 1.   Defective barrel assembly (1).

              Evacuate to support maintenance.

      Step 2.   Barrel assembly out of alignment with rear sight assembly (2) on upper receiver.

              Evacuate to support maintenance.

      Step 3.   Defective front sight (3).

              Remove front sight post (4), front sight detent (5), and helical spring (6). If damaged, replace.

      Step 4.   Defective rear sight assembly (2).

              Evacuate to support maintenance.

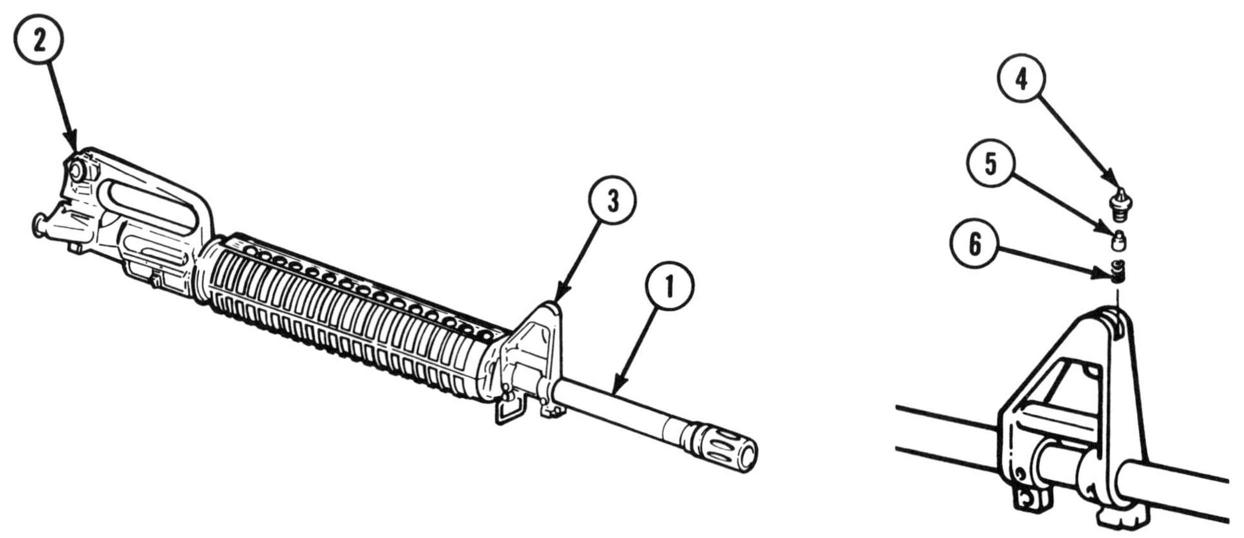

TROUBLESHOOTING (CONT)

MALFUNCTION
    TEST OR INSPECTION
        CORRECTIVE ACTION

---

12. **FAILURE TO CYCLE WITH SELECTOR LEVER SET ON BURST.**

    Faulty selector lever or broken cam, cam clutch spring, or burst lever.

    Evacuate to support maintenance.

13. **FIRES TWO ROUNDS WITH ONE PULL OF TRIGGER WITH SELECTOR LEVER SET ON SEMI (DOUBLE FIRING).**

    Perform function test.

        If any part of function test (p 2-68) fails, evacuate to support maintenance.

14. **FIRES WITH SELECTOR LEVER ON SAFE OR WHEN TRIGGER IS RELEASED WITH SELECTOR LEVER ON SEMI.**

    Worn, broken, or missing parts of firing mechanism.

    Evacuate to support maintenance.

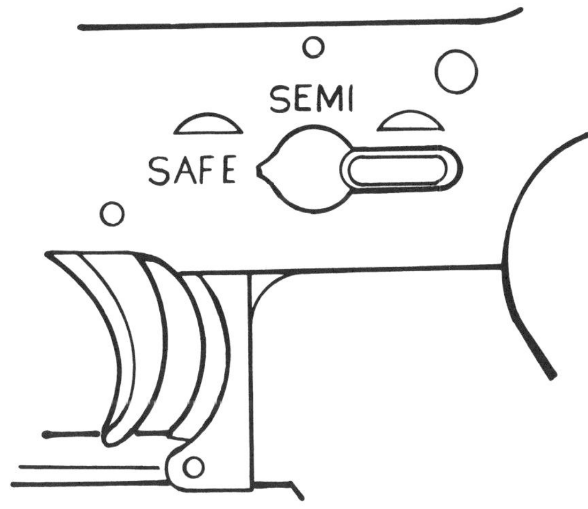

2-9. TROUBLESHOOTING PROCEDURES (CONT).

TROUBLESHOOTING (CONT)

MALFUNCTION
 TEST OR INSPECTION
  CORRECTIVE ACTION

---

15. BOLT ASSEMBLY FAILS TO LOCK TO REAR AFTER FIRING LAST ROUND.

Step 1. Magazine follower (1) worn or broken.

Replace magazine.

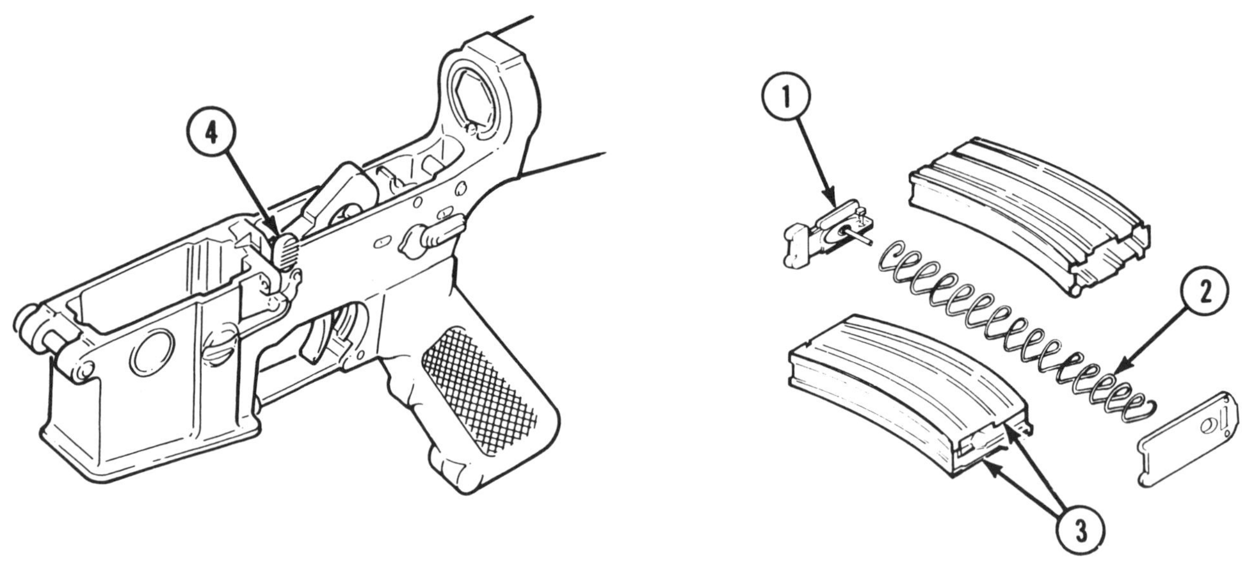

Step 2. Magazine catch spring (2) weak or broken.

Replace magazine.

Step 3. Magazine feeder lips (3) bent or broken.

Replace magazine.

Step 4. Magazine follower (1) binds during operation.

Replace magazine.

Step 5. Broken bolt catch (4) and/or spring.

Evacuate to support maintenance.

Section V.   MAINTENANCE PROCEDURES

2-10.   INITIAL SETUP. The following information will reduce the space required for the initial setup portion of the maintenance procedures.

   a.   Materials/Parts required are not listed unless they apply to the procedure.

   b.   Personnel Required is listed only if the task requires more than one person. If Personnel Required is not listed, it means one person can do the job.

   c.   The normal standard equipment condition is that the item is removed from the end item or next higher assembly and is in the assembled condition. Equipment Condition is not listed unless some other condition is required.

   d.   The approximate time required is listed on the applicable Maintenance Allocation Chart (MAC).

   e.   When the term evacuate to support maintenance is used, the entire rifle must be evacuated.

2-11.   LUBRICATION GENERAL.

   a.   Whenever the term ''cleaner, lubricant, and preservative (CLP)'' or the words ''lubricant'', ''lube'', ''LSA'', or ''LAW'' are cited in this TM, they are to be interpreted to mean CLP (item 9, app D), weapons lubricating oil (LSA) (item 23, app D), or weapons lubricating oil (LAW) (item 22, app D) can be utilized as applicable. The following constraints must be adhered to:

   b.   Under all but the coldest arctic conditions, LSA or CLP are the lubricants to use on the rifle. Either may be used at $-10\,°F$ $(-23\,°C)$ and above. However, do not use both on the same rifle at the same time.

   c.   LAW is the lubricant to use during cold arctic conditions, $+10\,°F$ $(-12\,°C)$ and below.

   d.   Any of the lubricants may be used from $-10\,°F$ to $+10\,°F$ $(-23\,°C$ to $-12\,°C)$.

   e.   **ARMY ONLY:** Do not mix lubricants on the same rifle. The rifle must be thoroughly cleaned during change from one lubricant to another. Dry Cleaning Solvent (SD) (item 16, app D) is recommended for cleaning during change from one lubricant to another.

   f.   Rifle Bore Cleaner (RBC) (item 12, app D) may be used to remove carbon buildup in the bore and other portions of the rifle.

2-12. MAJOR COMPONENTS OF M16A2 RIFLE.

This task covers disassembly.

INITIAL SETUP

References
  TM 9-1005-319-10 (operator's manual)

Equipment Conditions
  Rifle assembled

General Safety Instructions
  Before starting an inspection, be sure to clear the rifle. Do not keep live ammunition near the work area.

To avoid injury to your eyes, use care when removing and installing spring-loaded parts.

Below direct support maintenance, do not interchange bolt assemblies from one rifle to another. Doing so may result in injury to, or death of, personnel.

DISASSEMBLY

a. Refer to operator's manual.

b. Remove magazine (1), sling (2), bolt carrier assembly (3), charging handle assembly (4), and upper receiver and barrel assembly (5) from lower receiver and buttstock assembly (6).

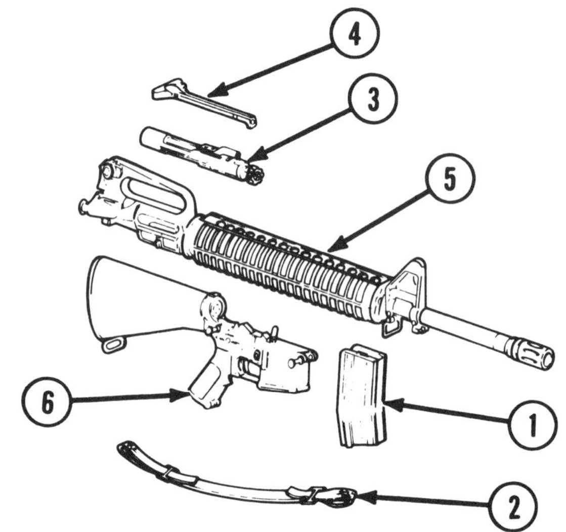

NOTE

Solid Film Lubricant (SFL) is the only authorized touchup for the M16A2 rifle and may be used on up to one third of the exterior finish of the rifle. **FOR ARMY CONUS USE ONLY AND AIR FORCE TRAINING RIFLES ONLY:** SFL may be used as touchup without limitation on the upper receiver and barrel assembly. This is to say that units which **DO NOT** fall under the category of Divisional Combat Units or rapid deployment type units may have up to 100 percent of the exterior surface of the upper receiver and barrel assembly protected with SFL if necessary.

2-13. BOLT CARRIER ASSEMBLY.

This task covers:

a. Disassembly
b. Cleaning
c. Inspection/Repair

d. Lubrication
e. Reassembly

INITIAL SETUP

Tools
  (ARMY) Small Arms Repairman Tool Kit
    (item 3, app B)

References
  TM 9-1005-319-10 (operator's manual)

Equipment Conditions
  2-34   Bolt carrier assembly removed

General Safety Instructions
  Bolt cam pin must be installed or rifle will
    blow up while firing the first round. If
    the bolt cam pin is not installed, injury
    to, or death of, personnel may result.

  Do not interchange bolt assemblies from
    one rifle to another. Doing so may
    result in injury to, or death of,
    personnel.

a. DISASSEMBLY

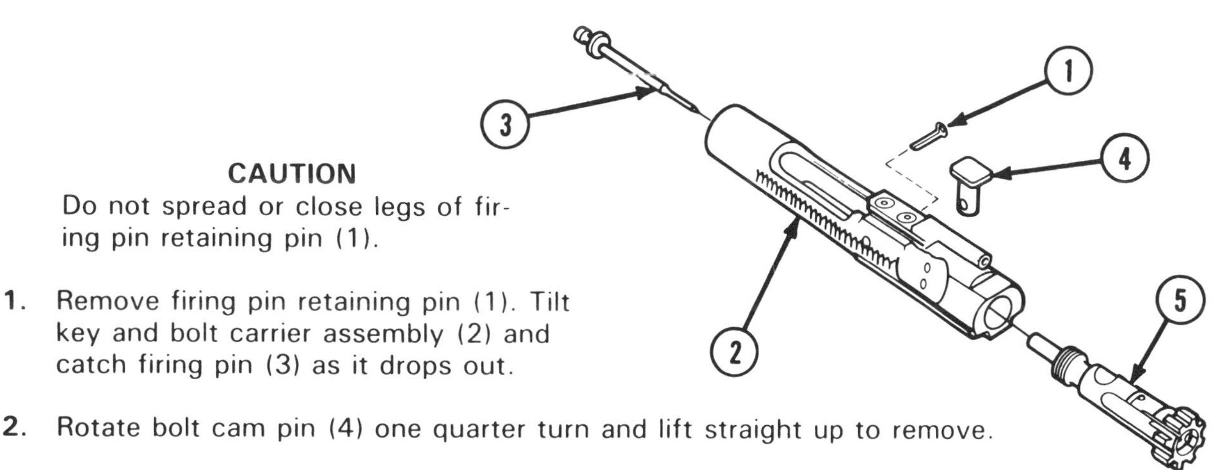

CAUTION
Do not spread or close legs of firing pin retaining pin (1).

1. Remove firing pin retaining pin (1). Tilt
    key and bolt carrier assembly (2) and
    catch firing pin (3) as it drops out.

2. Rotate bolt cam pin (4) one quarter turn and lift straight up to remove.

3. Remove bolt assembly (5) from key and bolt carrier assembly (2).

NOTE
For disassembly of bolt assembly (5), see page 2-38.

2-13. BOLT CARRIER ASSEMBLY (CONT).

b. CLEANING

Clean all items (operator's manual). Remove carbon deposits.

c. INSPECTION/REPAIR

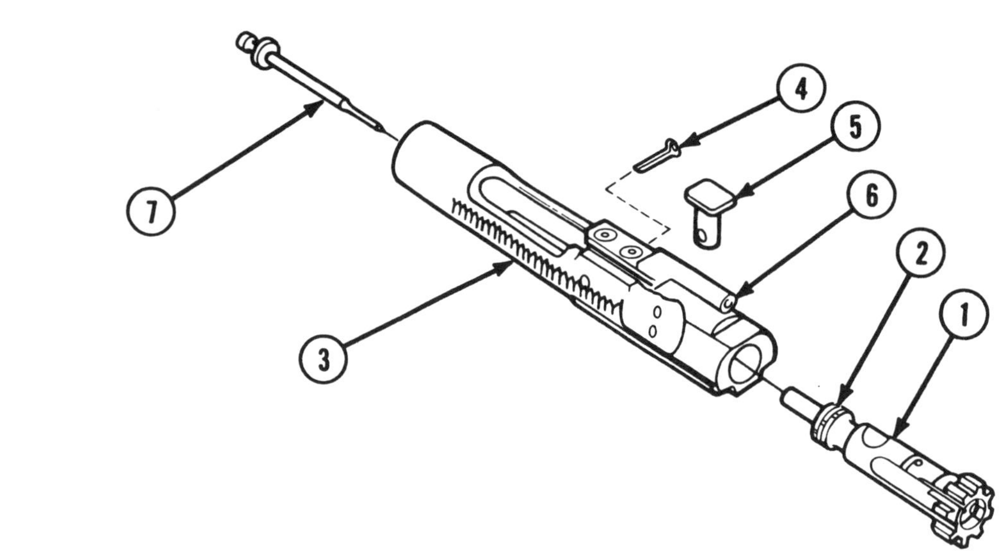

1. Inspect bolt assembly (1) as follows:

    a. Check to see that the bolt ring (2) gaps are staggered.

    b. Install bolt assembly (1) into key and bolt carrier assembly (3). Check for worn rings
        by holding the key and bolt carrier assembly (3) with the bolt assembly facing
        down. If bolt assembly falls out after firing pin retaining pin (4) and bolt cam pin (5)
        are removed, the rings are worn. Evacuate to support maintenance.

2. Inspect carrier key (6) for dents, distortion, or looseness. If dented or loose, evacuate
    to support maintenance.

3. Inspect firing pin retaining pin (4) and bolt cam pin (5) for cracks, damage, or excessive
    wear. Replace if unserviceable.

4. Inspect firing pin (7) for damage or if tip is chipped. Evacuate rifle to support
    maintenance if unserviceable.

5. Inspect key and bolt carrier assembly (3) for damage or wear. If unserviceable,
    evacuate to support maintenance.

d. LUBRICATION

Lubricate all items (p 2-33) (operator's manual).

e. REASSEMBLY

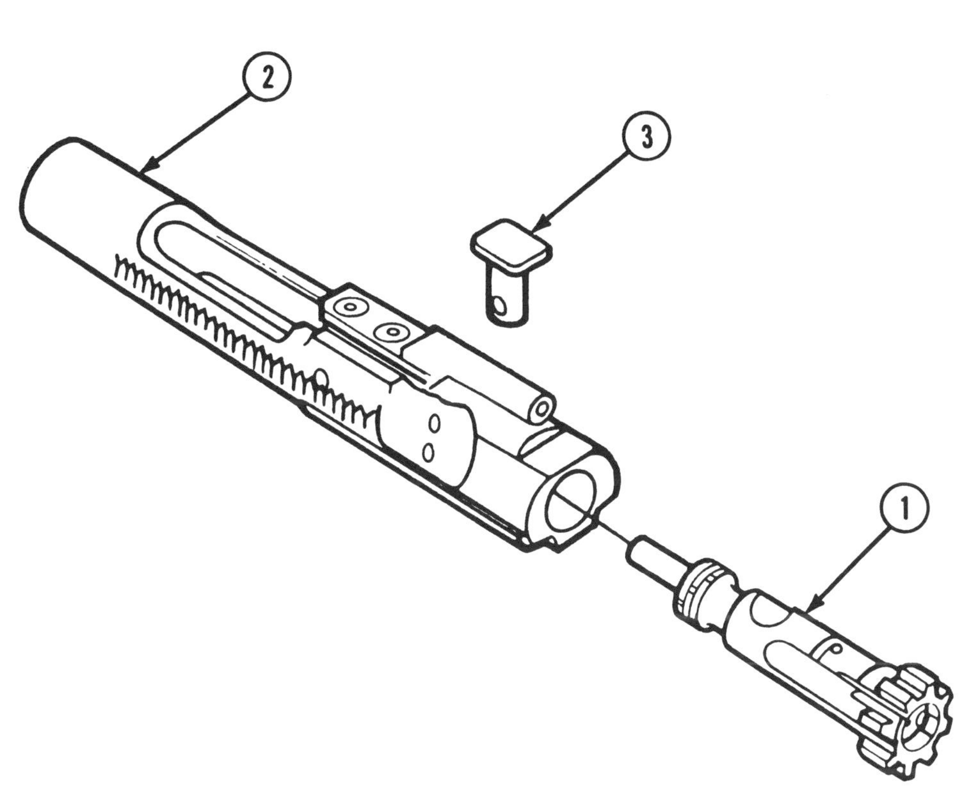

WARNING
Do not interchange bolt assemblies from one rifle to another. Doing so may result in injury, or death of, personnel.

Bolt cam pin must be installed or rifle will blow up while firing the first round. If the bolt cam pin is not installed, injury to, or death of, personnel may result.

NOTE
Before installing bolt assembly, check to see that the bolt ring gaps are staggered to prevent loss of gas pressure.

1. Install bolt assembly (1) in key and bolt carrier assembly (2).

2. Install bolt cam pin (3) and rotate one quarter turn.

2-13. BOLT CARRIER ASSEMBLY (CONT).

e. REASSEMBLY (CONT)

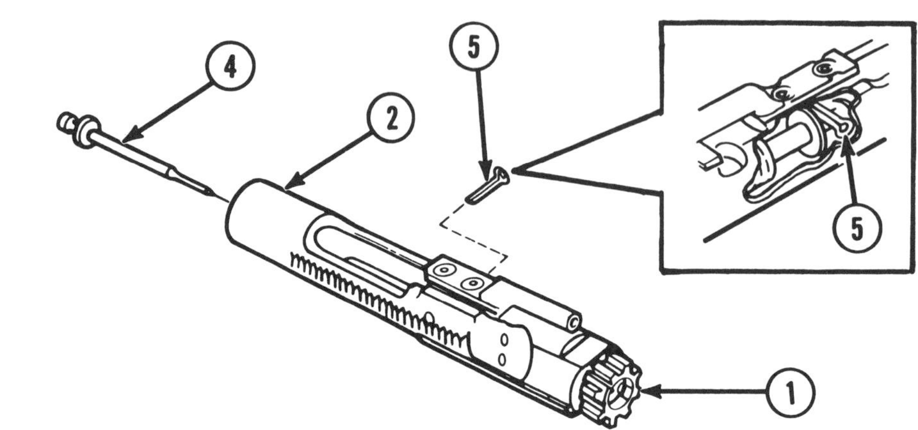

3. Hold key and bolt carrier assembly (2) with bolt assembly (1) down and drop in firing pin (4).

4. Install firing pin retaining pin (5) from left side only to ensure proper installation. Check for proper installation by holding key and bolt carrier assembly (2) with bolt assembly (1) up, then attempt to shake out firing pin.

5. Reassemble rifle, refer to page 2-68.

2-14. BOLT ASSEMBLY.

This task covers:

a. Disassembly	**d.** Lubrication
b. Cleaning	**e.** Reassembly
c. Inspection/Repair	

INITIAL SETUP

Tools
(ARMY) Small Arms Repairman Tool Kit
(item 3 app B)

Equipment Conditions
2-35 Bolt assembly removed

General Safety Instructions
Do not interchange bolt assemblies from one rifle to another. Doing so may result in injury to, or death of, personnel.

To avoid injury to your eyes, use care when removing and installing spring-loaded parts.

| a. | **DISASSEMBLY** |

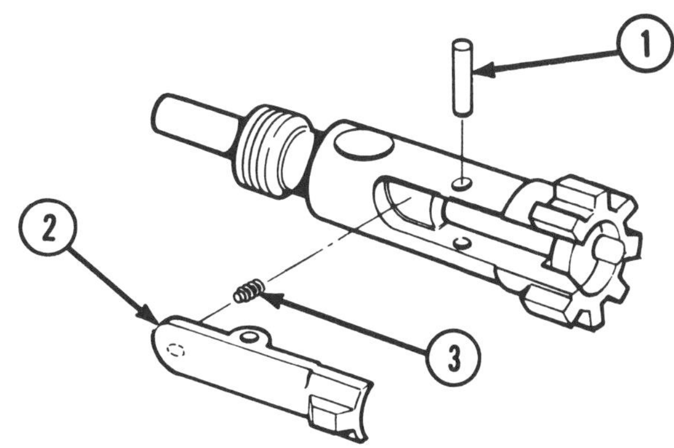

NOTE

Do not separate cartridge extractor and extractor spring assembly unless replacement of either or both is required.

Do not remove the rubber insert from the extractor spring assembly.

1. Push out extractor pin (1) and remove cartridge extractor (2) and extractor spring assembly (3) as a unit.

2. If required, twist extractor spring assembly (3) counterclockwise to remove from cartridge extractor (2).

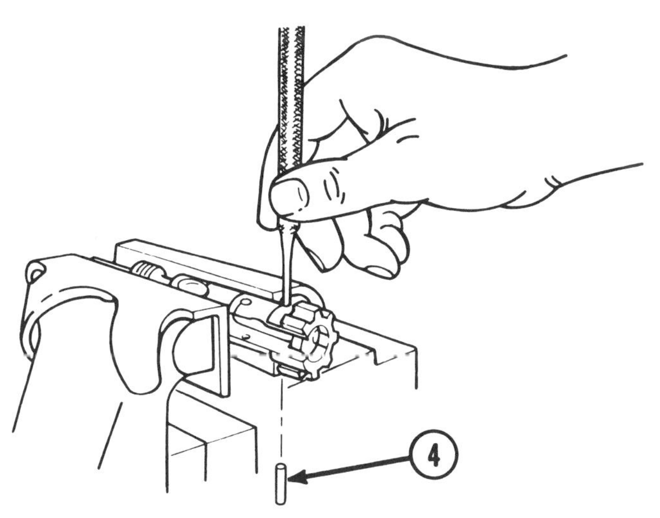

CAUTION
Be sure to use vise jaw protective caps.

3. Hold bolt body in vise and remove spring pin (4) using 1/16 inch punch and hammer.

2-14. BOLT ASSEMBLY (CONT).

a. DISASSEMBLY (CONT)

WARNING

To avoid injury to your eyes, use care when removing and installing spring-loaded parts.

4. Remove punch, be sure to catch cartridge ejector (5) and ejector spring (6) to prevent loss.

b. CLEANING

CAUTION

Do not distort extractor spring assembly during cleaning.

Clean all items (operator's manual). Remove carbon deposits.

c. INSPECTION/REPAIR

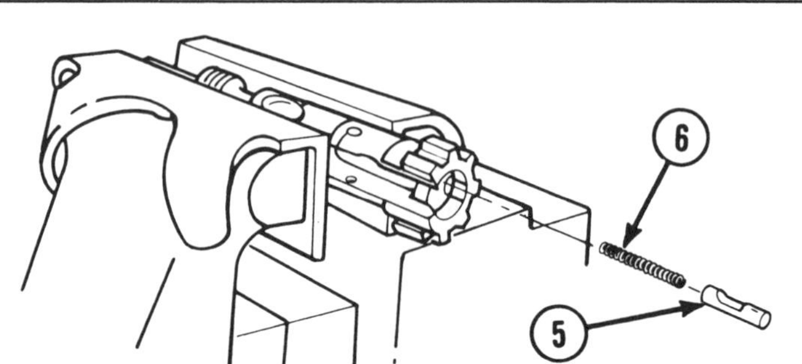

1. Check bolt rings (1) for damage and proper stagger (gaps approximately ⅓ turn apart).

2. Inspect for cracks or damage, especially around bolt cam pin hole (2) and locking lugs (3). If cracked or damaged, evacuate to support maintenance.

3. Inspect cartridge extractor (4), extractor spring assembly (5), and extractor pin (6) for cracks, breaks, chips, and other damage. Pay close attention to cartridge extractor lip (7). If damaged, replace.

4. Inspect cartridge ejector (8) and ejector spring (9) for cracks, breaks, and chips. If damaged, replace.

d. LUBRICATION

Lightly lubricate all items (p 2-33) (operator's manual).

e. REASSEMBLY

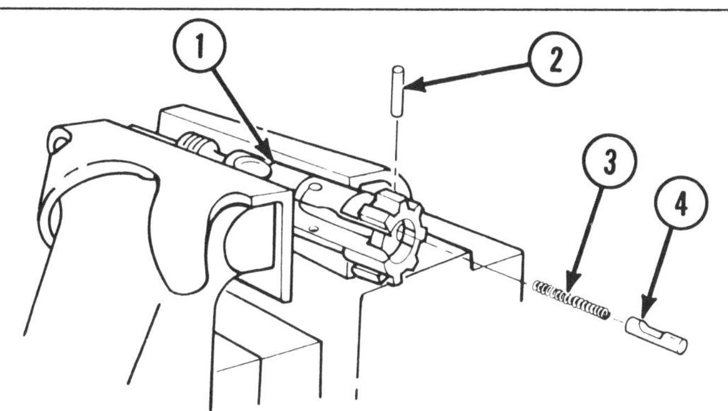

WARNING

To avoid injury to your eyes, use care when removing and installing spring-loaded parts.

Do not interchange bolt assemblies from one rifle to another. Doing so may result in injury to, or death of, personnel.

CAUTION

Be sure to use vise jaw protective caps.

1. Place bolt body (1) in a vise and start spring pin (2) in hole.

2. Install ejector spring (3) and cartridge ejector (4). Align groove on cartridge ejector (4) so that spring pin (2) can be installed.

3. Compress and hold ejector spring and cartridge ejector in place with a 3/8 inch punch.

4. Using hammer and 1/16 inch punch, complete installation of spring pin (2) so that the ends are flush with the outside of bolt body (1).

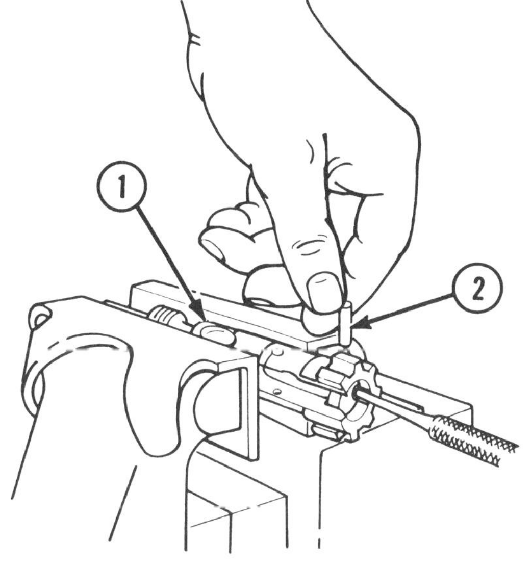

2-14. BOLT ASSEMBLY (CONT).

e. REASSEMBLY (CONT)

NOTE
Do not disassemble rubber insert from extractor spring assembly.

5. If removed, insert large end of extractor spring assembly (5) into cartridge extractor (6) and seat by pushing and turning clockwise.

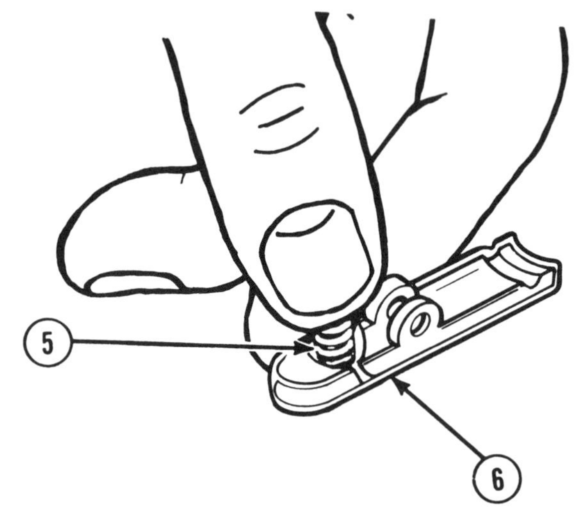

6. Position cartridge extractor (6) and extractor spring assembly (5) on bolt body (1).

7. Compress extractor spring assembly (5) and cartridge extractor (6) to align holes.

8. Install extractor pin (7) by hand.

9. Reassemble rifle, refer to page 2-68.

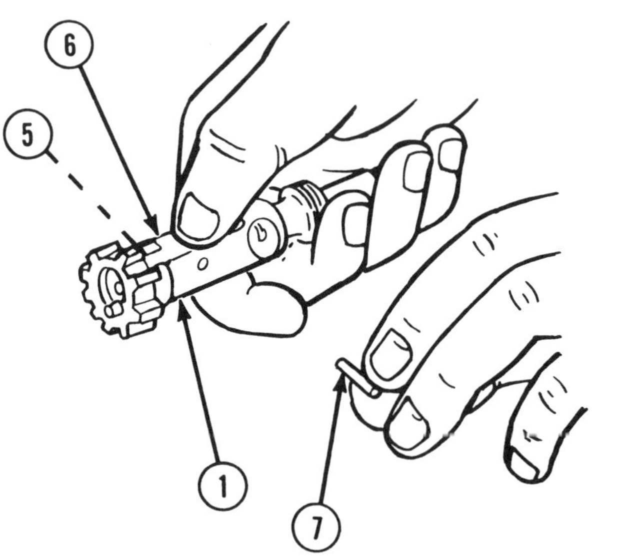

2-15. CHARGING HANDLE ASSEMBLY.

This task covers:

a. Disassembly
b. Cleaning
c. Inspection/Repair

d. Lubrication
e. Reassembly

INITIAL SETUP

Tools
   (ARMY) Small Arms Repairman Tool Kit
   (item 3, app B)

References
   TM 9-1005-319-10 (operator's manual)

Equipment Conditions
   2-34   Charging handle assembly re-
          moved

a. DISASSEMBLY

WARNING
To avoid injury to eyes, use care when removing and installing spring-loaded
parts.

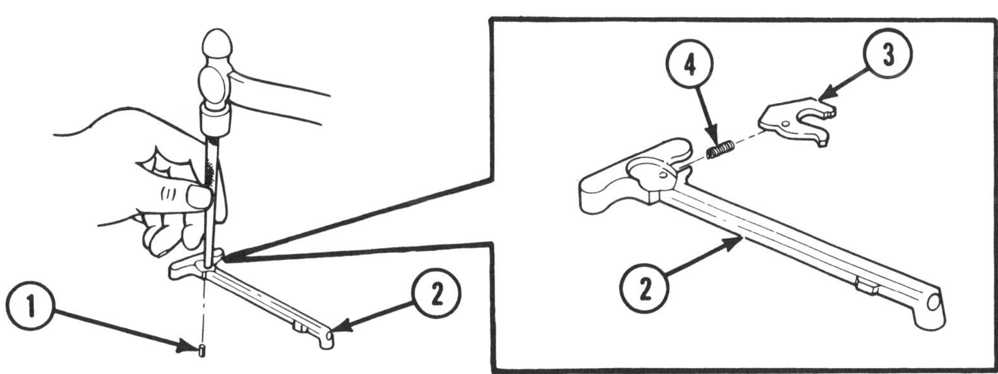

1. Remove spring pin (1) from charging handle (2) using a hammer and 1/16 inch punch.

2. As punch is withdrawn, catch charging handle latch (3) and helical spring (4) to prevent loss.

2-15. CHARGING HANDLE ASSEMBLY (CONT).

b. CLEANING

Clean all items (operator's manual). Remove carbon deposits.

c. INSPECTION/REPAIR

Inspect all items for breaks, cracks, or damage. Replace all unserviceable items.

d. LUBRICATION

Lightly lubricate all items (p 2-33) (operator's manual).

e. REASSEMBLY

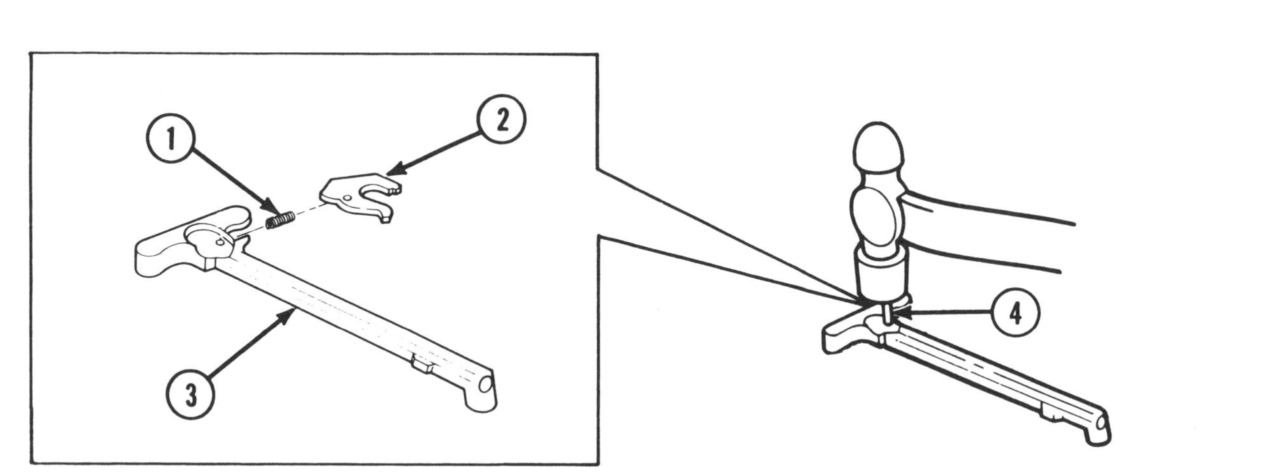

1. Position helical spring (1) and charging handle latch (2) in charging handle (3). Align holes and hold in position.

2. Install spring pin (4) using a hammer. Make sure spring pin is flush.

3. Reassemble rifle, refer to page 2-68.

2-16. UPPER RECEIVER AND BARREL ASSEMBLY, RIFLE BARREL ASSEMBLY, AND UPPER RECEIVER ASSEMBLY.

This task covers:

a. Disassembly
b. Cleaning
c. Inspection
d. Repair

e. Lubrication
f. Reassembly

INITIAL SETUP

Tools
   (ARMY) Small Arms Repairman Tool Kit
      (item 3, app B)
   Front sight post removal and installation
      tool (fig. E-2, app E)
   Front sight detent depressor (fig. E-1,
      app E)

Materials/Parts
   Lubricants (app D)

References
   TM 9-1005-319-10 (operator's manual)

Equipment Conditions
   2-34   Upper receiver and barrel assembly
             removed from lower receiver

General Safety Instructions
   To avoid injury to your eyes, use care
      when removing and installing spring-
      loaded parts.

---

a. DISASSEMBLY

CAUTION

Do not use a screwdriver or any other tool when removing the handguard assemblies. Doing so may damage the handguard assemblies and/or slip ring.

Do not remove heat shield for any reason. Doing so will damage the heat
shield and the handguard assemblies will have to be replaced.

NOTE

Refer to operator's manual for ''buddy system'' procedure on removing handguard assemblies.

2-16. UPPER RECEIVER AND BARREL ASSEMBLY, RIFLE BARREL ASSEMBLY, AND UPPER RECEIVER ASSEMBLY (CONT).

---

a. DISASSEMBLY (CONT)

1. Push down on handguard slip ring (1) and lift upper handguard assembly (2) up and out.

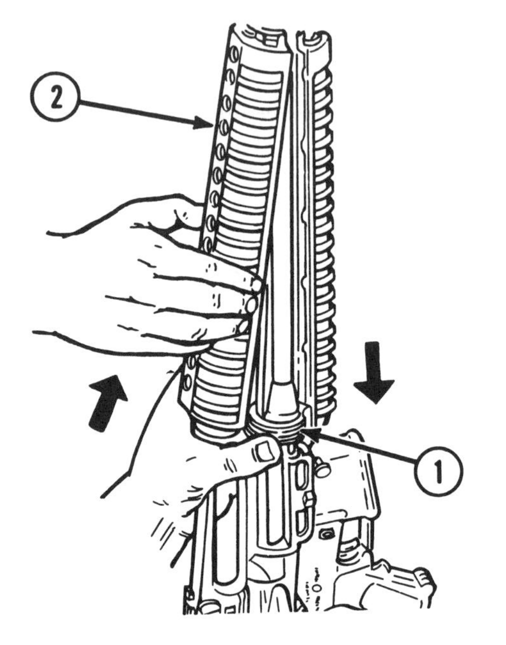

2. Push down on handguard slip ring (1) and lift the lower handguard assembly (3) up and out.

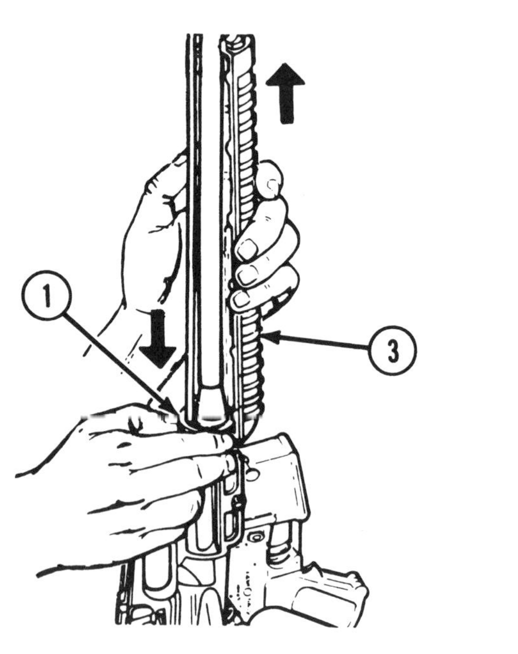

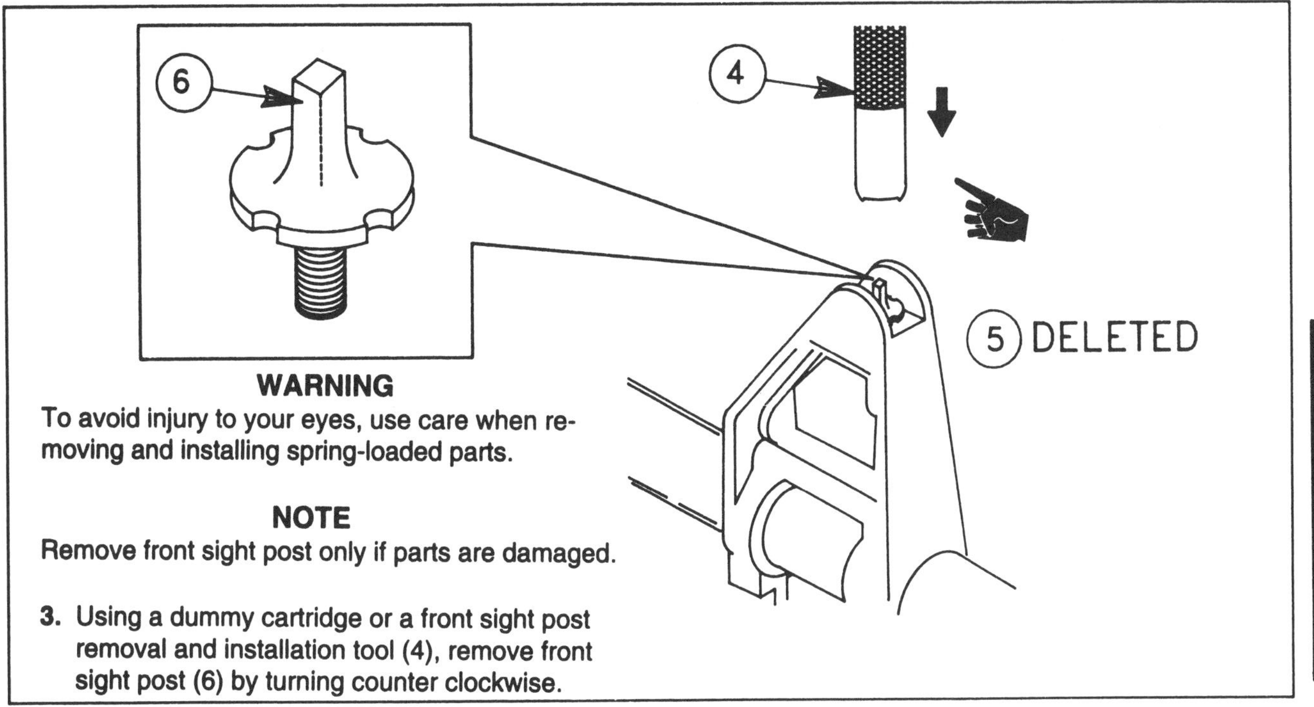

WARNING

To avoid injury to your eyes, use care when re-moving and installing spring-loaded parts.

NOTE

Remove front sight post only if parts are damaged.

3. Using a dummy cartridge or a front sight post removal and installation tool (4), remove front sight post (6) by turning counter clockwise.

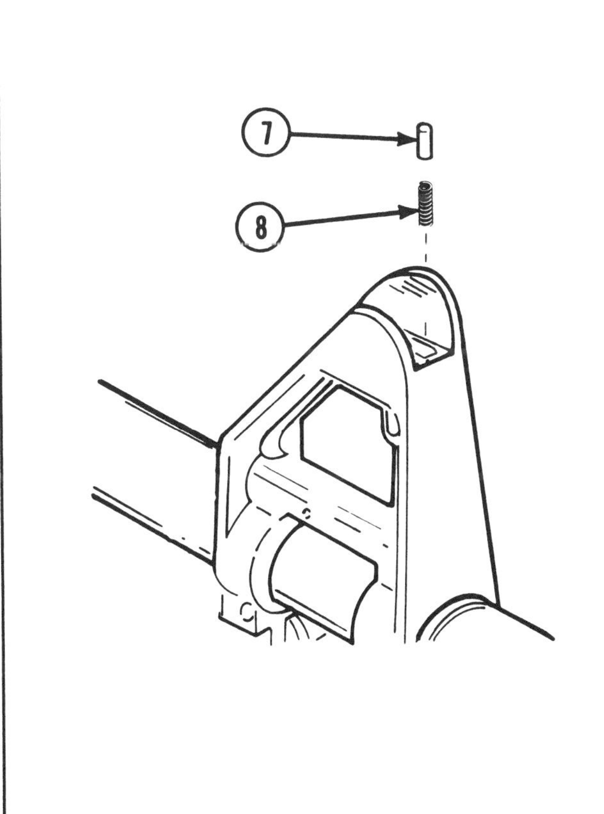

4. Catch front sight detent (7) and heli-cal spring (8) to prevent loss.

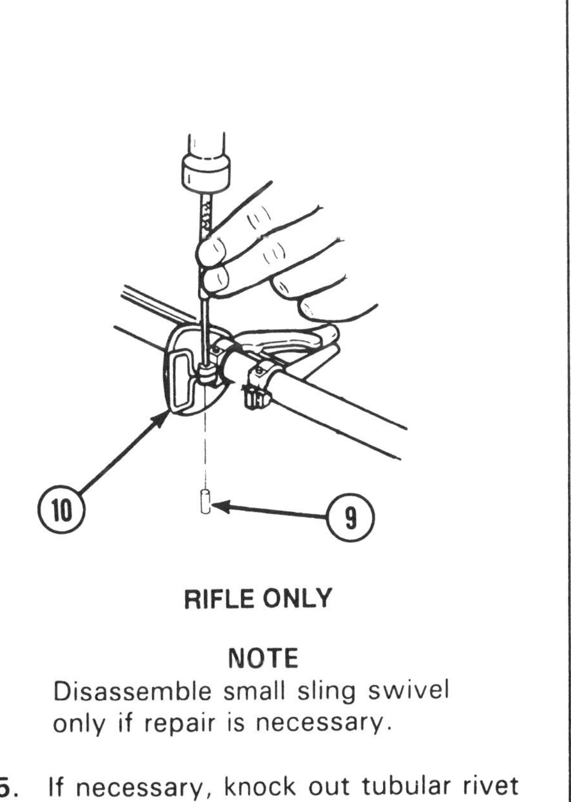

RIFLE ONLY

NOTE

Disassemble small sling swivel only if repair is necessary.

5. If necessary, knock out tubular rivet (9) with a hammer and punch and remove small sling swivel (10). Discard tubular rivet (9).

2-16. UPPER RECEIVE AND BARREL ASSEMBLY, RIFLE BARREL ASSEMBLY, AND UPPER RECEIVER ASSEMBLY (CONT).

a. DISASSEMBLY (CONT)

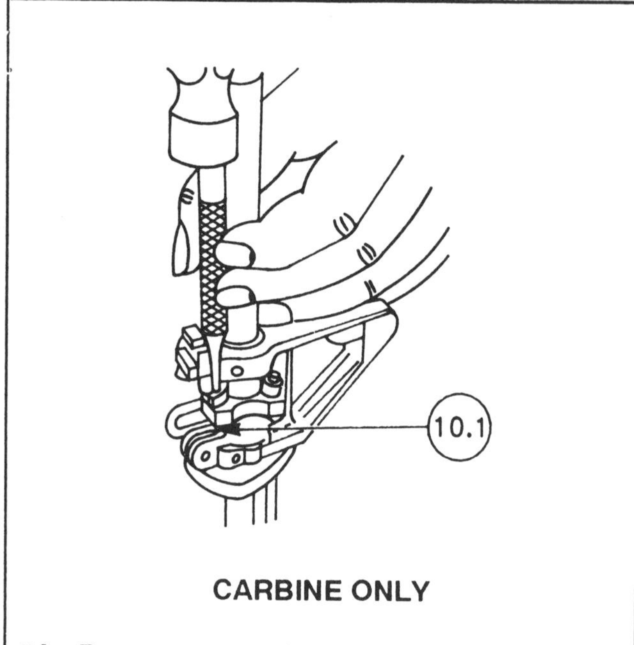

CARBINE ONLY

5A. Remove two spring pins (10.1).

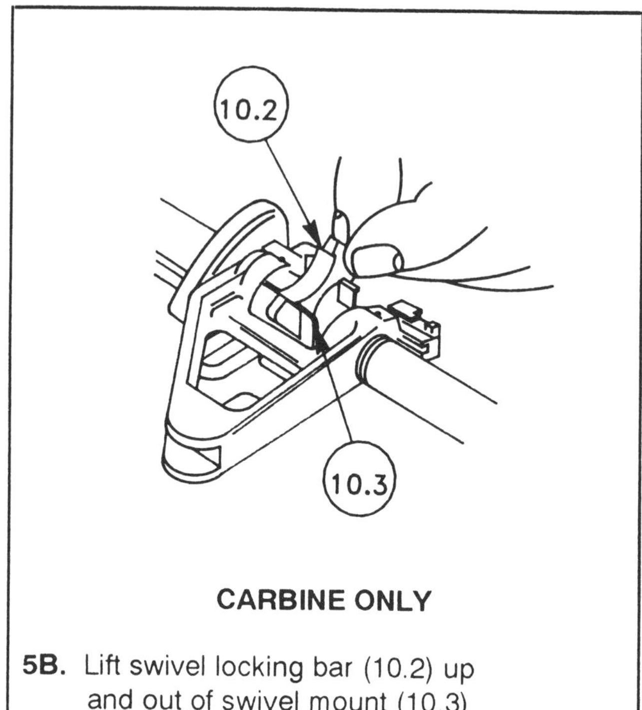

CARBINE ONLY

5B. Lift swivel locking bar (10.2) up and out of swivel mount (10.3)

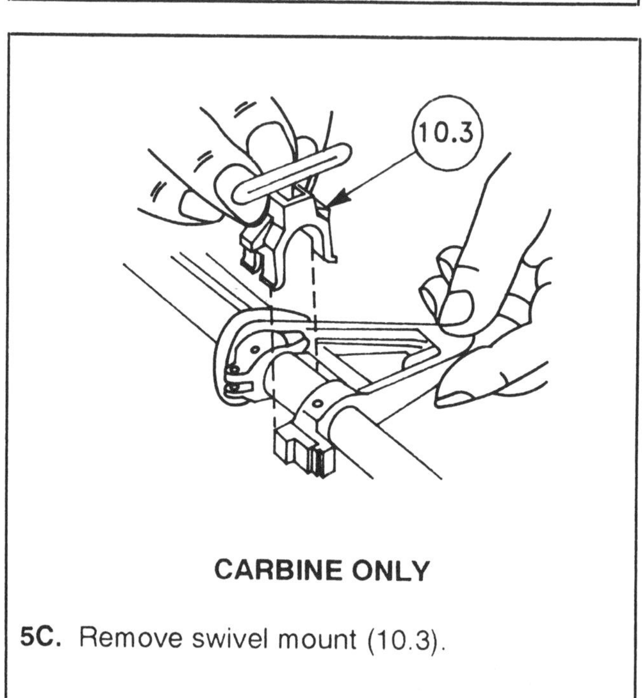

CARBINE ONLY

5C. Remove swivel mount (10.3).

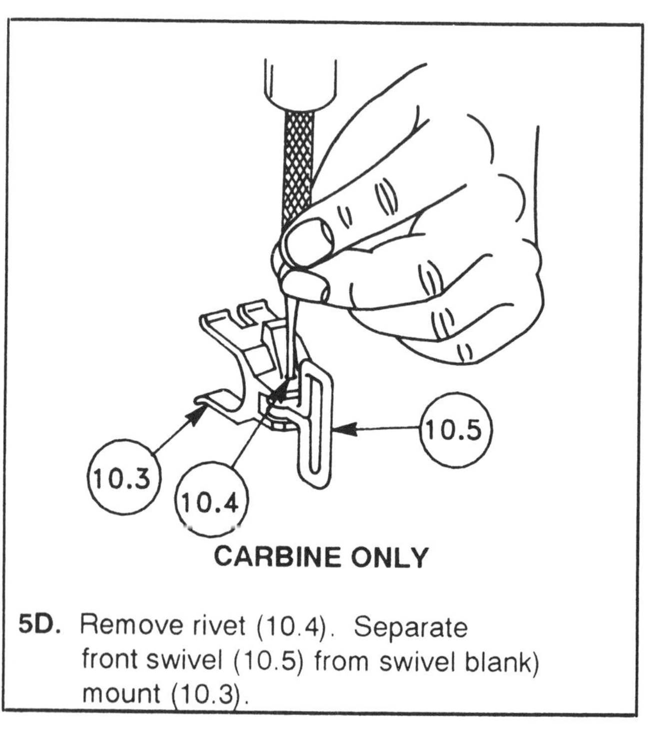

CARBINE ONLY

5D. Remove rivet (10.4). Separate front swivel (10.5) from swivel blank) mount (10.3).

2-16. UPPER RECEIVER AND BARREL ASSEMBLY, RIFLE BARREL ASSEMBLY, AND UPPER RECEIVER ASSEMBLY (CONT).

---

a. DISASSEMBLY (CONT)

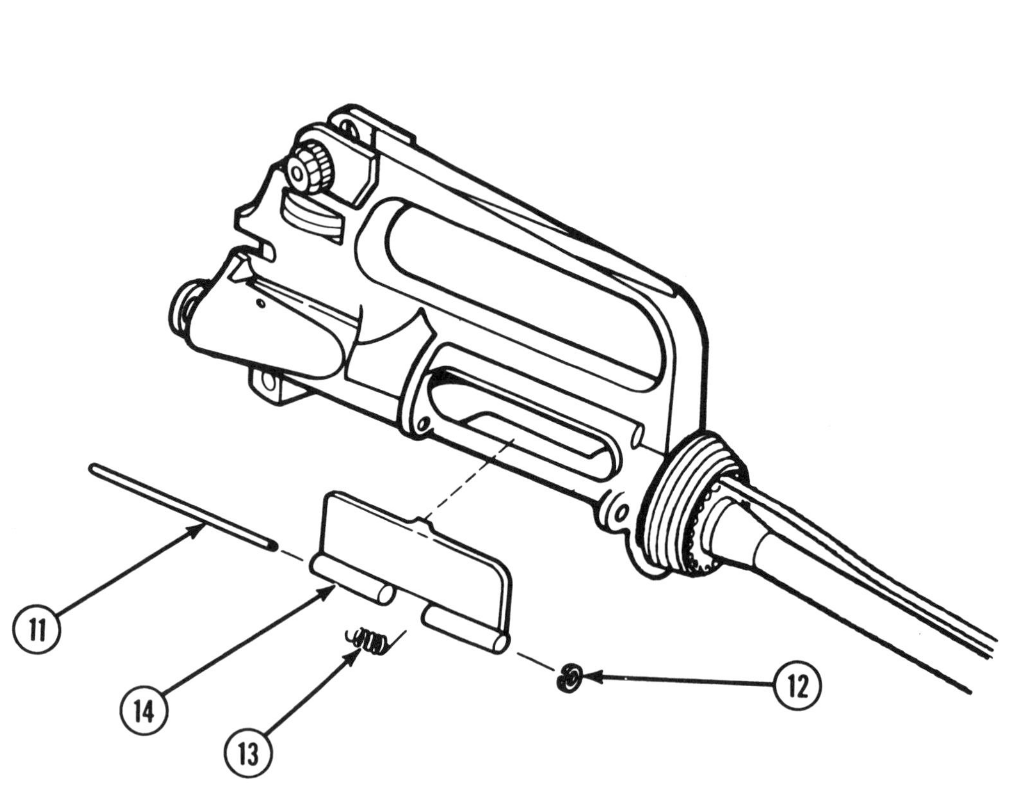

NOTE

Do not disassemble further unless repair is necessary. Headless grooved pin (11) may bind against the forward assist housing on the M16A2 rifle and re-quire some additional force to remove.

6. Remove retaining ring (12) and slide headless grooved pin (11) out to the rear.

7. Catch cover spring (13) and ejection port cover (14) to prevent loss as headless grooved pin is withdrawn.

---

b. CLEANING

Clean all items (operator's manual).

c. **INSPECTION**

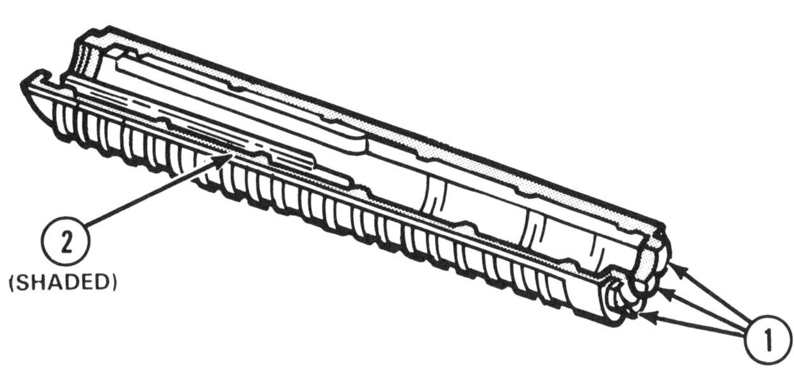

1. Inspect handguard assemblies for breaks, separation, and cracks using the following guidelines:

   a. Breaks and separations of material which prevent proper retention or interfere with functioning of the rifle will be cause for handguard assembly rejection and replacement.

   b. Each handguard assembly may have up to two of the three front retaining tabs (1) missing. If all three front retaining tabs are missing, handguard assembly must be replaced.

   c. Cracks are acceptable provided they do not extend into the retaining flange (CRITICAL AREA) (2).

   d. Replace severely cracked or damaged handguard assemblies. Handguard assemblies which have a heat shield which is loose enough to rattle when installed on the rifle must be replaced.

2. Inspect front sight assembly for chips, breaks, and cracks. Evacuate to support maintenance if broken, cracked, or bent.

3. Inspect front sight area for evidence of gas leakage around gas tube. Evacuate to support maintenance if short recoil results from gas leakage.

4. Inspect front sight post, front sight detent, and helical spring for damage. If damaged, replace.

2-16. UPPER RECEIVER AND BARREL ASSEMBLY, RIFLE BARREL ASSEMBLY, AND UPPER RECEIVER ASSEMBLY (CONT).

---

c. INSPECTION (CONT)

---

5. Inspect barrel for pits in bore, burrs, broken or worn locking lugs, and surface cracks and defects.

   a. Pits no wider than a land or groove and 3/8 inch (0.953 cm) or less in length are allowable in the bore.

   b. Uniformly fine pits in a densely pitted area of the bore are allowable.

   c. Lands that appear dark due to coating of gliding metal from projectiles are allowable.

   d. Striping of lands and grooves shall not be cause for rejection unless support maintenance determines by use of the barrel erosion gage.

   e. For pits other than mentioned above, broken or burred locking lugs, or surface cracks, evacuate to support maintenance.

6. Inspect bore for ringing. Definitely ringed bores or bores ringed sufficiently to bulge the outside surface of the barrel are causes for rejection. Evacuate to support maintenance.

7. Inspect chamber for pitting. Fine pits, or fine pits in a densely pitted area, are allowable. Pits 1/8 inch (0.318 cm) in length are cause for rejection. Evacuate to support maintenance.

8. Hand check compensator (3) for looseness on barrel. The third (middle) slot (4) must be straight up at top dead center (TDC). The alignment may vary as much as one half the width of the slot either direction. If loose or out of alignment, evacuate to support maintenance.

9. Inspect all items for serviceability in and tightness of latch assembly on ejection port cover. If items are damaged or nonfunctional, they are unserviceable.

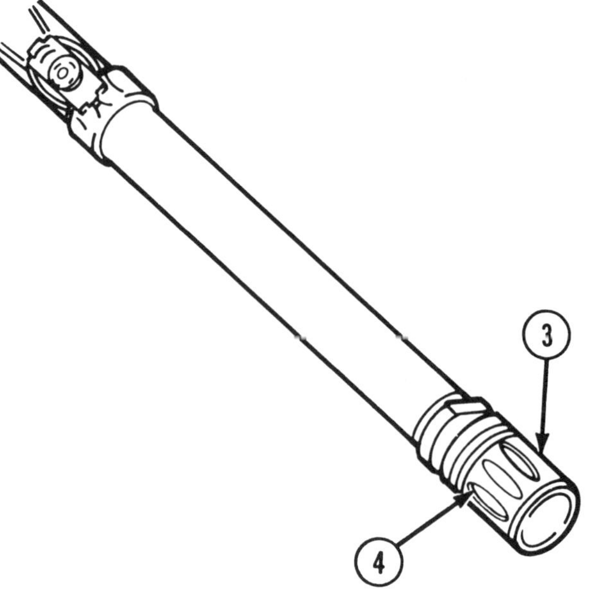

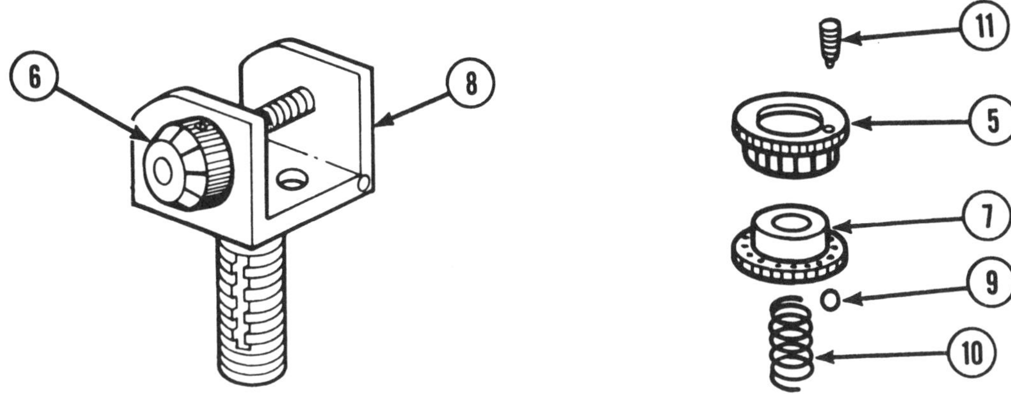

10. Rotate and test elevation index (5) and windage knob (6) for ease of functioning and legibility of markings.

11. Inspect elevation knob zero as follows:

   a. Rotate elevation knob (7) counterclockwise until the rear sight assembly is all the way down. If a whole click is not felt as the rear sight assembly stops, the rear sight assembly has bottomed out and will not pivot freely.

   b. Position elevation knob back slightly to its last whole click so the rear sight assembly base (8) is under tension of the ball bearing (9) and helical spring (10). The 300 meter mark should align with the mark on the receiver.

   c. If the 300 meter mark is not aligned with the mark on receiver, slip the range scale in the following manner:

      (1) Position the 300 meter mark with the mark on the receiver.

      (2) Insert a 1/16 inch allen wrench through the access hole of the rear sight assembly base and into the index screw (11).

      (3) Loosen the index screw three turns and leave the wrench in place.

      (4) Rotate lower portion of elevation knob counterclockwise until it stops (range scale should not have moved). Elevation knob should be positioned on its last whole click.

      (5) Tighten index screw and remove wrench.

      (6) Check for proper setting.

2-16. UPPER RECEIVER AND BARREL ASSEMBLY, RIFLE BARREL ASSEMBLY, AND UPPER RECEIVER ASSEMBLY (CONT).

---

d. REPAIR

1. Replace all authorized unserviceable parts.

2. Evacuate to support maintenance.

e. LUBRICATION

Lightly lubricate all items (p 2-31) (operator's manual).

f. REASSEMBLY

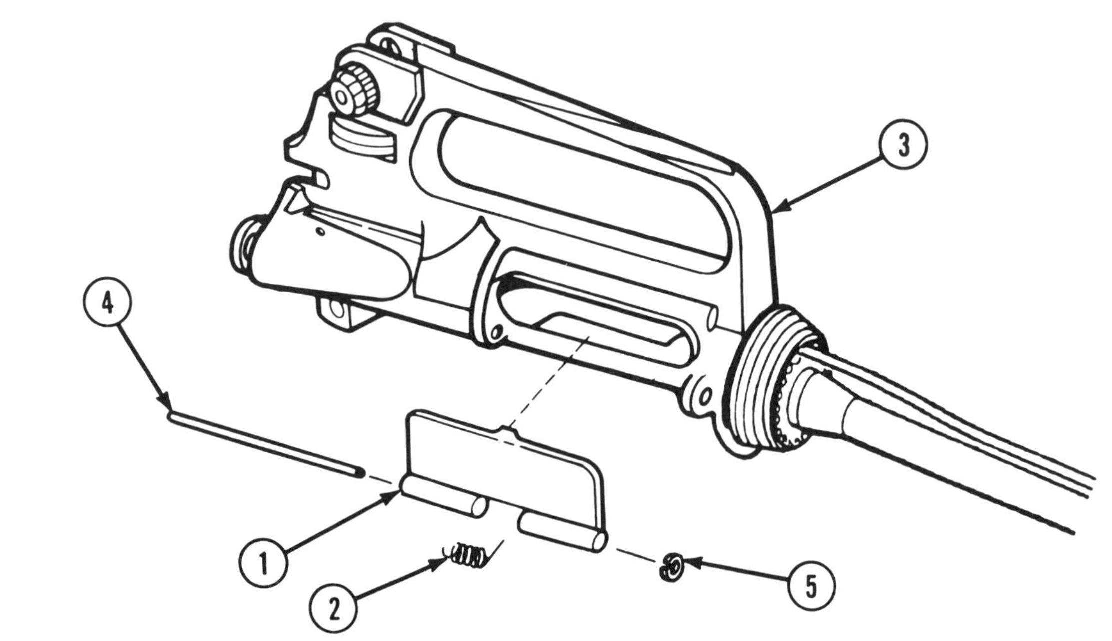

1. If previously disassembled, position ejection port cover (1) and helical spring (2) on upper receiver (3) with short leg of helical spring to the rear on inside of ejection port cover.

NOTE
Long legs of helical spring must be positioned and pretensioned before the headless grooved pin is installed.

2. Hold helical spring (2) short leg in this position and turn long leg one half turn (180 degrees) with fingers of right hand.

3. Position long leg of helical spring (2) against ejection port cover (1). Hold helical spring and ejection port cover in this position and install headless grooved pin (4). Check for proper spring tension during installation of retaining ring (5).

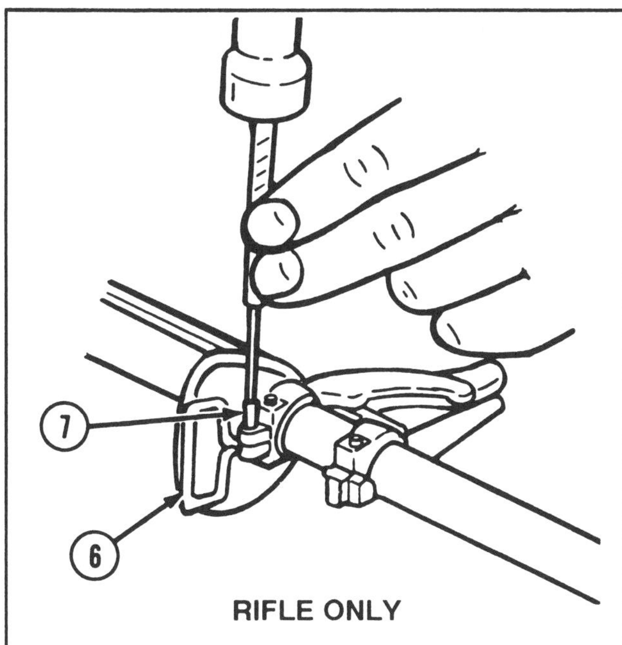

RIFLE ONLY

4. If previously disassembled, position small sling swivel (6) and install new tubular rivet (7) using center punch and hammer to spread and flare the hollow head of the tubular rivet.

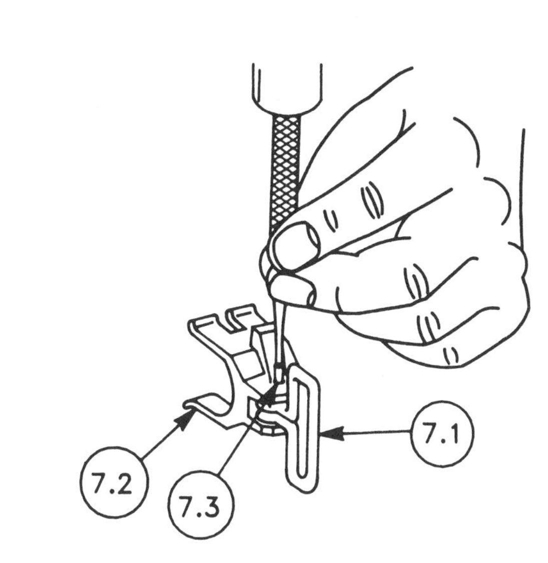

CARBINE ONLY

4A. Install front swivel (7.1) to swivel mount (7.2) with new rivet (7.3) using center punch and hammer to spread and flare the hollow head of the tubular rivet.

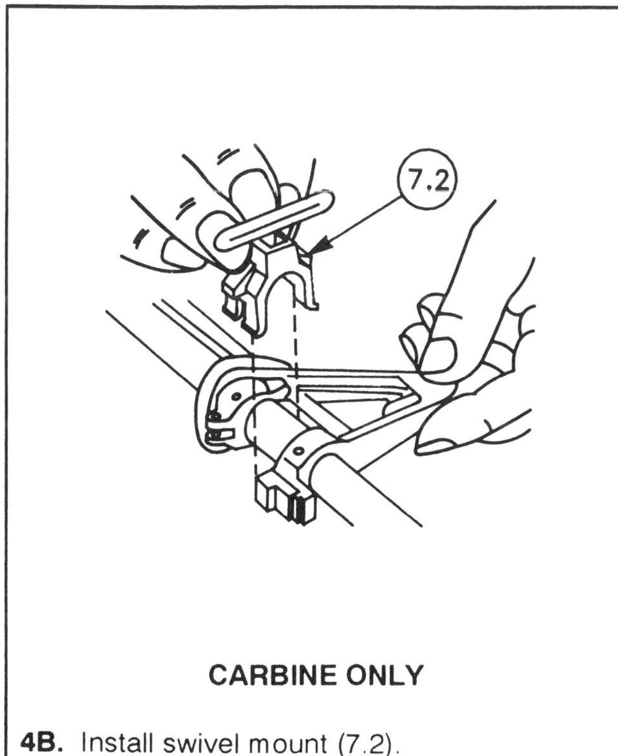

CARBINE ONLY

4B. Install swivel mount (7.2).

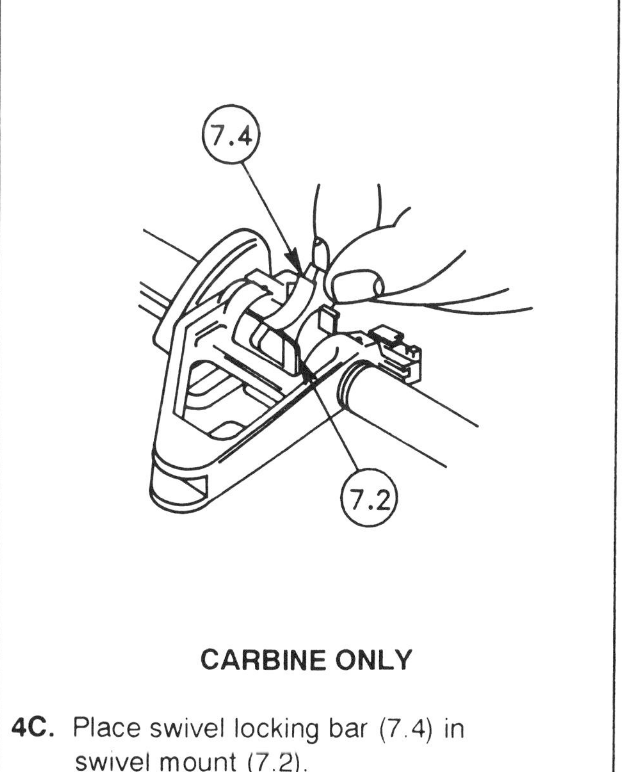

CARBINE ONLY

4C. Place swivel locking bar (7.4) in swivel mount (7.2).

2-16. UPPER RECEIVER AND BARREL ASSEMBLY, RIFLE BARREL ASSEMBLY, AND UPPER RECEIVER ASSEMBLY (CONT).

f. REASSEMBLY (CONT)

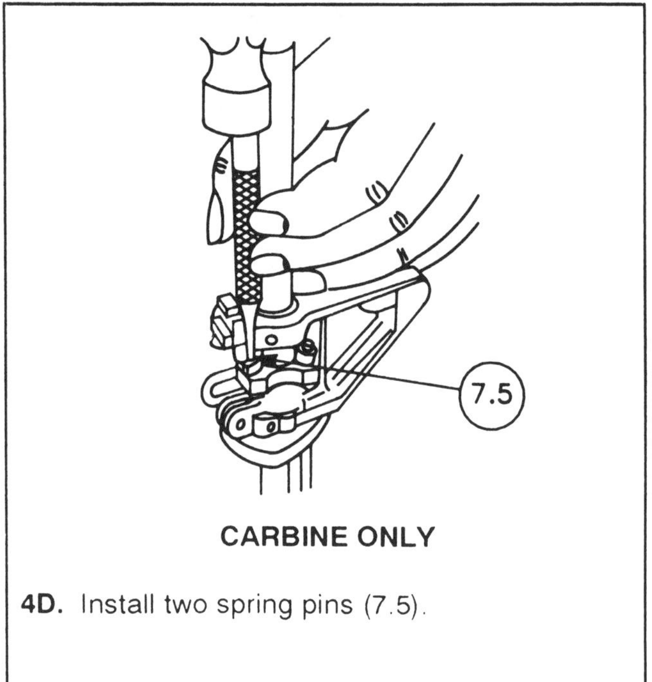

CARBINE ONLY

4D. Install two spring pins (7.5).

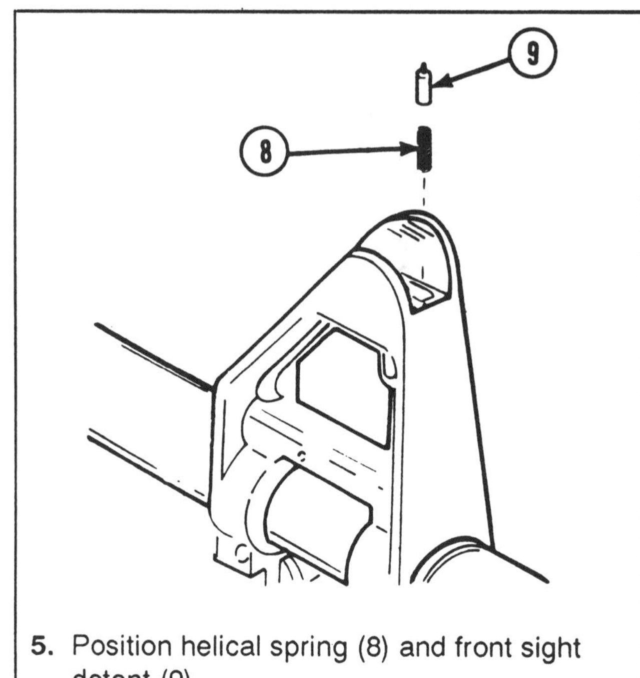

5. Position helical spring (8) and front sight detent (9).

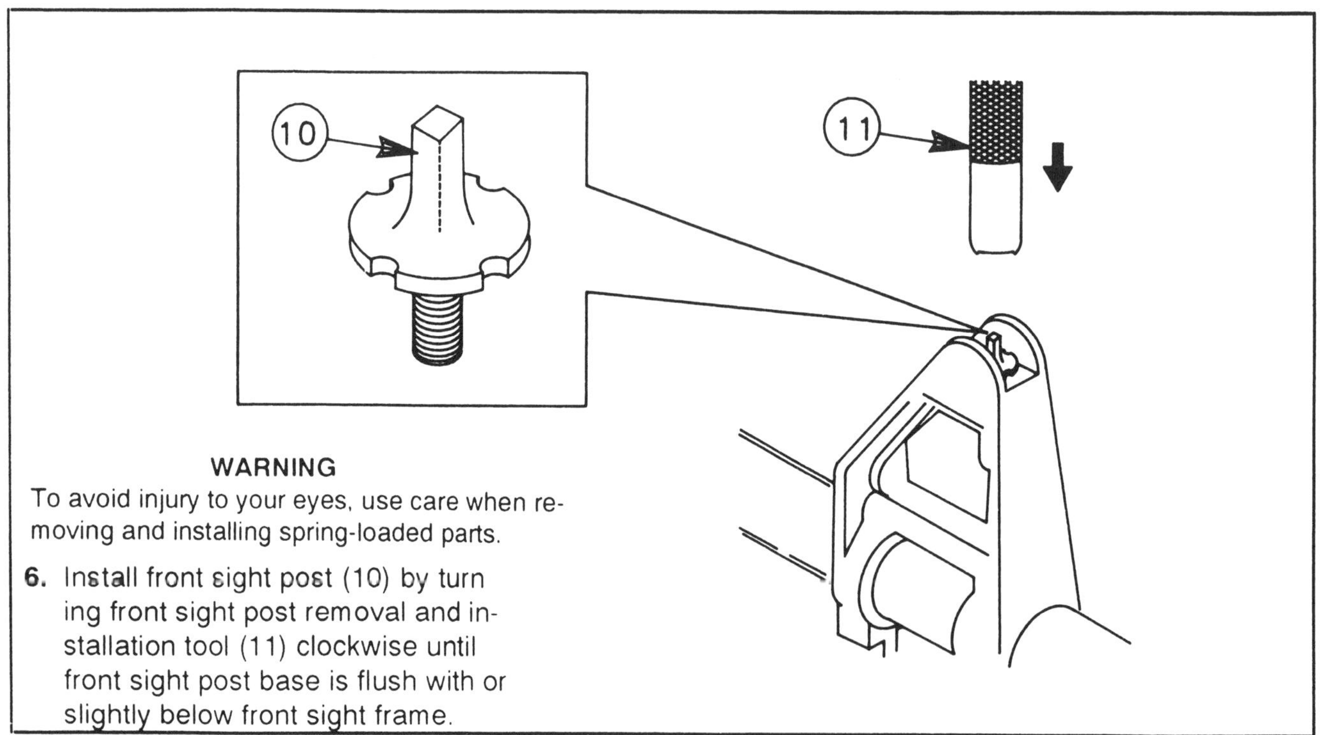

WARNING

To avoid injury to your eyes, use care when removing and installing spring-loaded parts.

6. Install front sight post (10) by turning front sight post removal and installation tool (11) clockwise until front sight post base is flush with or slightly below front sight frame.

2-16. UPPER RECEIVER AND BARREL ASSEMBLY, RIFLE BARREL ASSEMBLY, AND UPPER RECEIVER ASSEMBLY (CONT).

f. REASSEMBLY (CONT)

7. Mechanical Zero Procedures (A.F. Only).

a. Mark a piece of plastic card stock or rigid paper with lines from 1 to 5mm in 1mm increments. Set the card on the front sight frame and check the height of the top of the front sight post.

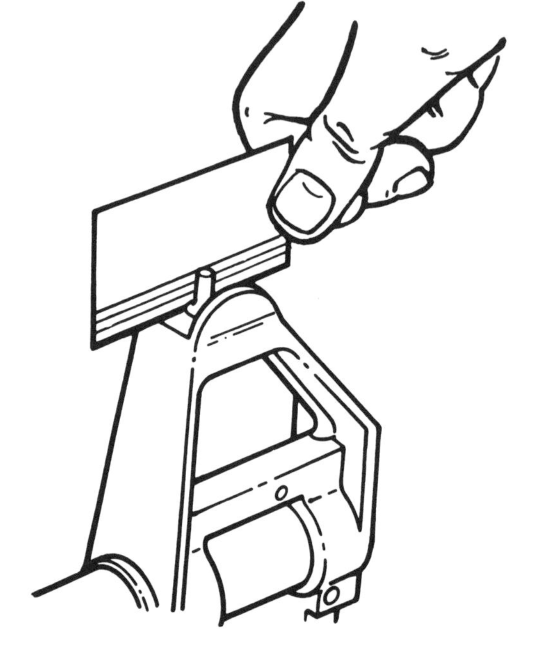

b. Adjust the front sight so the top of the front sight post is 5mm above the machined surfaces of the front sight frame.

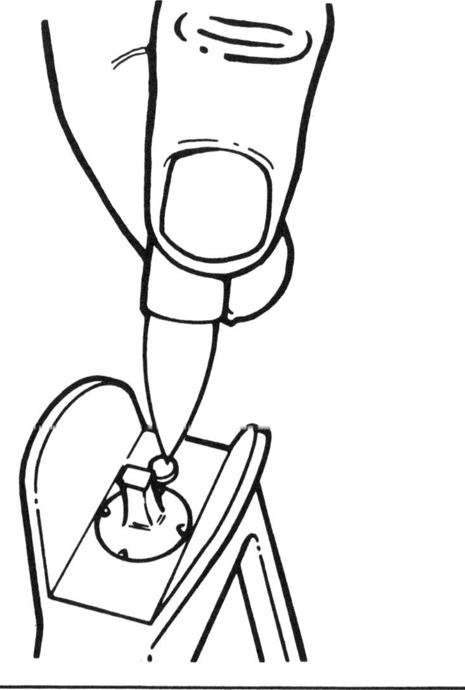

c. Visually check the front sight post top height by using the marked plastic or paper card. Card must set level on the machined surfaces of the front sight frame to obtain an accurate reading.

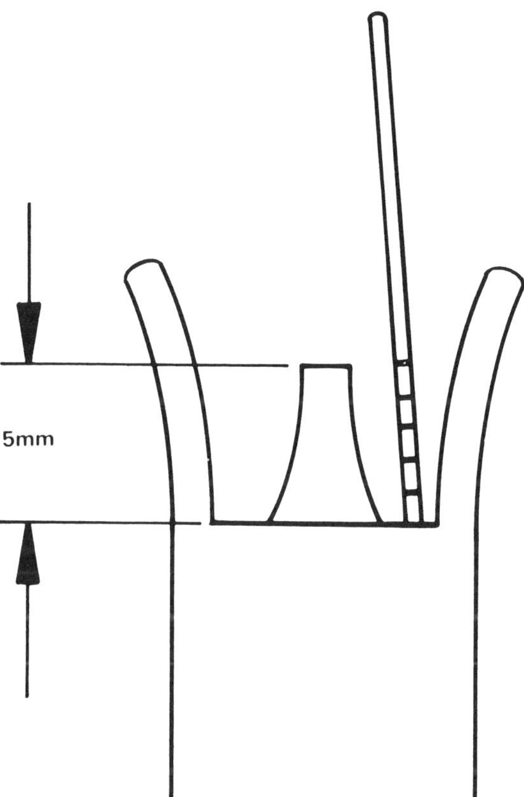

NOTE

This procedure, when used in conjunction with rear sight mechanical zero adjustment, will give an approximate battle sight zero to most M16A2 rifles. The above steps can also be used before firing a new or newly assigned rifle. Use the procedures to check rifles stored in preferred packaging during routine inspections. This will help ensure people armed with the rifles will stand a better chance of hitting an enemy if the rifles must be used before a live fire zero can be made. Whenever possible, zeroing of the rifle should be accomplished using ball ammunition on a 25 meter zeroing target using the ''L'' aperture.

2-16. UPPER RECEIVER AND BARREL ASSEMBLY, RIFLE BARREL ASSEMBLY, AND UPPER RECEIVER ASSEMBLY (CONT).

---

f. REASSEMBLY (CONT)

NOTE

Refer to operator's manual for "buddy system" procedure on installing hand-guard assemblies.

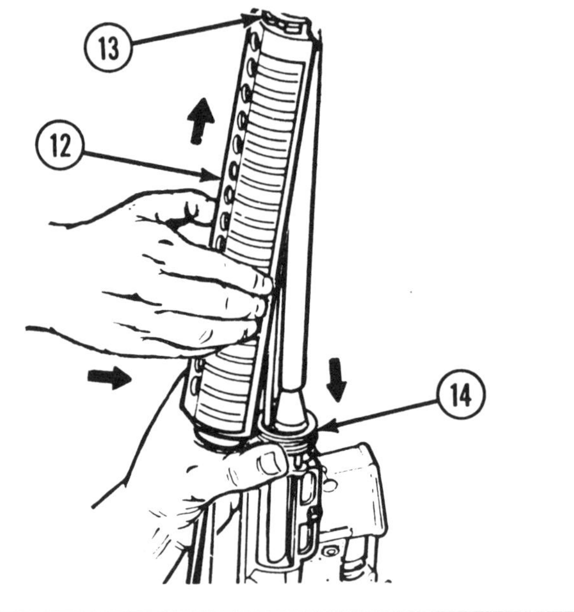

8. Install top of upper handguard assembly (12) in tube cap (13) while pushing down on handguard slip ring (14). Push bottom of upper handguard assembly (12) in place and release handguard slip ring (14) to lock handguard assembly in place.

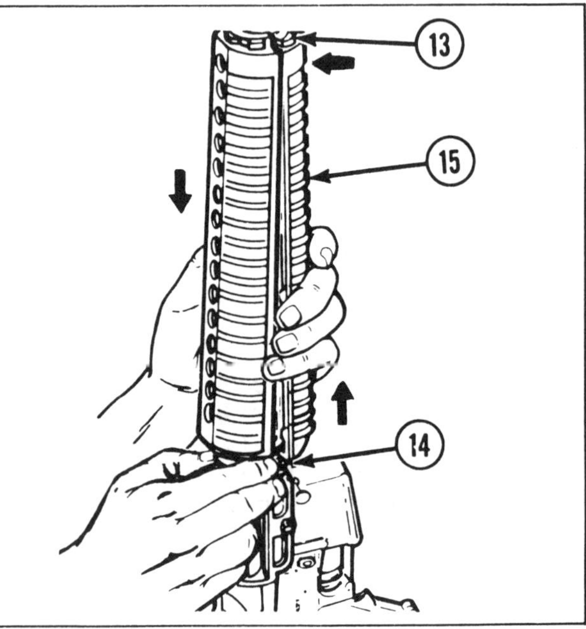

9. Install top of lower handguard assembly (15) in tube cap (13) while pushing down on handguard slip ring (14). Push bottom of lower handguard assembly (15) in place and release handguard slip ring (14) to lock both handguard assemblies in place.

10. Reassemble rifle, refer to page 2-68.

2-17. LOWER RECEIVER AND BUTTSTOCK ASSEMBLY.

This task covers:

a. Disassembly
b. Cleaning
c. Inspection

d. Repair
e. Lubrication
f. Reassembly

INITIAL SETUP

Tools
(ARMY) Small Arms Repairman Tool Kit
(item 3, app B)
Pivot Pin Removal Tool (fig. E-3, app E)
Pivot Pin Installation Tool (fig. E-6, app E)

Materials/Parts
Lubricant, solid film (SFL) (item 21,
app D)
Screw, self-locking (item 6, p C-11)

References
TM 9-1005-319-10 (operator's manual)

Equipment Conditions
2-34   Lower receiver and buttstock assembly removed

General Safety Instructions
To avoid injury to your eyes, use care when removing and installing spring-loaded parts.
When using solid film lubricant or dichloromethane, be sure the area is well ventilated.

a. DISASSEMBLY

1. Remove screw (1) and lockwasher (2).

WARNING
To avoid injury to your eyes, use care when removing and installing spring-loaded parts.

2. Carefully remove pistol grip (3) and catch helical spring (4) and safety detent (5) to prevent loss.

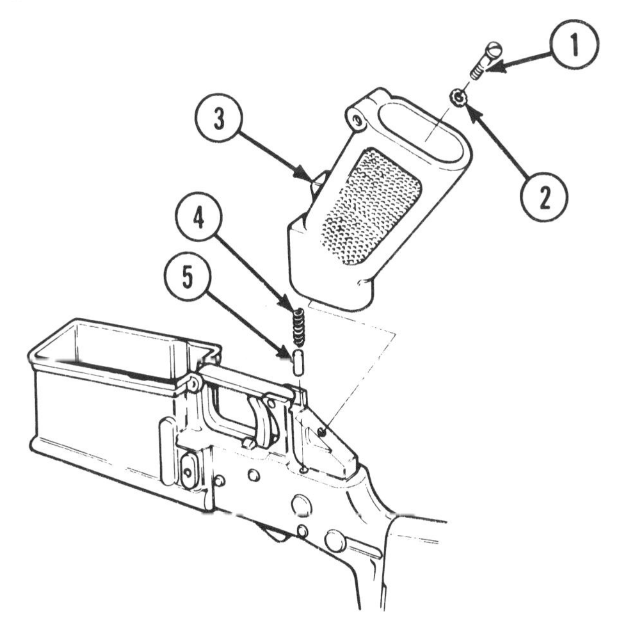

2-17. LOWER RECEIVER AND BUTTSTOCK ASSEMBLY (CONT).

a. DISASSEMBLY (CONT)

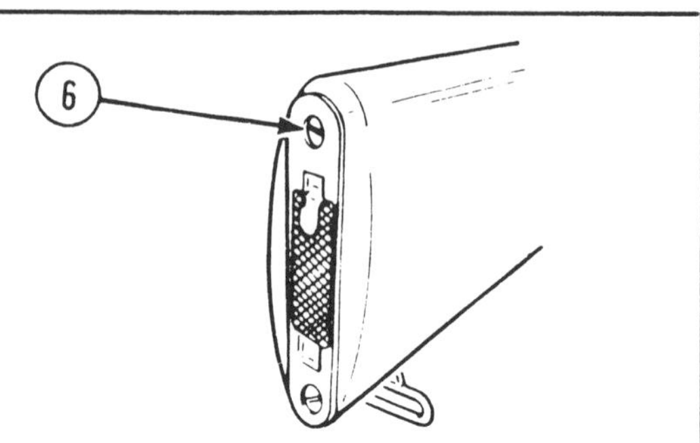

RIFLE ONLY

NOTE
If the self-locking screw is removed, it must be discared and
replaced with a new one.

3. Remove self-locking screw (6).

RIFLE ONLY

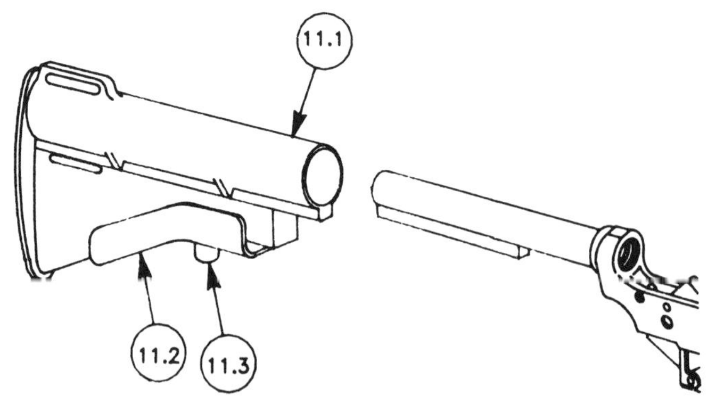

WARNING
To avoid injury to your eyes, use
care when removing and installing
spring-loaded parts.

4. Remove buttstock assembly (7) carefully and catch helical spring (8), takedown pin detent (9), takedown pin
(10), and stepped spacer (11) to prevent loss.

NOTE
If detent (9) will not come
out, use a wire to push It out.

CARBINE ONLY

4A. Extend buttstock assembly (11.1).

4B. Grasp the lock release lever (11.2) in
the area of the retaining nut (11.3), pull
downward, and slide buttstock to the
rear to separate the buttstock assembly
from the lower receiver extension.

WARNING

To avoid injury to your eyes, use care when removing and installing spring-loaded parts.

NOTE

Catch pivot pin detent and helical spring as pivot pin is removed (see step 6 on next page).

5. Depress pivot pin detent and spring and remove pin or insert fabricated pivot pin removal tool (12) to compress pivot pin detent. Turn pivot pin (13) a quarter turn. Remove tool and pivot pin.

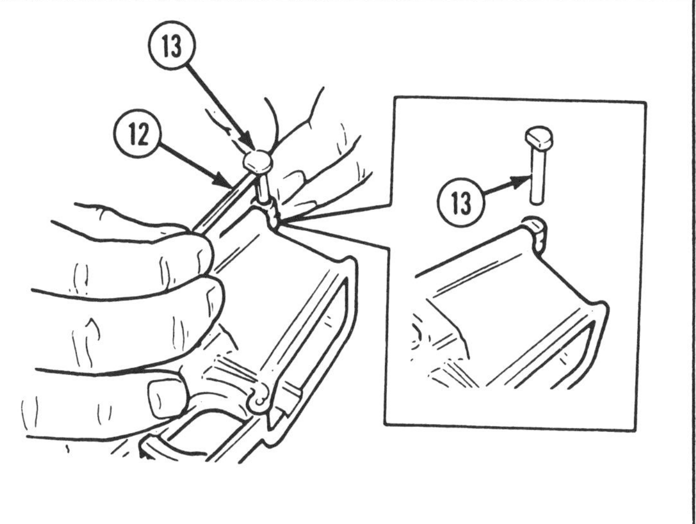

6. Be sure to hold cupped hand in front of pivot pin detent (14) and helical spring (15) to prevent loss of pivot pin detent and helical spring.

NOTE

If helical spring (15) will not come out, use a wire to pull it out.

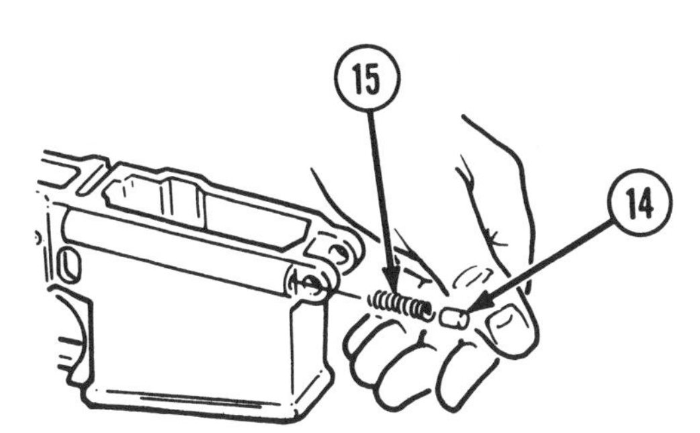

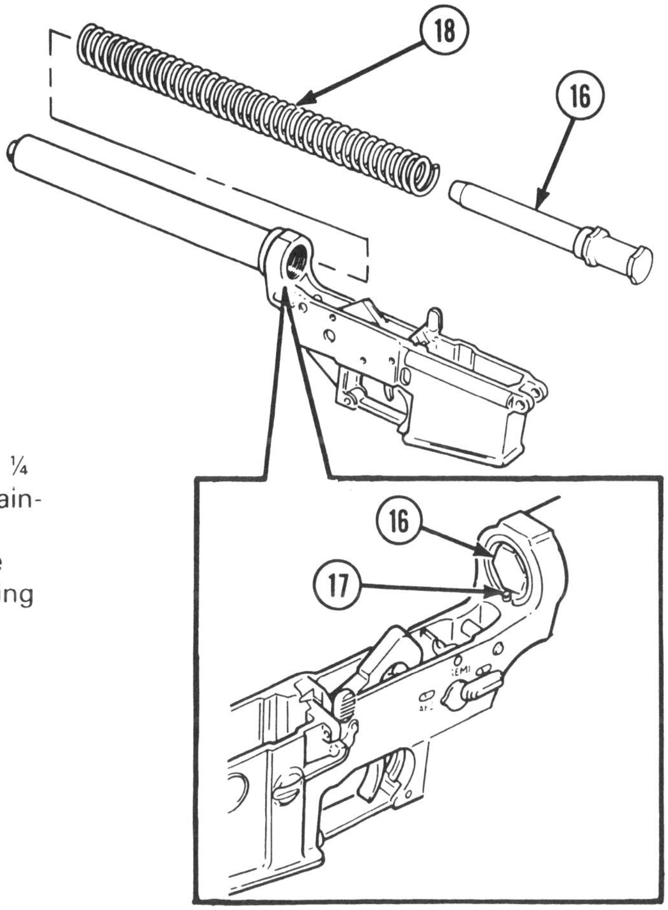

NOTE

Make sure the hammer is cocked and the selector lever is not set on BURST before removing the buffer assembly.

7. Press buffer assembly (16) in about ¼ inch (0.635 cm). Depress buffer retainer (17) and release buffer assembly (16) and action spring (18). Remove buffer assembly (16) and action spring (18) from receiver while depressing buffer retainer (17).

---

b. CLEANING

Clean all items (operator's manual). Remove carbon deposits.

2-17. LOWER RECEIVER AND BUTTSTOCK ASSEMBLY (CONT).

c. INSPECTION

1. Inspect buffer assembly (1, 2, 3, or 3.1) for cracks or damage.

   RIFLE ONLY
   a. Some old buffers (1) have a hole with pin installed which protrudes equally on each side approximately 1/32 inch (0.08 cm).

   b. Some buffers (2) have a hole in the housing but no pin.

   c. New buffers (3) do not have a hole in buffer body or a pin.

   BOTH WEAPONS
2. If cracked or damaged, replace.

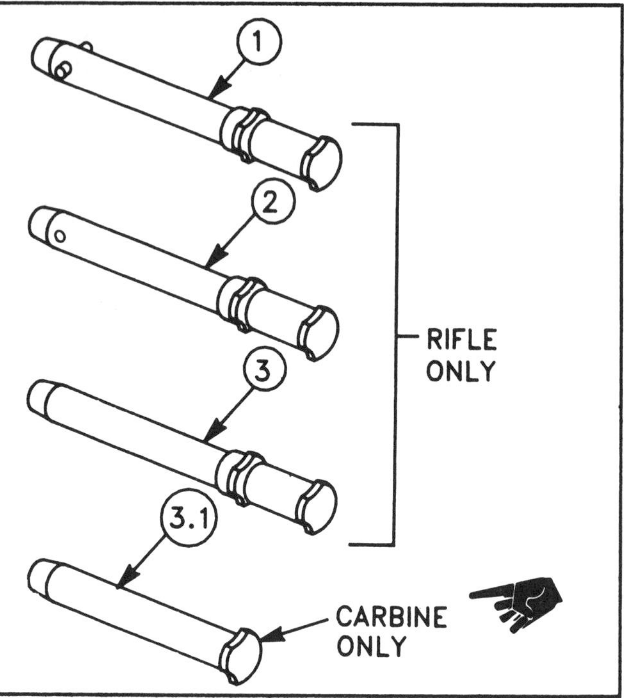

3. Check free length of action spring (4). The free length **FOR RIFLE ONLY** must be between 11 3/4 inches (29.85 cm) minimum and 13 1/2 inches (34.29 cm) maximum; **FOR CARBINE ONLY** must be between 10 1/16 inches (25.56 cm) minimum and 11 1/4 inches (28.58 cm) maximum; if not, replace. Do not attempt to adjust the length by stretching the action spring.

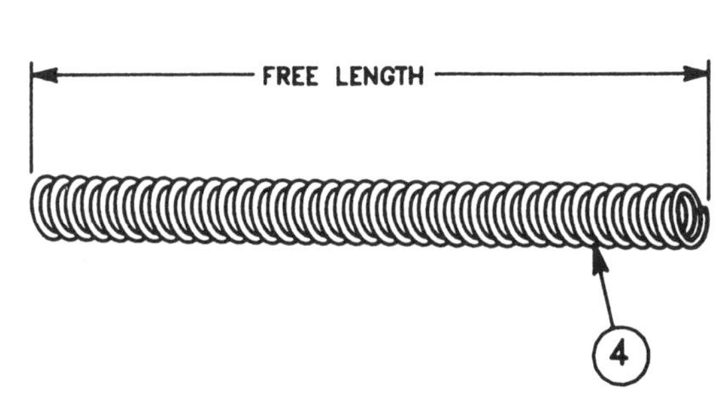

4. Inspect lower receiver (5) (without further disassembly) for legibility of serial number. **ARMY ONLY:** If the serial number is hard to read, evacuate to direct support maintenance. **AIR FORCE ONLY:** If the serial number is hard to read, evacuate to depot maintenance.

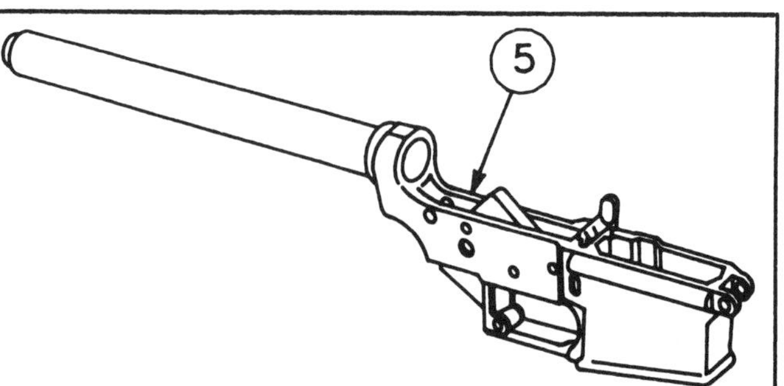

2-61

NOTE

AIR FORCE ONLY: Only depot maintenance is authorized to restamp the serial number.

5. Inspect for missing or damaged parts. Inspect finish of lower receiver for shiny spots. Touch up with solid film lubricant as required (p 2-34).

NOTE

If a M16A2 rifle lower receiver is missing one third or more of its exterior protective finish, resulting in an unprotected, light reflecting surface, it is candidate for overhaul. This missing finish will be considered a shortcoming. This shortcoming requires action to obtain a replacement rifle. Once a replacement has been received, evacuate the original rifle to depot maintenance for overhaul.

d. REPAIR

Replace all authorized unserviceable parts. If repair is not authorized at this level, evacuate to support maintenance.

e. LUBRICATION

Lightly lubricate all metal components (p 2-33) (operator's manual).

2-17. LOWER RECEIVER AND BUTTSTOCK ASSEMBLY (CONT).

f. REASSEMBLY

WARNING

To avoid injury to your eyes, use care when removing and installing spring-loaded parts.

NOTE

Make sure the hammer is cocked and the selector lever is not set on BURST before installing the buffer assembly.

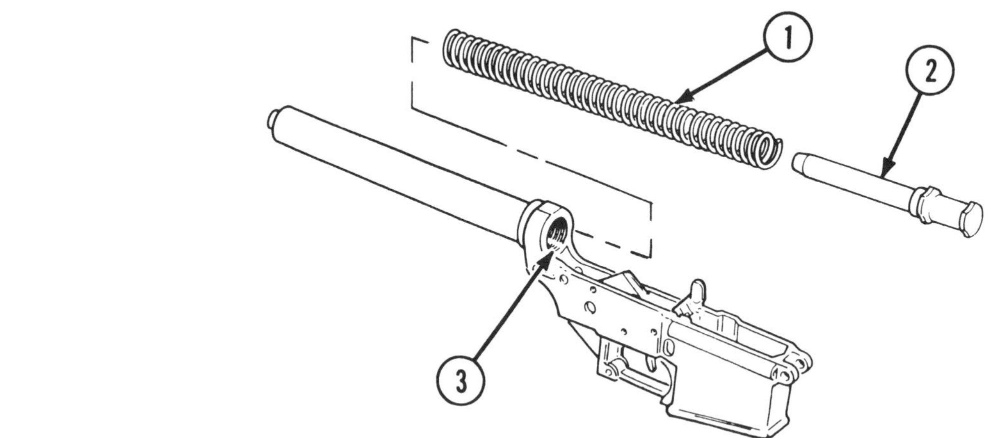

1. Press action spring (1) and buffer assembly (2) in until buffer retainer (3) snaps up and holds them in place.

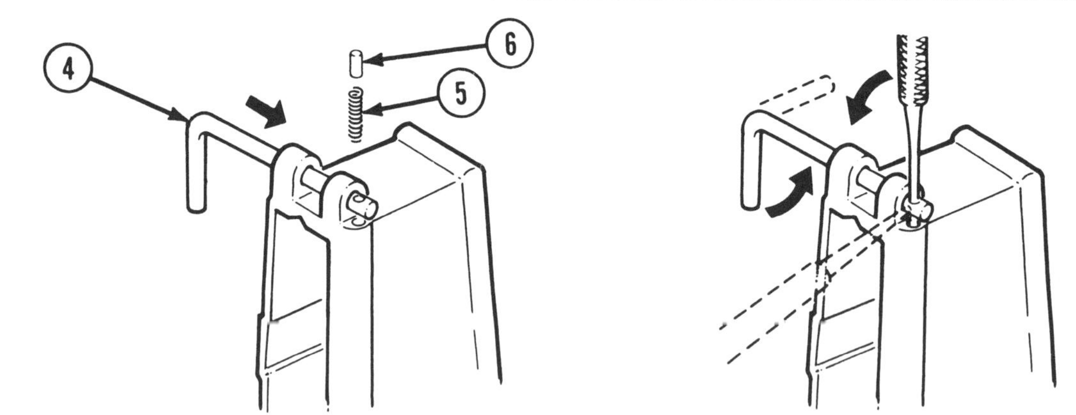

2. Install fabricated pivot pin installation tool (4). Insert helical spring (5) and pivot pin detent (6). Compress pivot pin detent in recess with 3/32 inch punch and rotate tool. Remove 3/32 inch punch.

NOTE
Rounded end of pivot pin detent must be in the groove of the pivot pin (7) when assembly is complete.

3. Install pivot pin (7) while removing fabricated pivot pin installation tool (4). Maintain pressure while sliding pivot pin (7) into hole. Rotate pivot pin until pivot pin detent is inserted into pivot pin groove.

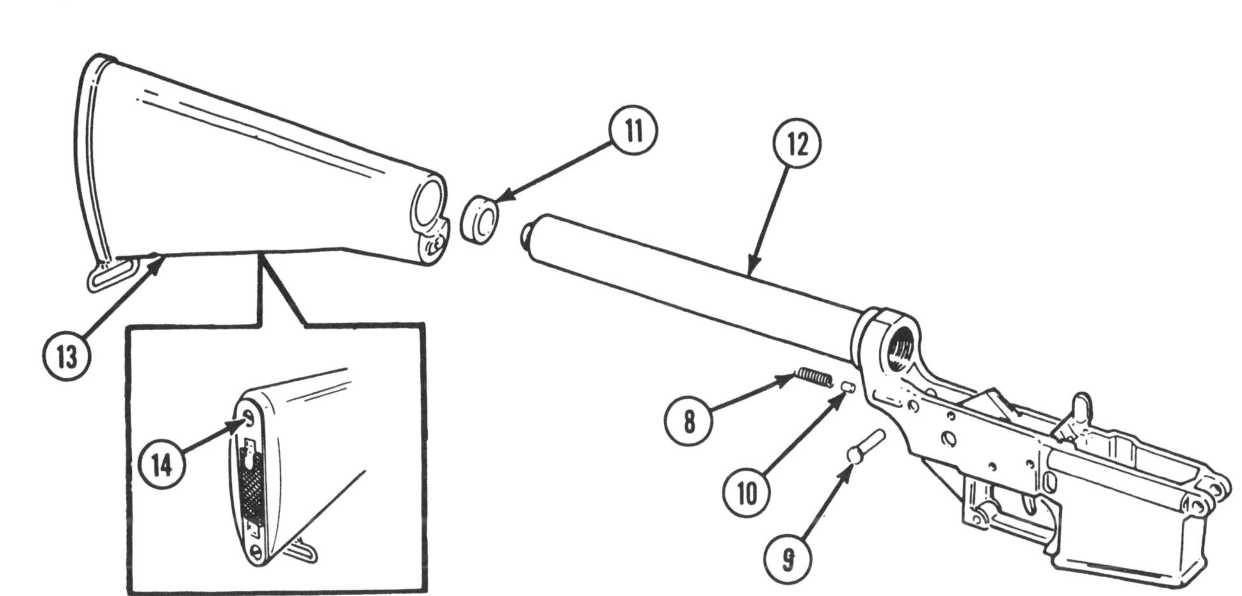

RIFLE ONLY

CAUTION
Do not kink helical spring (8) during assembly.

4. Install takedown pin (9) with groove toward the rear. Install takedown pin detent (10) and helical spring (8) from the rear.

5. Install stepped spacer (11) on receiver extension (12) and carefully slide buttstock assembly (13) on to compress helical spring (8).

NOTE
Self-locking screw (14), if removed, must be discarded and replaced with a new one.

6. Install self-locking screw (14) to secure buttstock assembly (13).

2-17. LOWER RECEIVER AND BUTTSTOCK ASSEMBLY (CONT).

f. REASSEMBLY (CONT)

CARBINE ONLY

6A. Grasp the lock release lever (14.1) in
the area of the retaining nut (14.2) and
pull to reinstall the buttstock assembly
(14.3) onto the lower receiver extension
(14.3) onto the lower receiver extension
(14.4).

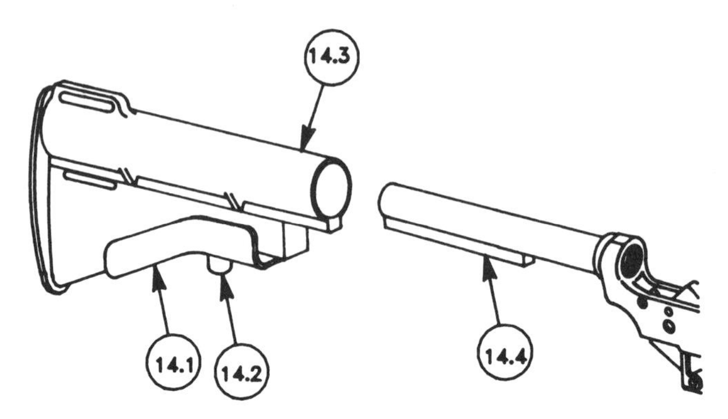

2-17. LOWER RECEIVER AND BUTTSTOCK ASSEMBLY (CONT).

f. REASSEMBLY (CONT)

WARNING

When utilizing the enhanced rifle grip (it has a bump between the second and third finger for a better grip) PN 9349127, rifle grip screw, PN AN 501D416-18 (1-1/8 in.) or AN501D416-16 (1 in.), is authorized to be used with the enhanced grip. Any screw longer than 1-1/8 in. used with 9349127 could cause a hazardous situation. Also, ensure the washer is in place.

CAUTION

Do not kink helical spring (15) during assembly.

7. Install safety detent (16), pointed end first, and helical spring (15) into bottom of lower receiver (17).

NOTE

A portion of the helical spring will fit in a hole in the pistol grip.

8. Carefully install pistol grip (18) to compress helical spring (15). Secure pistol grip (18) in place with lockwasher (19) and screw (20).

9. Reassemble rifle, refer to page 2-68.

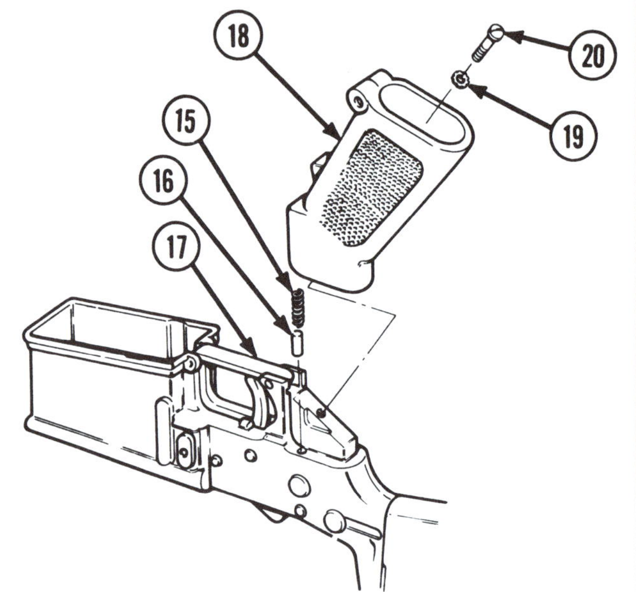

2-18. BUTTSTOCK ASSEMBLY.

This task covers:

a. Disassembly	**d.** Repair
b. Cleaning	**e.** Lubrication
c. Inspection	**f.** Reassembly

INITIAL SETUP

Tools
   (ARMY) Small Arms Repairman Tool Kit
   (item 3, app B)

Equipment Conditions
   2-57 Buttstock assembly removed from lower receiver and buttstock assembly

a. **DISASSEMBLY**

RIFLE ONLY

1. Remove self-locking screw (1), small sling swivel (2), and buttplate group (3) from buttstock (4).

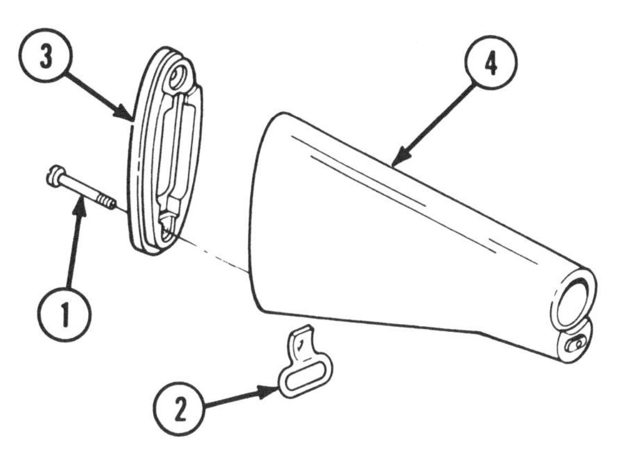

RIFLE ONLY

2. Push down on plunger (5) and lift door assembly (6) out of buttplate (7).

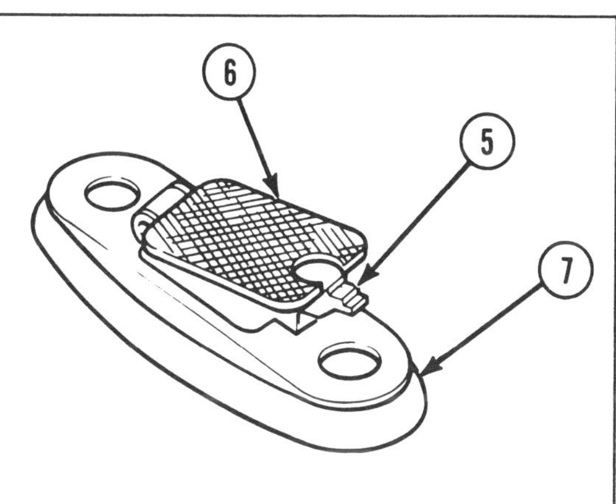

RIFLE ONLY

3. Remove straight pin (8) and separate hinge (9) and door assembly (6).

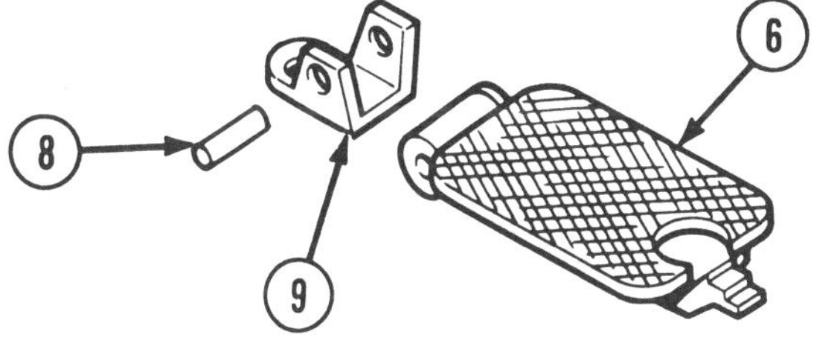

2-18. BUTTSTOCK ASSEMBLY (CONT).

a. DISASSEMBLY (CONT)

CARBINE ONLY

4. Disassemble the buttstock assembly
    by tapping out the spring pin (10) located in
    the oval slot of the retaining nut (11). This
    is done using a 1/16 inch punch.

5. Insert your index finger into the forward
    end of the buttstock (12) and push down on
    the locking pin (13). Unscrew the retaining nut
    (11) and remove the release lever (14), locking
    pin (13) and spring (15).

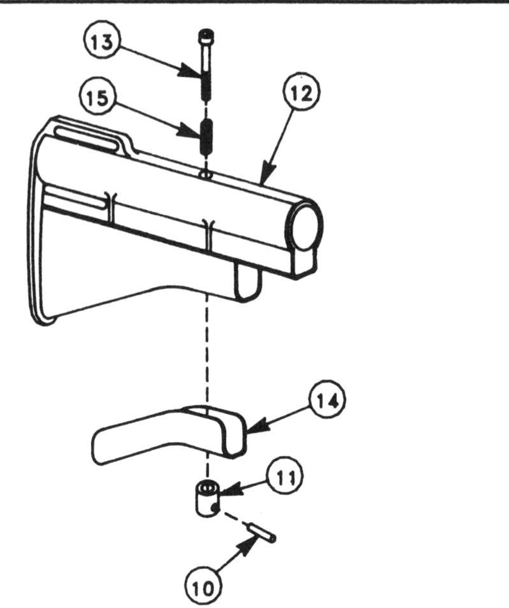

b. CLEANING (RIFLE ONLY)

Clean all parts with CLP (operator's manual). Use brush to clean knurled surface of door assembly.

2-18. BUTTSTOCK ASSEMBLY (CONT).

c. INSPECTION

NOTE

M16A2 buttstocks, PN 9349121, with unauthorized markings may be used under the following conditions:

a. The only authorized markings are those which are temporary in nature; i.e., paint, tape, etc.

b. When marking a buttstock, only use temporary markings.

c. Buttstocks with unauthorized markings that have been stamped into the surface of the buttstock will not be used.

d. Unauthorized markings that have previously been scratched, etched, carved, etc. may continue in use if the marks do not extend into the fiber of the buttstock. Cutting into the fiber of the buttstock may weaken it.

e. These marks may be at any location on the buttstock. Unauthorized markings are not desirable. However, if previously applied, they will be allowed to continue in use due to the cost of the buttstock.

1. Inspect buttstock for cracks using the following guidelines:

a. Under the following conditions, hairline cracks (no chipped away material allowed) originating from buttplate end of buttstock are acceptable.

(1) One hairline crack, not to exceed 1 in. (2.54 cm) in length, per side of buttstock.

(2) Two additional hairline cracks up to 0.25 in. (0.64 cm) in length, per side of buttstock.

(3) A total of three cracks per side of buttstock, originating from buttplate end, are allowable.

b. Cracks in the critical area at the front end of the buttstock are not acceptable and these buttstocks must be replaced.

2. While buttplate is installed on rifle, inspect for cracks around the mounting holes. Check for cracks in excess of 0.25 in. (0.64 cm) in length which extend through the buttplate. Replace if cracked.

3. Inspect door assembly for cracks, corrosion, stuck plunger, separations on outer face, or other damage. Replace if defective.

d. REPAIR

Replace all authorized unserviceable items. Unserviceable items are those items which are damaged.

e. LUBRICATION

Lubricate all metal components (p 2-33) (operator's manual).

f. REASSEMBLY

RIFLE ONLY

1. Position hinge (1) on door assembly (2) and install straight pin (3).

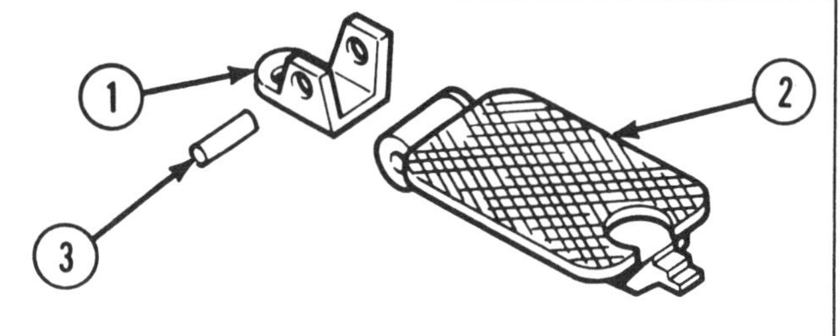

RIFLE ONLY

2. Install door assembly (2) into buttplate (4) and press plunger (5) to lock.

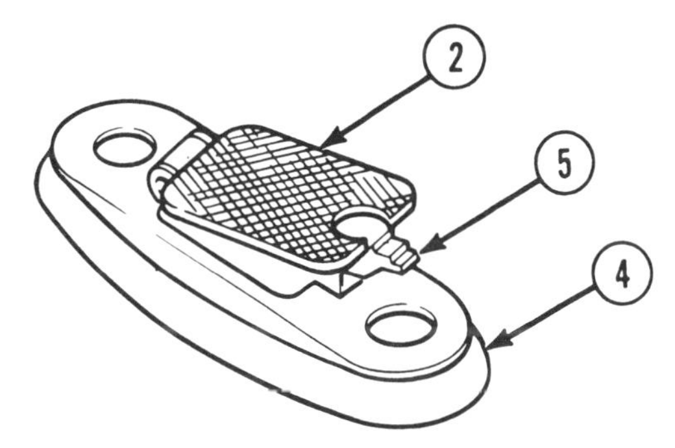

RIFLE ONLY

3. Position buttplate group (6) and small sling swivel (7) to the buttstock (8) and secure with self-locking screw (9).

NOTE
See page 2-62, reassembly, for reassembly of buttstock assembly to lower receiver.

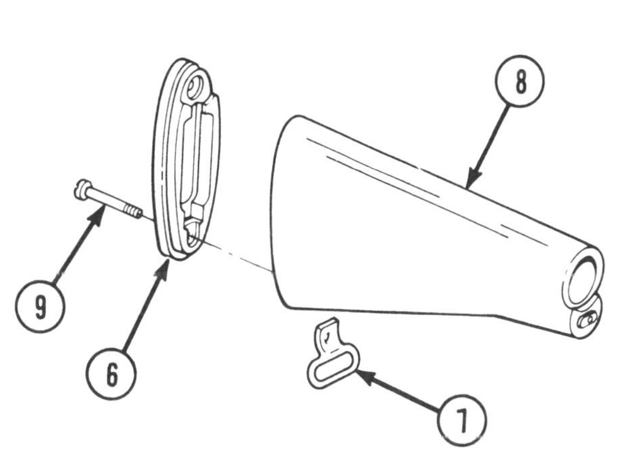

2-18. BUTTSTOCK ASSEMBLY (CONT).

f. REASSEMBLY (CONT)

CARBINE ONLY

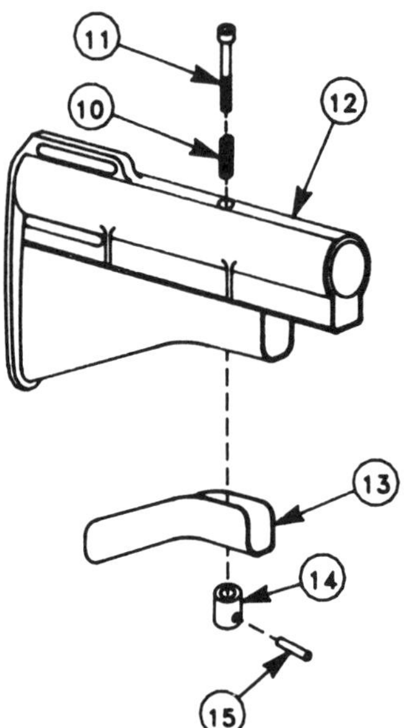

4. Insert locking pin spring (10) onto locking pin (11).

5. Insert locking pin (11) and spring (10) into the hole on top of the buttstock (12), threaded end first.

6. Insert your index finger into the forward end of the buttstock (12) and push down on locking pin (11).

7. Install the release lever (13) onto the threaded portion of the locking pin (11) protruding through the bottom of the buttstock (12).

8. Screw on the locking nut (14) until flush with locking pin (11). Align the slot in locking nut (14) with the spring pin hole in the locking pin (11).

9. Lightly tap in spring pin (15) until flush on both sides of the retaining nut (14).

2-19. MAJOR COMPONENTS OF M16A2 RIFLE.

This task covers:

a. Reassembly   **c.** Stowage
b. Inspection

INITIAL SETUP

References
  TM 9-1005-319-10

Equipment Conditions
  Rifle disassembled into major components

General Safety Instructions
  To avoid injury to eyes, use care when removing and installing spring-loaded parts.

Do not interchange bolt assemblies from one rifle to another. Doing so may result in injury to, or death of, personnel.
Do not keep live ammunition near the work area.

a. REASSEMBLY

Refer to operator's manual and assemble lower receiver and buttstock assembly (1), upper receiver and barrel assembly (2), charging handle assembly (3), bolt carrier assembly (4), sling (5), and magazine (6).

b. **INSPECTION**

Perform the following function checks on assembled rifle:

1. Remove magazine if installed. Pull charging handle assembly to rear. Check that chamber is clear. Let bolt and bolt carrier close. Do not pull trigger. Leave hammer in cocked position.

WARNING

If rifle fails any of the following tests, continued use of the rifle could result in injury to, or death of, personnel.

2. Place selector lever in SAFE position and pull trigger. Hammer should not fall.

3. Place selector lever in SEMI position. Pull trigger. Hammer should fall.

NOTE

For the purpose of the following check "SLOW" is defined as one fourth to one half the normal rate of trigger release.

4. Hold trigger to the rear, charge rifle, and release the trigger with a slow, smooth motion, without hesitations or stops, until the trigger is fully forward; an audible click should be heard. Hammer should not fall.

5. Repeat the SEMI position test five times. The rifle must not malfunction during any of these five tests. If the rifle malfunctions during any of these five tests, evacuate rifle to support maintenance for repair.

6. Place selector lever in BURST position. Charge rifle and pull trigger. Hammer should fall.

7. Hold trigger to the rear, pull charging handle assembly to rear and release three times. Release trigger. Hammer should not fall. The burst disconnector should have held the hammer to the rear while the trigger was in the pulled position.

8. Pull trigger. Hammer should fall. This should be the first round of a three round burst.

9. With hammer in forward position, attempt to place the selector lever in the SAFE position. If selector lever can be placed on SAFE, evacuate the rifle to support maintenance.

2-19. MAJOR COMPONENTS OF M16A2 RIFLE (CONT).

c. STOWAGE

Prior to stowing the rifle in arms room, perform the following procedures:

NOTE
Rifle passed into arms room issue window should be passed butt first with the bolt locked to the rear.

1. Clear. Refer to operator's manual.

2. Place selector lever in SEMI position.

3. Pull trigger. Hammer should fall.

4. Close ejection port (dust) cover.

5. Place rifle in rack.

6. **M4 Carbine Only**, use M12 Arms Rack.

    a. The M12 arms rack is the correct arms rack in which to store the M4 Carbine. When storing the M4 Carbine in the M12 arms rack a mounting bracket, NSN 5210-01-230-3181, must be used for each M4 Carbine being stored. This requirement is for the safety of the person that opens the arms rack as well as to prevent damage to the carbine when the rack is opened. Without the mounting bracket the M4 Carbine can fall out of the rack when it is opened.

    b. To install the mounting bracket on the M12 arms rack, for use with the M4 Carbine, install the bracket with the hooks of the bracket facing toward the carbine, so that the lower receiver extension will contact the bent end of the bracket. The bent end of the bracket will hold the carbine upright when the arms rack is opened.

Section VI. PREPARATION FOR STORAGE OR SHIPMENT

2-20. PREPARATION FOR STORAGE OR SHIPMENT.

a. Packaging of the M16A2 Rifle and the M4 Carbine shall be In accordance with the following:

ARMY ONLY: Army users shall package the rifle and the carbine in accordance with each respective Packaging Data Sheet (PDS) for shipment or storage which may exceed 90 days. The PDS is part of the Army Master Data File Retrieval Microform System (ARMS) Packaging File.

AIR FORCE ONLY: Air Force users shall package the rifle in accordance with each respective Special Packaging Instruction (SPI) 00-856-6885 for shipment or storage which may exceed 90 days. The SPIs are part of the Army Master Data File Retrieval Microform System (ARMS) Packaging File.

b. Packaging, if required, for shipping/storage which will not exceed 90 days shall be as follows:

(1) Clean in accordance with operator's manual.

(2) Wrap with MIL-B-121 waterproof material.

(3) Place in barrier bag MIL-B-117, Type I, Class C, or wrap with MIL-B-121, Type I, Grade A, and seal with tape, PPP-T-76.

(4) Place one or more of item in minimum size container. Block and brace in accordance with MIL-STD-1186. Cushion the M16 and similar weight items with PPP-C-843, and use PPP-F-320 as filler, to create a tight pack.

    (a) Fiber board containers shall be in accordance with PPP-B-636 and may be Class Domestic. Gross weight and size of material shall determine grade of fiberboard container. PPP-B-640 may also be used.

    (b) Wood containers shall be in accordance with PPP-B-601 or PPP-B-621.

(5) Equivalent materials may be used.

c. NSNs are not assigned to all the specified material. If it is necessary to specify an NSN in the TMs, the packing materials will have to be spared and part numbers and NSNs assigned.

d. The specifications used are:

(1) MIL-B-121   Barrier Material, Greaseproofed, Waterproofed, Flexible (NSN 8135-00-753-4661)

2-20. PREPARATION FOR STORAGE OR SHIPMENT (CONT).

(2) MIL-B-117 — Bag, Sleeve and Tubing - Interior Packaging (NSN 8135-00-543-6574)

(3) PPP-B-636 — Boxes, Shipping, Fiberboard

(4) PPP-B-601 — Boxes, Wood, Cleated Plywood

(5) MIL-STD-129 — qMarking for Shipment and Storage

(6) PPP-T-76 — Tape, Packaging, Paper

(7) MIL-STD-1186 — Cushioning, Anchoring, Bracing, Blocking and Waterproofing; with Appropriated Test Methods

(8) PPP-C-843 — Cushioning Material, Cellulosic

(9) PPP-F-320 — Fiberboard, Corrugated and Solid Sheet Stock (Container Grade), and Cut Shapes

(10) PPP-B-640 — Boxes, Fiberboard, Corrugated, Triple-Wall

(11) PPP-B-621 — Boxes, Wood, Nailed and Locked - Corner

CHAPTER 3
DIRECT SUPPORT MAINTENANCE INSTRUCTIONS

CHAPTER OVERVIEW

This chapter provides information and instructions to keep the weapon in good repair and contains the following sections:

.

a. Repair Parts and Special Tools
b. Direct Support Troubleshooting
c. Direct Support Maintenance Procedures for the M16A2 Rifle
d. Preparation for Storage or Shipment
e. Preembarkation Inspection of Material in Units Slated for Overseas Movement

Section I. REPAIR PARTS, SPECIAL TOOLS, TMDE, AND SUPPORT EQUIPMENT

3-1. COMMON TOOLS AND EQUIPMENT. For authorized common tools and equipment, refer to the Modified Table of Organization and Equipment (MTOE) applicable to your unit. Air Force users must maintain the following common tools:

Flat tip screwdriver	Torque wrench	5/64-inch drive pin
Socket wrench handle	Retaining ring pliers	punch
and socket head screw	(Two) 8-inch	1/8-inch drive pin punch
socket wrench	adjustable wrenches	1/16-inch drive pin
Vise jaw caps	Flat file	punch
Machinist's vise	Ball-peen hammer	3/32-inch drive pin
Solid center punch	Trigger pull test	punch
Hammer	fixture rod and	
Combination wrench	weights	

3-2. SPECIAL TOOLS, TMDE, AND SUPPORT EQUIPMENT. Special tools required for direct support maintenance are listed in appendixes B and C. Fabricated tools are listed and illustrated in appendix E.

3-3. REPAIR PARTS. Repair parts are listed and illustrated in appendix C of this manual.

NOTE

Bolt assemblies, and/or barrel assemblies may be interchanged, at the Direct Support Maintenance level, from one rifle to another, under the provisions of the note on page C-3: If these parts are interchanged the rifle must be checked/inspected as depicted on pages 3-17, 3-21, and 3-33. While performing these checks and inspections, pay special attention to the headspace requirements on page 3-47.

Section II. TROUBLESHOOTING

3-4. GENERAL.

a. This section contains direct support level troubleshooting information for locating and correcting most of the operating troubles which may develop in the M16A2 rifle. Each malfunction for the individual part or assembly is followed by a list of tests or inspections which will help you to determine the corrective action to take. You should perform the tests/inspections and corrective actions in the order listed.

b. This manual cannot list all malfunctions that may occur, nor all tests or inspections and corrective actions. If a malfunction is not listed or is not corrected by listed corrective actions, see individual repair sections in the maintenance procedures on each major assembly.

3-5. TROUBLESHOOTING PROCEDURES. Refer to troubleshooting table for malfunctions, tests, and corrective actions. The symptom index is provided for a quick reference of the malfunctions covered in the table.

SYMPTOM INDEX

TROUBLESHOOTING

MALFUNCTION
    TEST OR INSPECTION
        CORRECTIVE ACTION

---

1. FAILURE OF MAGAZINE TO LOCK IN RIFLE.

    Step 1. Dirty or corroded magazine catch (1).

        Disassemble and clean.

    Step 2. Defective magazine catch spring (2).

        Replace magazine catch spring (2) (p 3-62).

    Step 3. Worn or broken magazine catch (1).

        Replace magazine catch (1) (p 3-62).

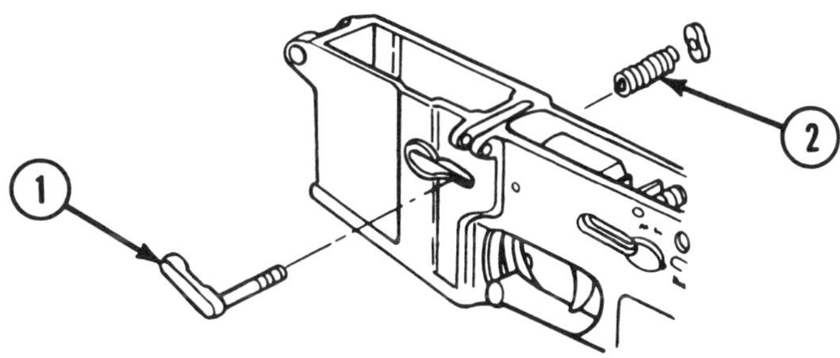

2. FAILURE TO FEED.

    Step 1. Magazine catch spring (1) weak or broken.

        Replace magazine catch spring (1) (p 3-62).

    Step 2. Short recoil.

        Refer to page 3-9.

3. FAILURE TO CHAMBER.

    Short recoil.

        Refer to page 3-9.

3-5. TROUBLESHOOTING PROCEDURES (CONT).

TROUBLESHOOTING (CONT)

MALFUNCTION
    TEST OR INSPECTION
        CORRECTIVE ACTION

---

4. FAILURE TO LOCK.

    Step 1.   Damaged bolt carrier key (1).

        Repair or replace bolt carrier key (1) (p 3-25).

    Step 2.   Loose screws (2) on bolt carrier key (1).

        a.   Disassemble and repair (p 3-25).

        b.   Reassemble using new screws.

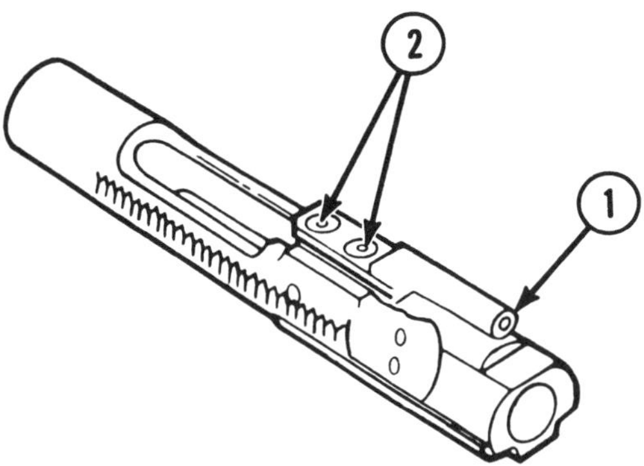

    Step 3.   Bent gas tube (3).

        a.   Adjust by bending gas tube (3) in area of handguards.

        b.   Replace gas tube (3) and check alignment (p 3-29).

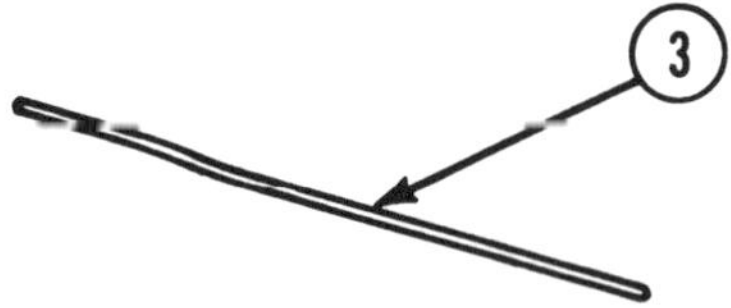

    Step 4.   Short recoil.

        Refer to page 3-9.

TROUBLESHOOTING (CONT)

MALFUNCTION
    TEST OR INSPECTION
        CORRECTIVE ACTION

---

5.   FAILURE TO FIRE.

      Step 1.   Broken hammer (1).

              Replace hammer (1) (p 3-73).

      Step 2.   Weak or broken hammer spring (2).

              Replace hammer spring (2) (p 3-73).

      Step 3.   Hammer spring (2) improperly assembled.

              Assemble properly.

      Step 4.   Burst cam (3) and/or cam spring (4) frozen or improperly assembled.

              Disassemble, clean, lubricate, and reassemble correctly (p 3-73).

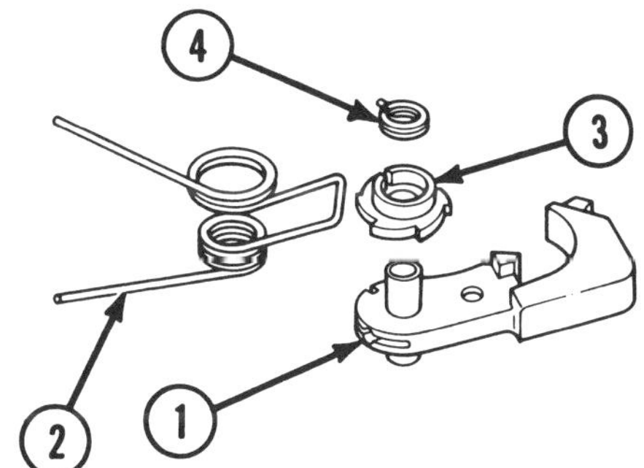

      Step 5.   Selector lever (5) frozen on SAFE position.

              Disassemble and clean (p 3-62).

      Step 6.   Broken firing pin (6) or firing pin does not meet gage protrusion requirement.

              Replace firing pin (6) (p 3-16).

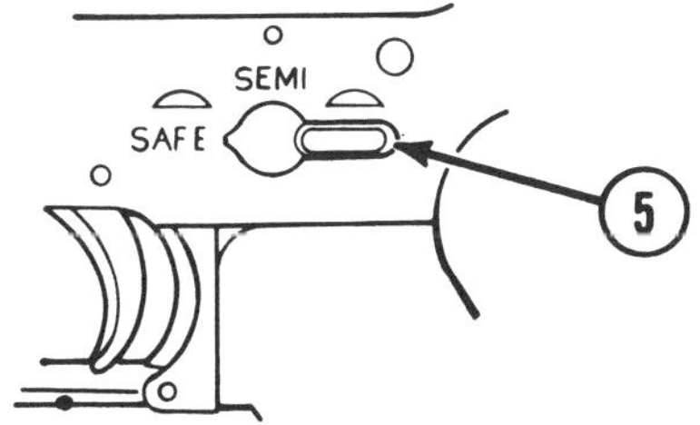

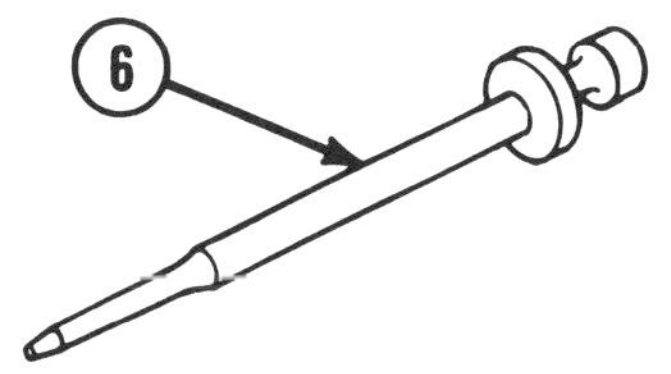

3-5. TROUBLESHOOTING PROCEDURES (CONT).

TROUBLESHOOTING (CONT)

MALFUNCTION
> **TEST OR INSPECTION**
> > **CORRECTIVE ACTION**

---

6. FAILURE TO UNLOCK.

> Step 1. Burred locking lugs (1) on bolt assembly.
>
> Remove burrs.
>
> Step 2. Burred lugs (2) on barrel extension.
>
> Remove burrs.
>
> Step 3. Short recoil.
>
> Refer to page 3-9.

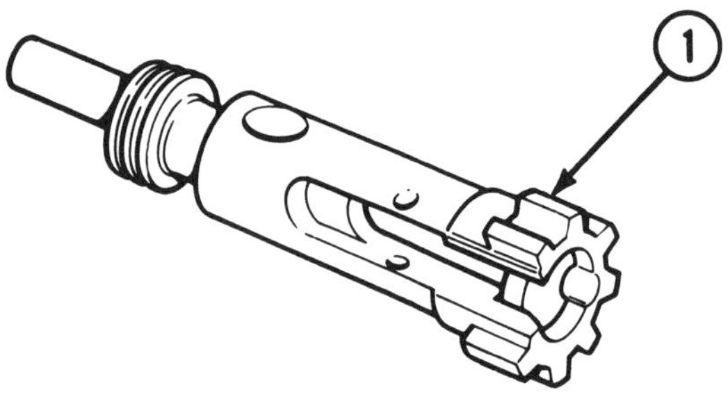
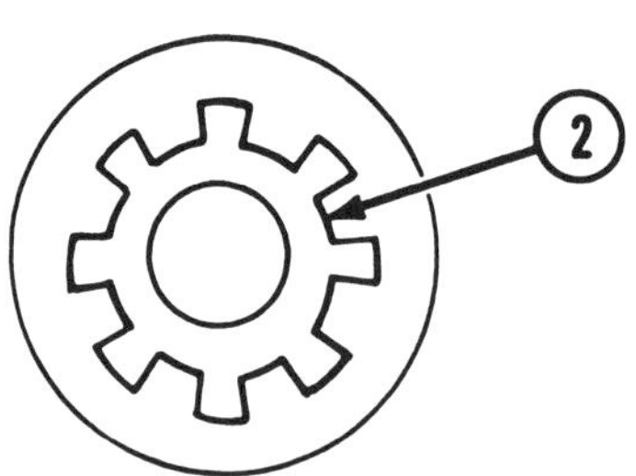

7. FAILURE TO EXTRACT.

> Step 1. Inspect cartridge extractor and extractor spring assembly.
>
> Replace if cracked or broken (p 2-35).
>
> Step 2. Inspect badly pitted chamber with reflector tool (item 2, fig. C-17).
>
> Replace barrel assembly if chamber is badly pitted (p 3-29).

8. FAILURE TO EJECT.

> Short recoil.
>
> Refer to page 3-9.

TROUBLESHOOTING (CONT)

MALFUNCTION
    TEST OR INSPECTION
        CORRECTIVE ACTION

9.   FAILURE TO COCK.

      Step 1.   Worn or broken trigger nose (1) or trigger spring (2).

           Replace trigger (3) or defective trigger spring (2) (p 3-62).

      Step 2.   Worn or broken hammer trigger notch (4).

           Replace hammer (5) (p 3-62).

      Step 3.   Worn or broken hammer disconnector hook (6).

           Replace hammer (5) (p 3-62).

      Step 4.   Worn or broken hammer automatic sear hook (7).

           Replace hammer (5) (p 3-62).

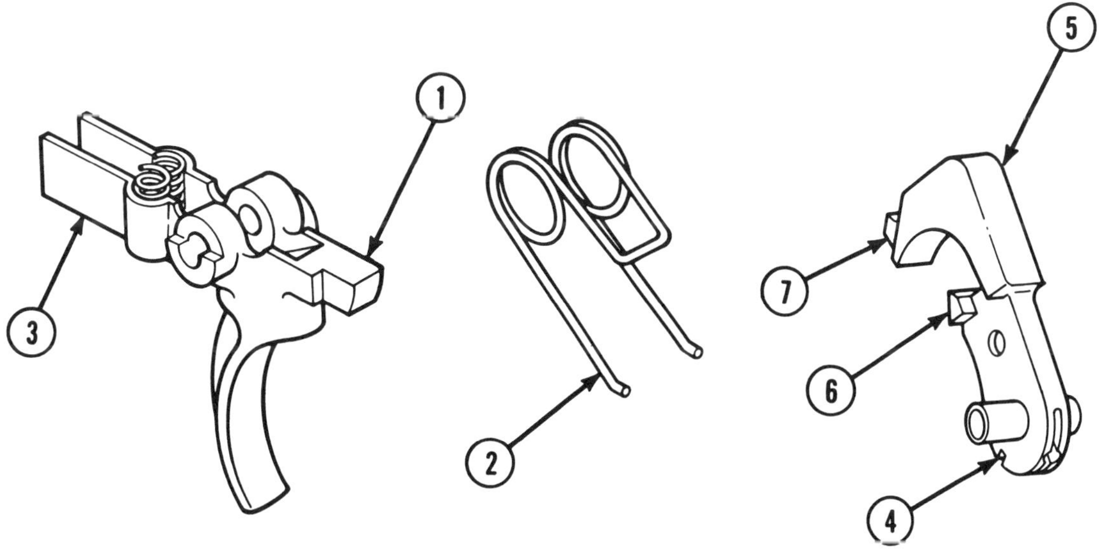

3-5. TROUBLESHOOTING PROCEDURES (CONT).

TROUBLESHOOTING (CONT)

MALFUNCTION
    TEST OR INSPECTION
        CORRECTIVE ACTION

---

9. FAILURE TO COCK (CONT).

    Step 5. Worn or broken disconnector hooks (8).

        Replace defective disconnectors (9) (p 3-62).

    Step 6. Weak, broken, or missing disconnector springs (10).

        Replace disconnector springs (10) (p 3-62).

    Step 7. Worn, broken, or missing automatic sear (11).

        Replace automatic sear (11) (p 3-62).

    Step 8. Weak or broken automatic sear spring (12).

        Replace automatic sear (11) (p 3-62).

    Step 9. Long leg (13) of automatic sear spring incorrectly assembled in receiver.

        Remove automatic sear assembly (11) and install correctly (p 3-62).

    Step 10. Burst cam (14) or clutch spring (15) frozen or improperly assembled.

        Disassemble, inspect, clean, lubricate, or replace as required (p 3-73).

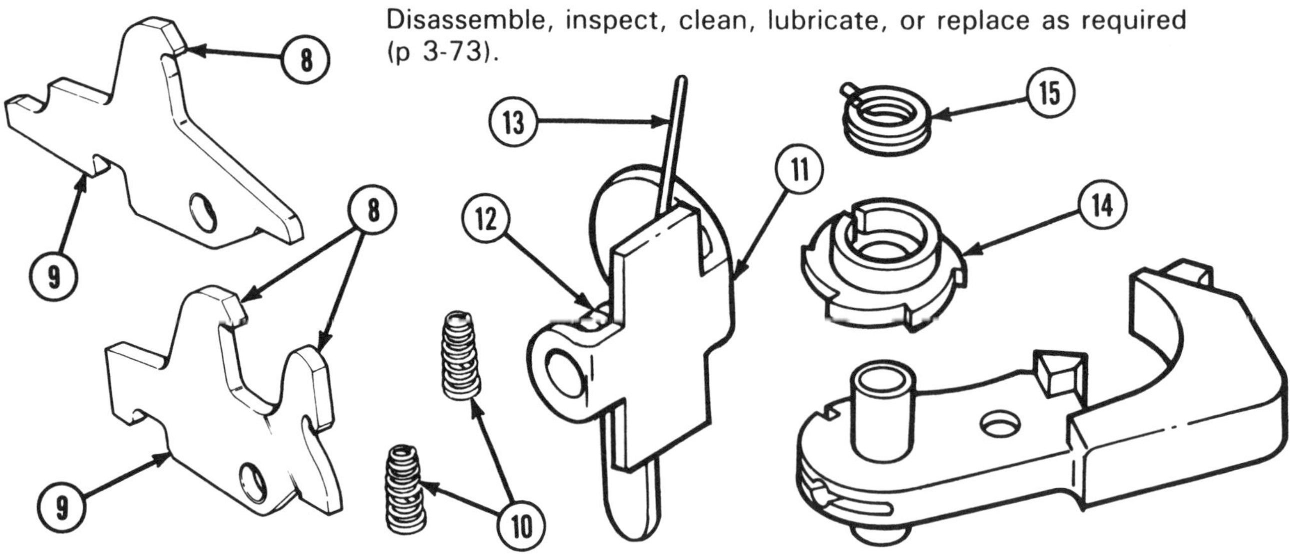

TROUBLESHOOTING (CONT)

MALFUNCTION
    TEST OR INSPECTION
        CORRECTIVE ACTION

---

10. SHORT RECOIL.

    Step 1. Improper gap space or worn, missing, or broken bolt rings (1).

        a. Stagger bolt ring gaps (p 3-21).

        b. Replace bolt rings (1) and stagger gaps (p 3-21).

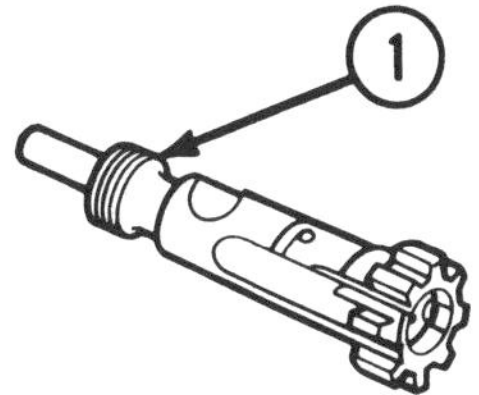

    Step 2. Broken or bent gas tube (2).

        Adjust by bending in area of handguards or replace gas tube (2) (p 3-29).

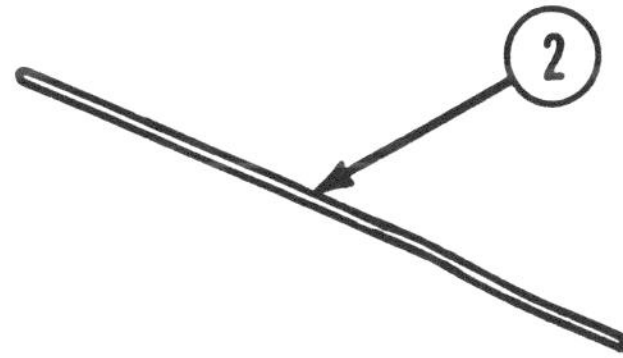

    Step 3. Gas tube spring pin (3) missing from front sight (4).

        Replace gas tube spring pin (3) (p 3-29).

    Step 4. Partially plugged gas system because of carbon build-up in the gas tube (2).

        Replace gas tube (2) (p 3-29).

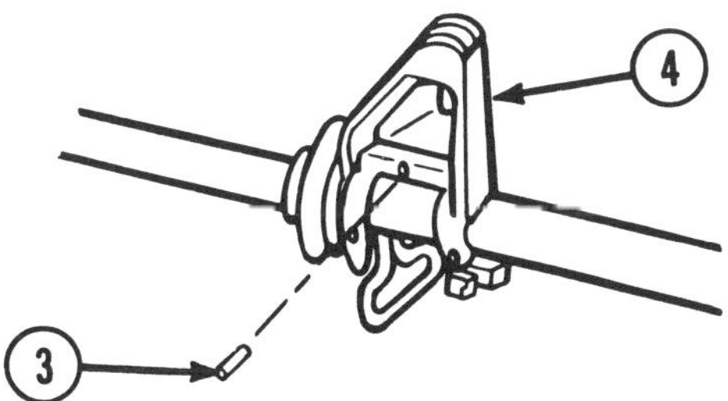

3-5. TROUBLESHOOTING PROCEDURES (CONT).

TROUBLESHOOTING (CONT)

MALFUNCTION
    TEST OR INSPECTION
        CORRECTIVE ACTION

---

10. SHORT RECOIL (CONT).

WARNING

When using carbon removing compound (item 8, app D), avoid skin contact. If it comes in contact with the skin, wash off thoroughly with running water. The use of a good lanolin base cream after exposure to compound is helpful. The use of gloves and protective equipment is required.

Step 5. Carbon build-up in barrel gas port (5).

Remove carbon build-up by soaking barrel in carbon removing compound (item 8, app D). Use rubber gloves (item 18, app D) with carbon removing compound. Use a bore small arms cleaning brush (item 4, app D).

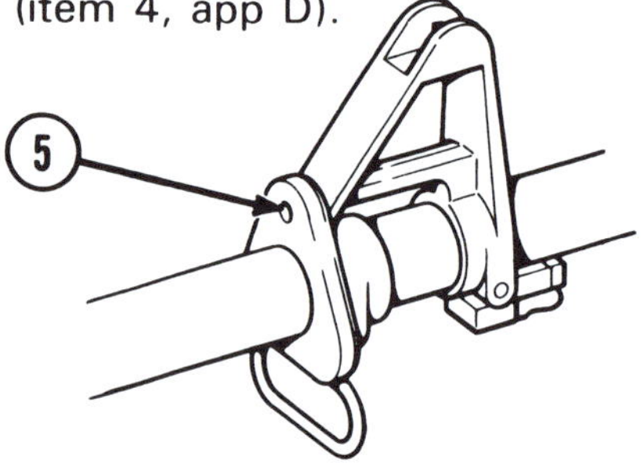

11. RIFLE CANNOT BE ZEROED.

Step 1. Inspect for defective or bent barrel assembly (1) (p 3-29).

Replace barrel assembly (1) (p 3-29).

Step 2. (For windage) barrel assembly (1) out of alignment with rear sight on upper receiver.

Align barrel assembly (1) and upper receiver (p 3-29).

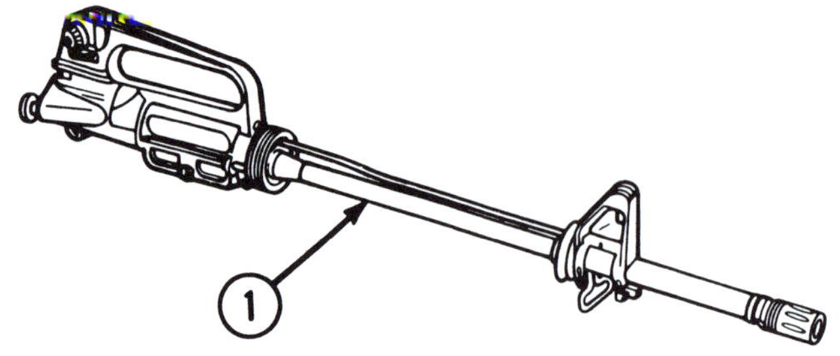

TROUBLESHOOTING (CONT)

MALFUNCTION
    TEST OR INSPECTION
        CORRECTIVE ACTION

Step 3.   (For elevation) defective front sight (2) or rear sight (3).

    Repair as required (p 3-48).

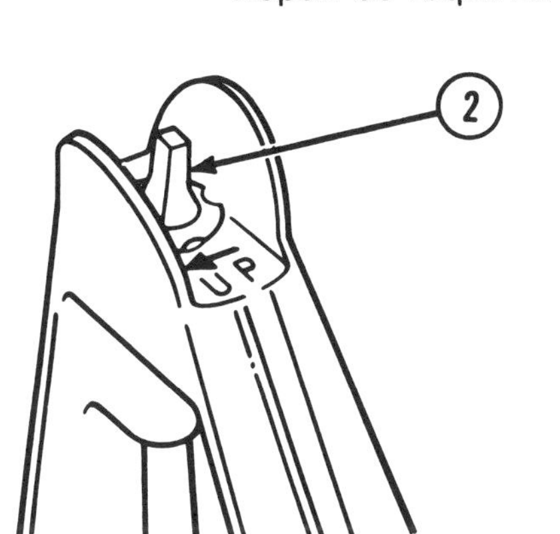
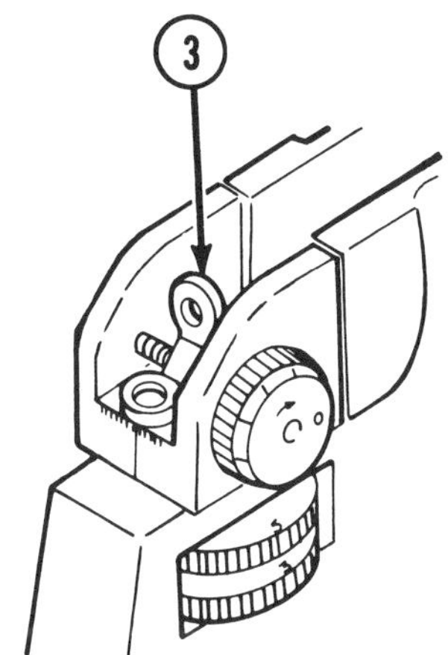

12.   FAILURE TO CYCLE WITH SELECTOR LEVER SET ON BURST.

    Step 1.   Broken automatic sear (1) or spring (2).

        Replace automatic sear assembly (1) (p 3-62).

    Step 2.   Faulty selector lever (3).

        Replace selector lever (3) (p 3-62).

    Step 3.   Broken tooth on burst cam (4).

        Replace burst cam (4) (p 3-73).

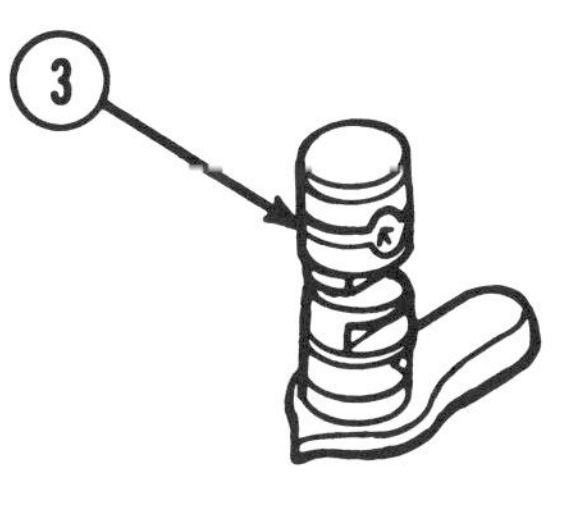
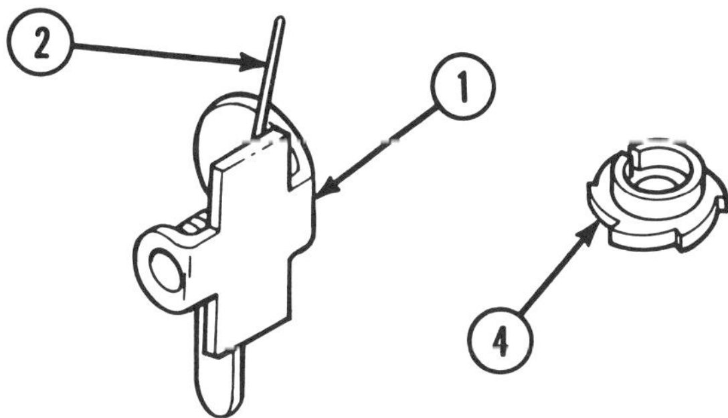

3-5. TROUBLESHOOTING PROCEDURES (CONT).

TROUBLESHOOTING (CONT)

MALFUNCTION
    TEST OR INSPECTION
        CORRECTIVE ACTION

---

12.   FAILURE TO CYCLE WITH SELECTOR LEVER SET ON BURST (CONT).

      Step 4.   Broken cam clutch spring (5). Cam clutch spring should be bent and properly formed without any sharp edges or corners.

           Inspect and replace if required.

      Step 5.   The bend in the cam clutch spring (5) installed backwards (toward outside).

           Install cam clutch spring (5) properly with the bend to the inside (p 3-73).

NOTE

When hammer is rotated back to cocked position, cam should rotate to allow the burst disconnector to latch in the next notch.

      Step 6.   Cam clutch spring (5) fails to "clutch" and burst cam (4) fails to rotate back with hammer (6).

           Replace cam clutch spring (5) (p 3-73). If problem continues, replace hammer (6) and cam (4) (p 3-73).

      Step 7.   Broken front hook (7) on burst disconnector (8).

           Replace burst disconnector (8) (p 3-62).

      Step 8.   Short recoil.

           Refer to page 3-9.

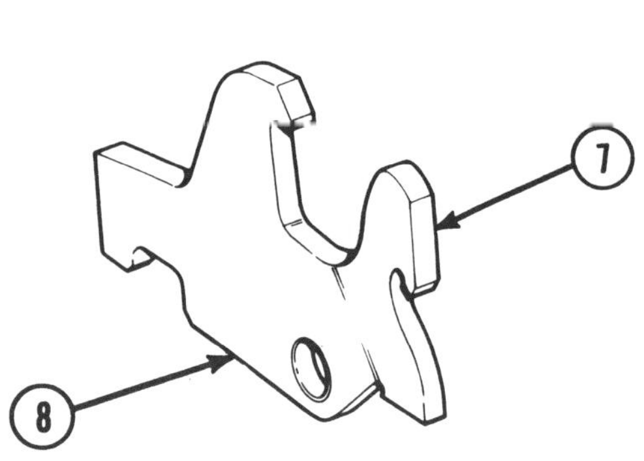

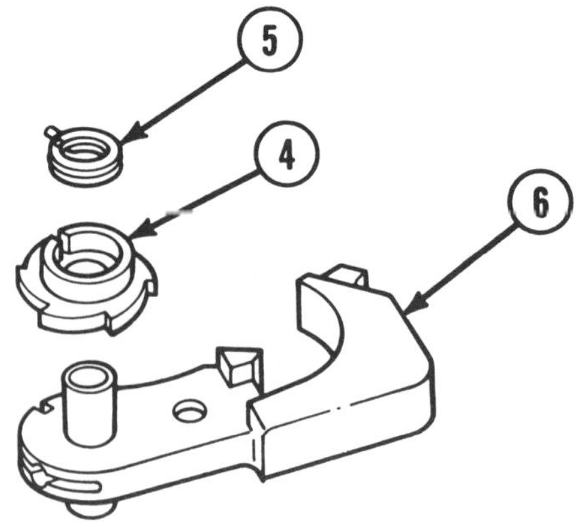

TROUBLESHOOTING (CONT)

MALFUNCTION
    TEST OR INSPECTION
        CORRECTIVE ACTION

---

13.   FIRES TWO ROUNDS WITH ONE PULL OF TRIGGER WITH SELECTOR LEVER SET ON SEMI (DOUBLE FIRING).

       Step 1.   Defective semiautomatic disconnector (1).

               Replace semiautomatic disconnector (1) (p 3-62).

       Step 2.   Worn or broken trigger notch (2) of hammer (3) (searing portion).

               Replace hammer (3) (p 3-62).

       Step 3.   Worn or broken disconnector notch (4) of hammer (3).

               Replace hammer (3) (p 3-62).

       Step 4.   Worn or broken trigger (5) (searing portion).

               Replace trigger (5) (p 3-62).

       Step 5.   Worn trigger or hammer pin holes (6).

               Gage trigger and hammer pin holes (p 3-62). If test fails, replace rifle.

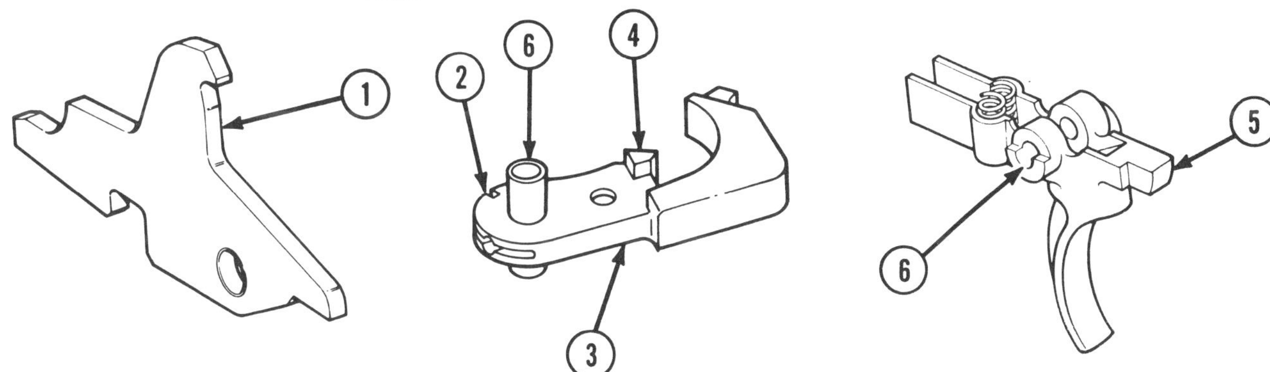

14.   FIRES WITH SELECTOR LEVER ON SAFE OR WHEN TRIGGER IS RELEASED WITH SELECTOR LEVER ON SEMI.

       Step 1.   Defective selector lever (1).

               Replace selector lever (1) (p 3-62).

       Step 2.   Worn or broken trigger (rear portion) (2).

               Replace trigger (3) (p 3-62).

3-5. TROUBLESHOOTING PROCEDURES (CONT).

TROUBLESHOOTING (CONT)

MALFUNCTION
>   **TEST OR INSPECTION**
>>   **CORRECTIVE ACTION**

---

15. HAMMER PIN "WALKS".

Hammer pin (1) "walks" or works loose during firing or hammer pin is very easy to push out of receiver when hammer is installed.

Replace hammer assembly (p 3-62).

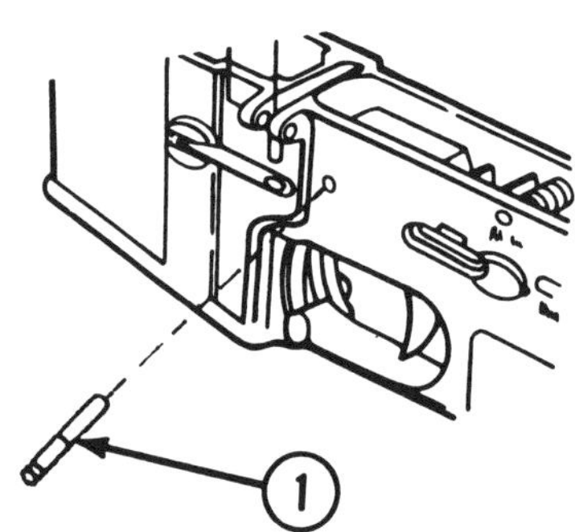

16. BOLT ASSEMBLY FAILS TO LOCK TO REAR AFTER FIRING LAST ROUND.

Step 1. Broken bolt catch (1).

Replace bolt catch (1) (p 3-62).

Step 2. Weak or broken bolt catch spring (2).

Replace bolt catch spring (2) (p 3-62).

Step 3. Restricted movement of bolt catch (1).

Disassemble and clean.

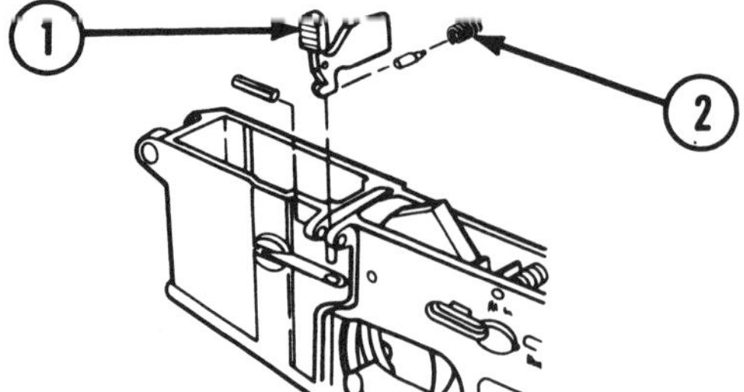

Section III.   MAINTENANCE PROCEDURES FOR THE M16A2 RIFLE

3-6.   MAJOR COMPONENTS OF M16A2 RIFLE.

This task covers disassembly.

INITIAL SETUP

Tools
  (ARMY) Small Arms Repairman Tool Kit
    (item 3, app B)

References
  TM 9-1005-319-10 (operator's manual)

Equipment Conditions
  Rifle assembled

General Safety Instructions
  Before starting an inspection, be sure to
  clear the rifle. Do not pull the trigger
  until the rifle has been cleared.
  Inspect the chamber and receiver to ensure no ammunition is present. Do not
  keep live ammunition near work area.

DISASSEMBLY

a. Refer to operator's manual.

b. Remove magazine (1), sling (2), bolt
carrier assembly (3), charging handle
assembly (4), upper receiver and barrel
assembly (5), and lower receiver and
buttstock assembly (6).

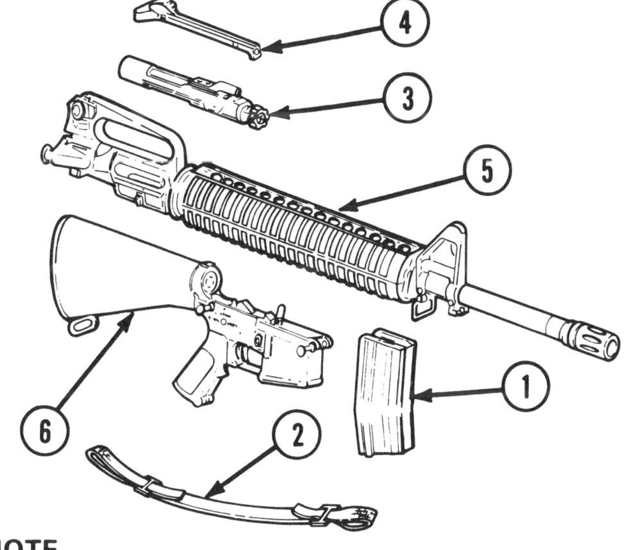

NOTE

Solid Film Lubricant (SFL) is the only authorized touchup for the M16A2 rifle
and may be used on up to one third of the exterior finish of the rifle. **FOR AR-
MY CONUS USE ONLY AND AIR FORCE TRAINING RIFLES ONLY:** SFL may
be used as touchup without limitation on the upper receiver and barrel assembly. This is to say that units which **DO NOT** fall under the category of Divisional Combat Units or rapid deployment type units may have up to 100 percent of the exterior surface of the upper receiver and barrel assembly protected with SFL if necessary.

3-7. BOLT CARRIER ASSEMBLY.

This task covers:

a. Disassembly   d. Test
b. Cleaning   e. Repair
c. Inspection   f. Reassembly

INITIAL SETUP

Test Equipment
  Tool and Gage Set (item 2, app B)

Tools
  (ARMY) Small Arms Repairman Tool Kit
    (item 3, app B)

Materials/Parts
  Cleaner, lubricant, and preservative (CLP)
    (item 9, app D)
  Cleaning compound, rifle bore (RBC)
    (item 12, app D)
  Pipe cleaner (item 11, app D)

Equipment Conditions
  3-15 Bolt carrier assembly removed

a. DISASSEMBLY

1. Remove firing pin retaining pin (1).

2. Tip key and bolt carrier assembly (2) allowing firing pin (3) to drop out. Catch the firing pin.

3. Rotate bolt cam pin (4) one quarter turn and lift straight up to remove.

4. Remove bolt assembly (5) from key and bolt carrier assembly (2).

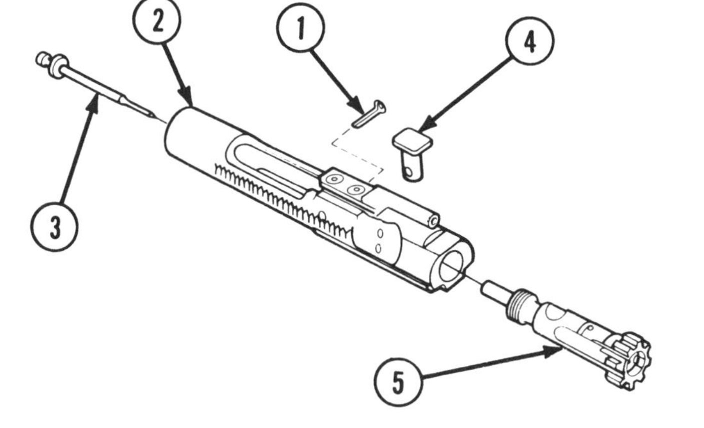

b. CLEANING

Clean carrier key and gas relief ports using CLP or RBC and pipe cleaner.

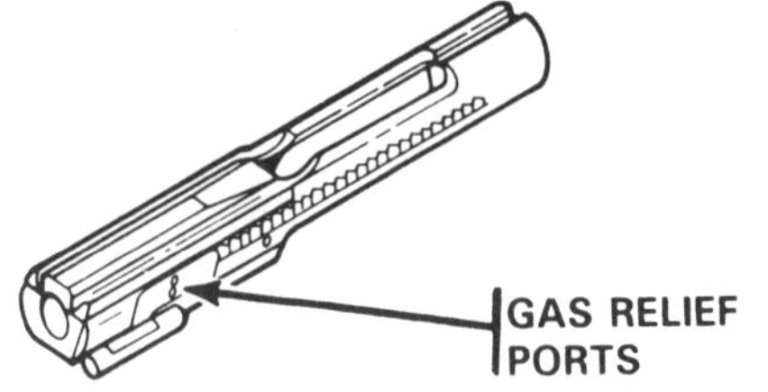

c. **INSPECTION**

1. Inspect bolt carrier assembly using the following guidelines.

   (a) Inspect bolt carrier assembly (1) for burrs, cracks, wear, and evidence of gas loss.

   (b) Visually inspect the carrier and key screws (2) for looseness and proper staking as shown below.

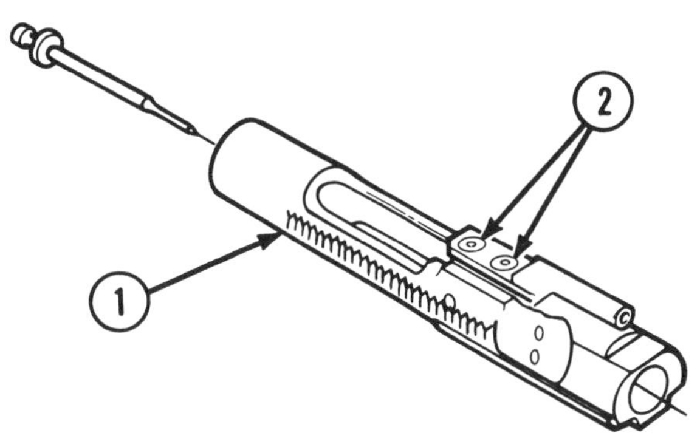

NOTE
Do not attempt to retorque if there is no loosening of the screws indicated by the staking marks.

Surface "A" must not indicate distortion or damage which impairs parallelism.

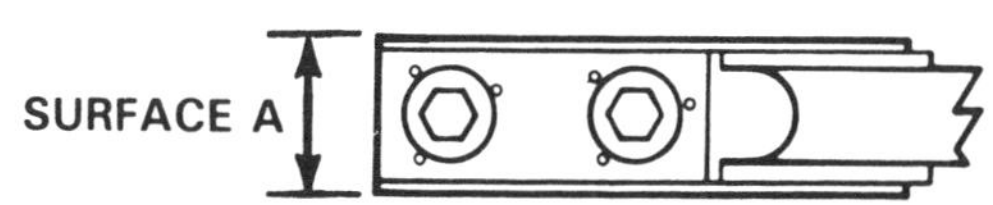

NOTE
A maximum of 0.025 in. (0.064 cm) protrusion in an upward direction is permissible.

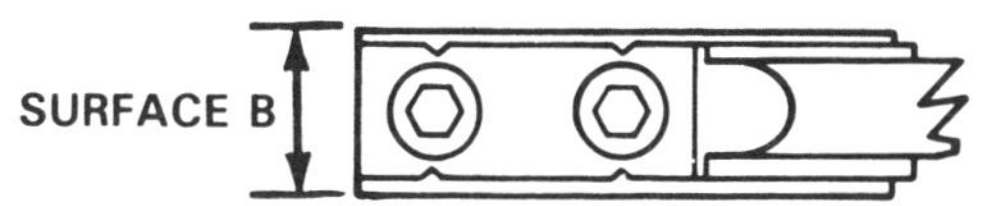

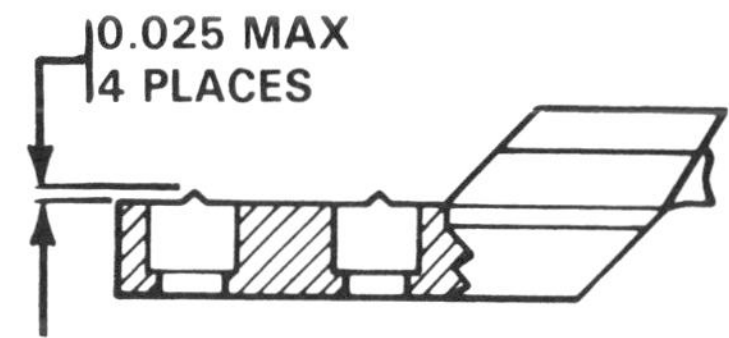

NOTE
There are bolts and bolt carriers on fielded rifles, some with chrome-plated exterior surface finishes and some with phosphate coating. Both finishes are acceptable under certain operational requirements and/or restrictions. Phosphate-coated bolt carriers are required for divisional combat units. Chrome-plated bolt carriers are acceptable for divisional noncombat units and training center units. Chrome-plated and phosphate-coated bolt assemblies, bolt carrier assemblies, and repair parts for these assemblies may be intermixed in any combination, with the following exception:

Phosphate-coated bolt carriers are required for all deployable and deploying units. Chrome-plated bolt carriers are acceptable for nondeployable and training center units.

3-7. BOLT CARRIER ASSEMBLY (CONT).

c. INSPECTION (CONT)

2. Inspect firing pin (3) tip for proper contour. Inspect for pitting, wear, and burrs. Pits or wear in area (4) is permissible. Replace firing pin if defective.

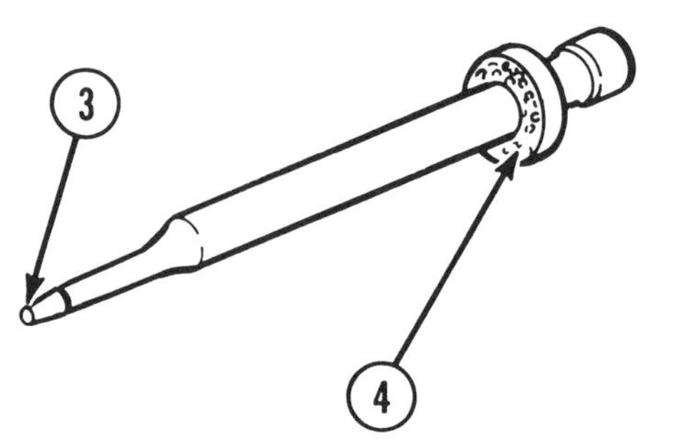

3. Prior to reassembly, insert bolt assembly (5) into key and bolt carrier assembly (6) (do not insert bolt cam pin) and exercise bolt assembly in and out of key and bolt carrier assembly. Check for binding.

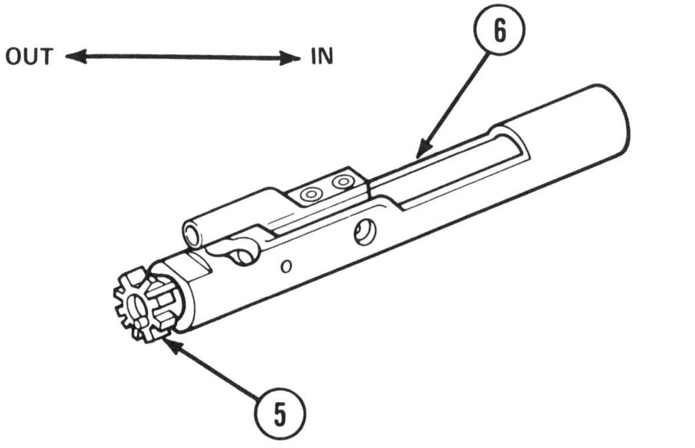

4. Check bolt assembly (5) for proper fit with bolt cam pin removed. Turn key and bolt carrier assembly (6) and suspend so the bolt assembly is pointed down.

NOTE

The bolt assembly must not drop out. If weight of bolt assembly allows it to drop out of key and bolt carrier assembly, replace bolt rings (p 3-21).

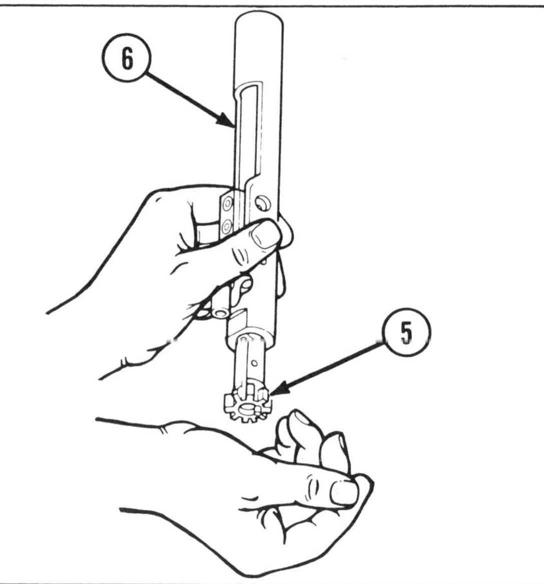

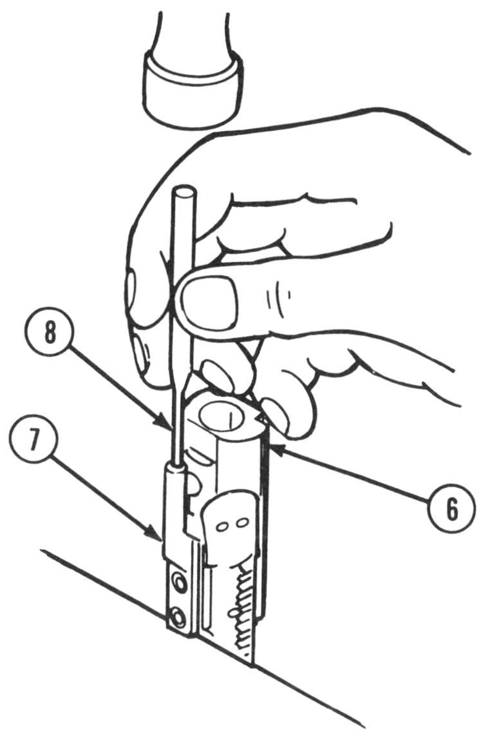

CAUTION

Extreme care must be exercised during the following procedure to assure that the striking force is not directed to the attaching screws and that the tube portion is not enlarged or flared beyond original requirement. Such enlargement would permit loss of gas pressure when the key and gas tube come together during functioning.

5. Repair small dents and/or distortions in carrier key (7) using fabricated key tool as follows:

   (a) Place the key and bolt carrier assembly (6) in a vertical position, supported in a manner that contact is made with the rear surface of the carrier key (7).

   (b) Insert the small end of the key tool (8) into the tube portion of the carrier key (7).

   (c) Strike the large end of the key tool (8) lightly with a 3 ounce, soft-brass hammer.

   (d) Repeat striking (gently) until carrier key (7) is reformed to original configuration.

   (e) If carrier key (7) cannot be reformed to original configuration, replace carrier key.

3-7. BOLT CARRIER ASSEMBLY (CONT).

d. TEST

1. Insert firing pin (1) through bolt (2).

2. Position firing pin protrusion gage (3) PN 7799735 to check for proper firing pin (1) protrusion (minimum 0.028 in. (0.07 cm) — maximum 0.036 in. (0.09 cm)).

   NOTE

   Firing pin should touch the gage on minimum but should not touch on maximum.

3. Replace a defective firing pin.

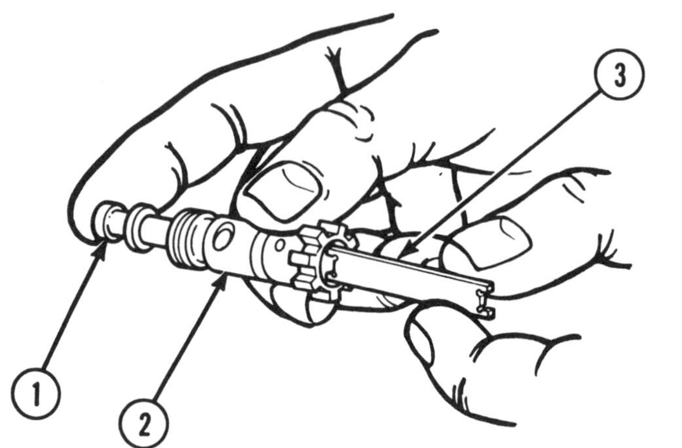

e. REPAIR

Replace all authorized unserviceable items. Retest all replaced parts.

f. REASSEMBLY

1. Install bolt assembly (1) into key and bolt carrier assembly (2).

2. Install bolt cam pin (3) and rotate one quarter turn to secure bolt assembly (1).

3. Hold key and bolt carrier assembly (2) with bolt assembly (1) down and drop in firing pin (4).

4. Install firing pin retaining pin (5) from left side only to ensure proper installation. Check installation by attempting to shake out firing pin.

5. Reassemble rifle, refer to page 3-77.

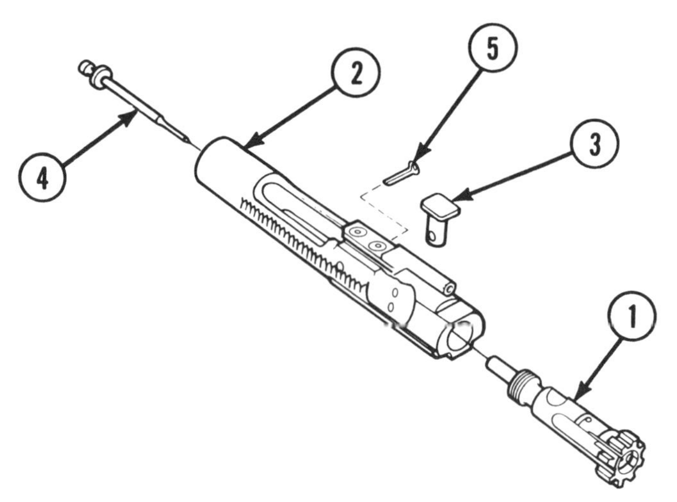

3-8. BOLT ASSEMBLY.

This task covers:

a. Disassembly
b. Inspection/Repair

c. Test
d. Reassembly

INITIAL SETUP

Test Equipment
  Tool and Gage Set (item 2, app B)

Tools
  (ARMY) Small Arms Repairman Tool Kit
  (item 3, app B)

Materials/Parts
  Penetrant kit (item 25, app D)
  Rag, wiping (item 26, app D)

Equipment Conditions
  3-16   Bolt assembly removed

a. DISASSEMBLY

NOTE
Do not remove bolt rings unless
they require replacement and
three new replacement bolt rings
are on hand.

Using small flat tip jeweler's screwdriver,
remove the three bolt rings (1) from the
bolt (2).

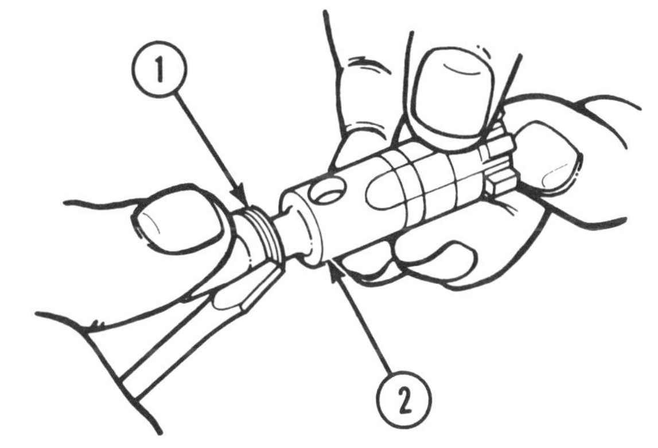

b. INSPECTION/REPAIR

1. Visually inspect bolt rings for cracks, kinks, and bends. Replace all three bolt rings if
    one or more bolt rings is damaged. See page 3-16 for bolt ring wear check.

3-8. BOLT ASSEMBLY (CONT).

b. INSPECTION/REPAIR (CONT)

2. Inspect bolt for pits, burrs, and wear as
    follows:

    (a) Bolt faces with a cluster of pits
         which are touching or tightly
         grouped, covering an area measur-
         ing approximately 1/8 inch across,
         will be rejected and replaced.

    (b) Bolts which contain individual pits
         or a scattered pattern will not be
         cause for rejection.

    (c) Bolts that contain pits extending
         into the firing pin hole will not be
         rejected unless firing pin hole gag-
         ing check determines excess wear.

    (d) Rings on the bolt face (machine
         tool marks), grooves, or ridges less
         than approximately 0.010 inch will
         not be cause for rejection.

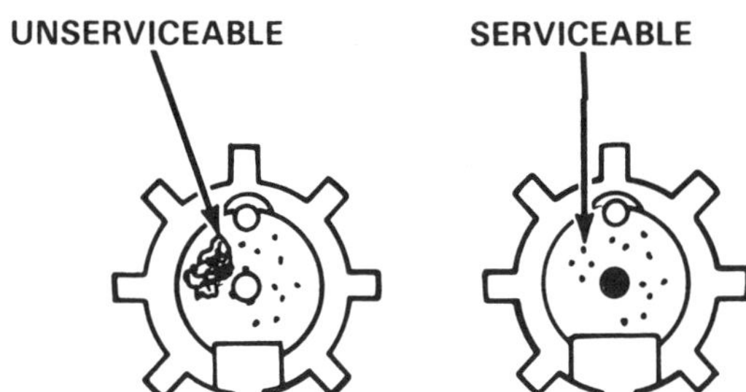

3. Inspect bolt for cracks in the locking lugs and the bolt cam pin hole area. Use black light if available; otherwise, use a glass of no more than 3X magnification or use a penetrant kit (item 25, app D). Pay close attention to the area where the locking lugs meet the body. Replace bolt assembly if bolt is defective.

WARNING

Dry cleaning solvent is flammable and toxic and should be used in a well-ventilated area. The use of rubber gloves is necessary to protect the skin when washing rifle parts.

4. Use penetrant kit (item 25, app D) to check for cracks in bolt as follows:

(a) The area to be inspected must be clean, free of oil, etc. Spray a small amount of remover on the area to be inspected, let dry and wipe off with a wiping rag.

(b) Spray penetrant (only enough to wet the area) on the area of the bolt (1) to be inspected.

(c) Spray developer over the penetrant and let the developer work. Cracks will be indicated by a change in color where there is a crack. If there are cracks, the component is unserviceable.

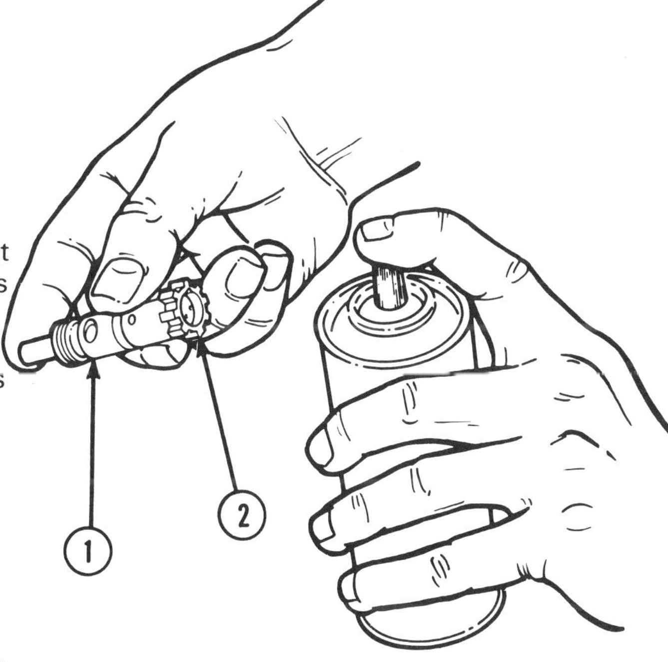

(d) Pay close attention to the area where the locking lugs (2) meet the body.

(e) If there are no cracks, spray remover on the area; let dry and wipe off with a wiping rag. Oil the area to prevent corrosion.

(f) Replace bolt assembly if bolt (1) is defective.

NOTE

Replacement of the bolt assembly will require that the headspace be tested (p 3-45, TEST).

3-8. BOLT ASSEMBLY (CONT).

c. TEST

Test bolt (1) for elongated or oversized firing pin hole using special no-go plug gage
(2) PN 12620101.

NOTE
Bolts with firing pin holes which
permit the special no-go plug gage
to fully penetrate at any position
on the circumference will be rejected and replaced.

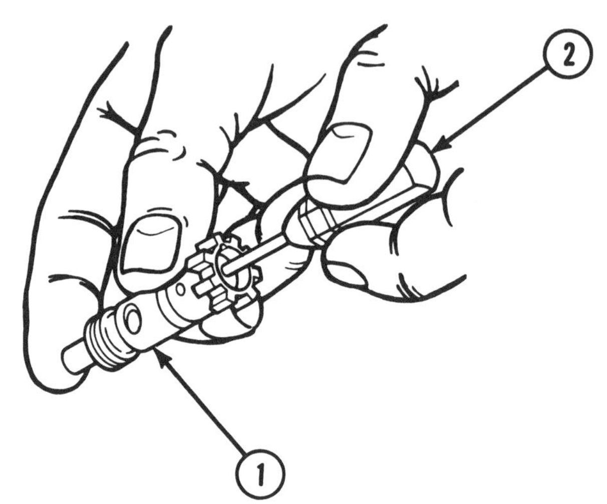

d. REASSEMBLY

NOTE
To install a bolt ring, carefully
place one end in the bolt ring
groove and hold in place with the
thumb of one hand. With the index finger of the other hand, gently guide and push the rest of the
bolt ring into the groove a little bit
at a time until the entire bolt ring
is in place.

1. Install the three bolt rings (1) one at a
    time onto the bolt (2) using care not to
    bend or "spring" new bolt rings. Stag-
    ger the bolt ring gaps (approximately ⅓
    turn apart).

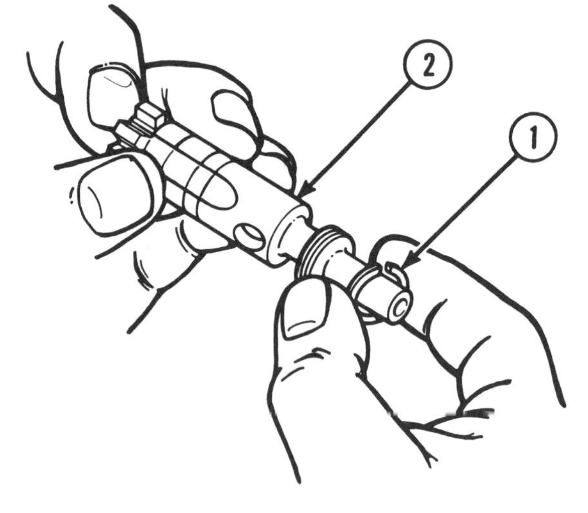

NOTE

Make certain bolt ring gaps are staggered to prevent loss of gas pressure. New bolt rings will make installing the bolt assembly difficult. Lubricate inside key and bolt carrier assembly and use gentle pressure when installing.

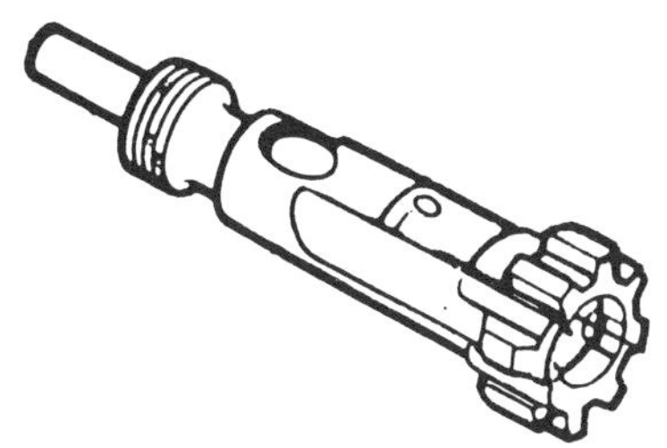

2. Reassemble rifle, refer to page 3-77.

3-9. KEY AND BOLT CARRIER ASSEMBLY.

This task covers:

a. Disassembly
b. Repair

c. Reassembly

INITIAL SETUP

Tools
  Field Maintenance Basic Less Power
    Small Arms Shop Set (item 1, app B)
  (ARMY) Small Arms Repairman Tool Kit
    (item 3, app B)

Materials/Parts
  Carrier and key screws (2) (8448508)

Equipment Conditions
  3-16 Key and bolt carrier assembly removed

3-9. KEY AND BOLT CARRIER ASSEMBLY (CONT).

a. DISASSEMBLY

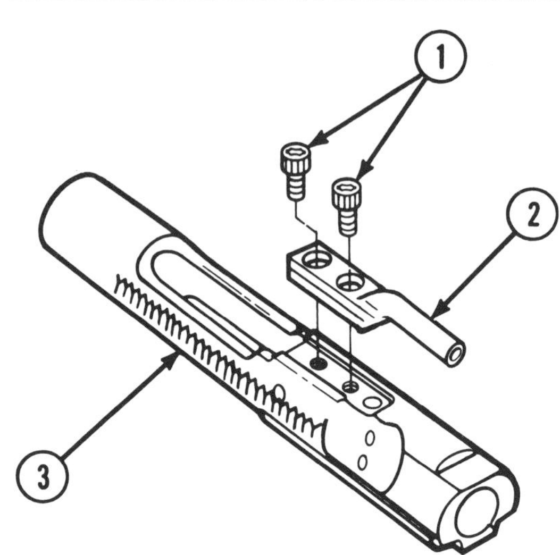

NOTE

Do not disassemble the key and bolt carrier assembly unless the bolt carrier key or bolt carrier is defective as determined by inspection procedures described in section III, preceding.

1. Using socket wrench handle and tight fitting 1/8 inch socket head screw socket wrench attachment, remove two carrier and key screws (1).

NOTE

The heads and part of the bolt carrier key may be ground off in order to remove bolt carrier key from bolt carrier if carrier and key screws cannot otherwise be removed.

2. Remove bolt carrier key (2) from bolt carrier (3).

b. REPAIR

NOTE

Do not retorque carrier and key screws if staking marks do not indicate loosening screws.

Repair by replacing, torquing, and restaking carrier and key screws. Refer to the following reassembly procedures.

c. REASSEMBLY

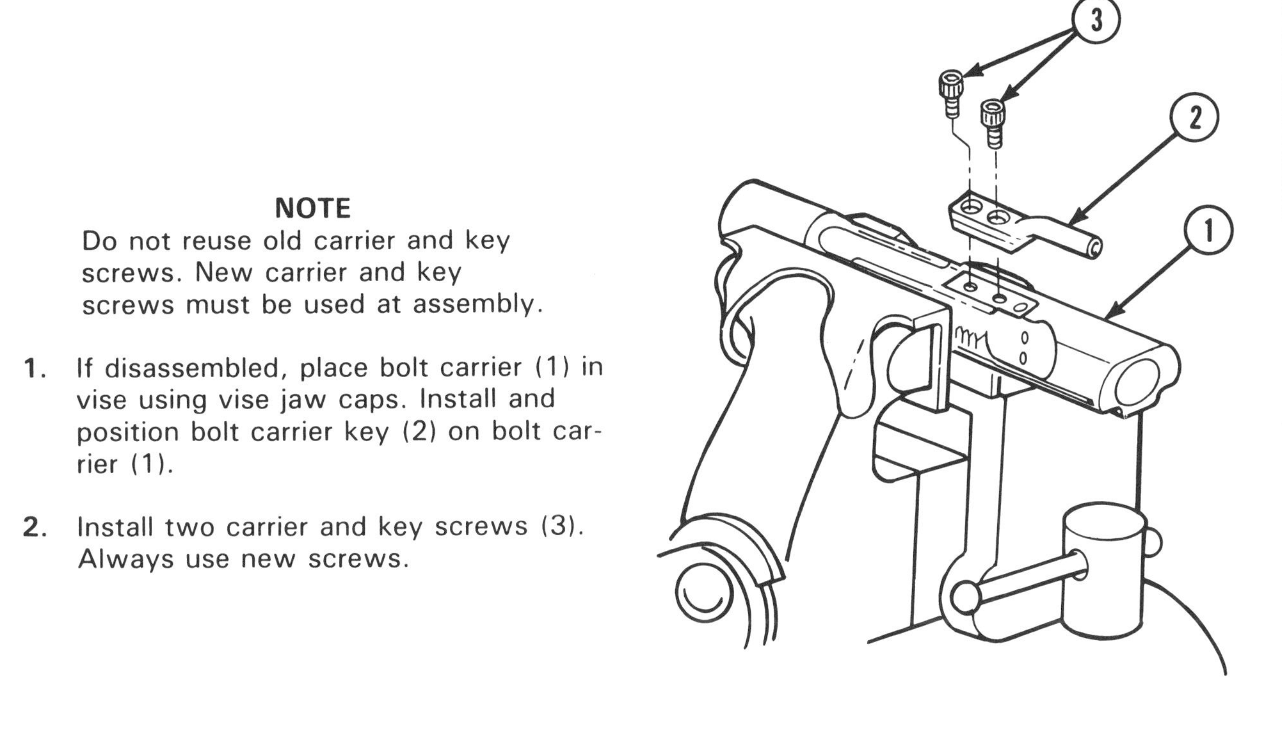

NOTE
Do not reuse old carrier and key screws. New carrier and key screws must be used at assembly.

1. If disassembled, place bolt carrier (1) in vise using vise jaw caps. Install and position bolt carrier key (2) on bolt carrier (1).

2. Install two carrier and key screws (3). Always use new screws.

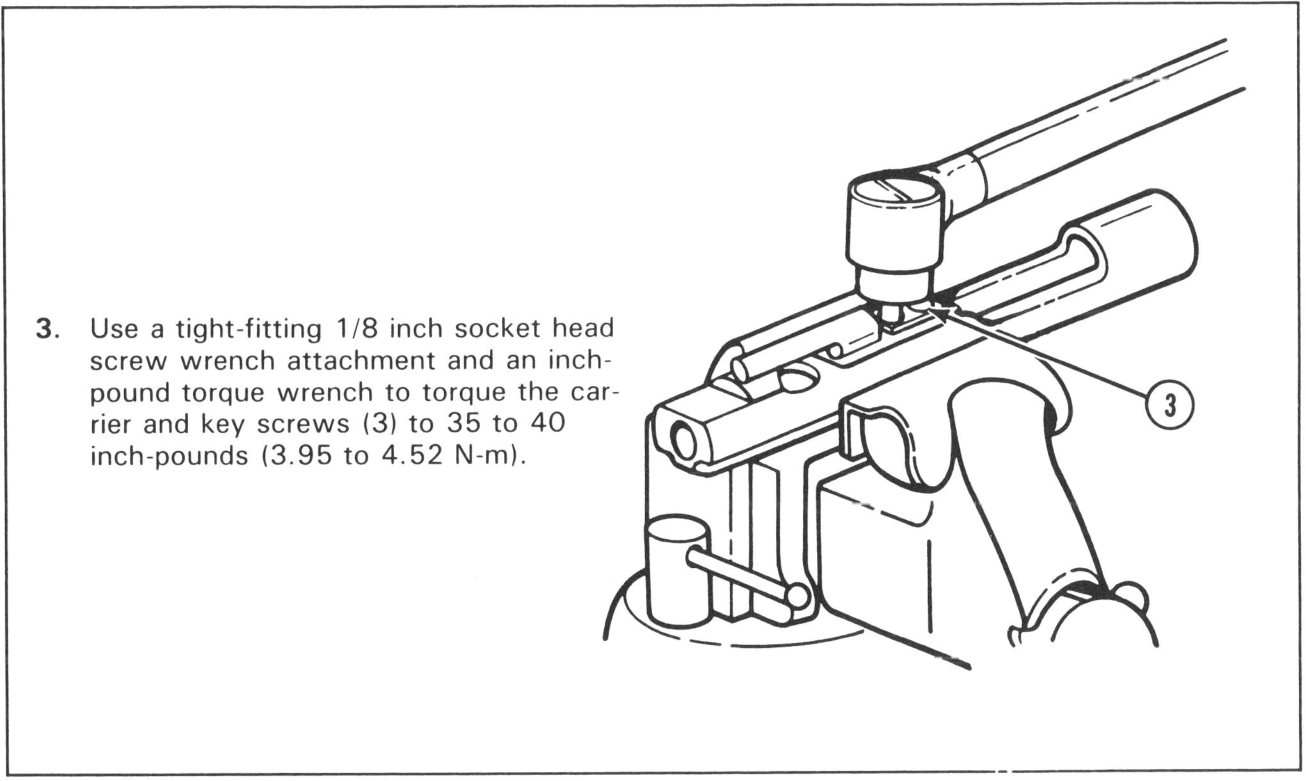

3. Use a tight-fitting 1/8 inch socket head screw wrench attachment and an inch-pound torque wrench to torque the carrier and key screws (3) to 35 to 40 inch-pounds (3.95 to 4.52 N-m).

3-9. KEY AND BOLT CARRIER ASSEMBLY (CONT).

c. REASSEMBLY (CONT)

NOTE
Field staking method will be used
by field units.

4. Use solid center punch and hand hammer to stake the two carrier and key screws (3) in three places.

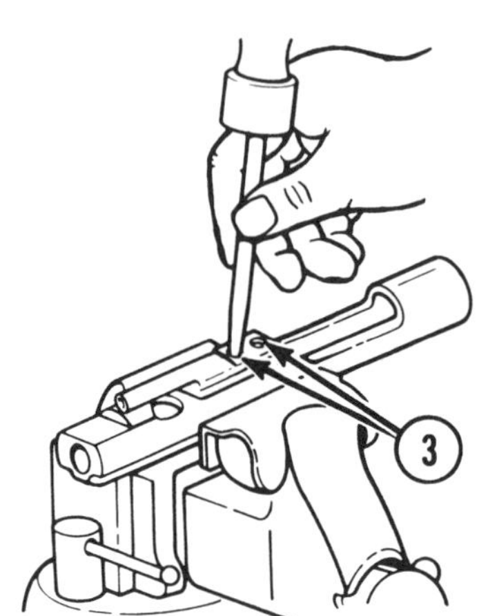

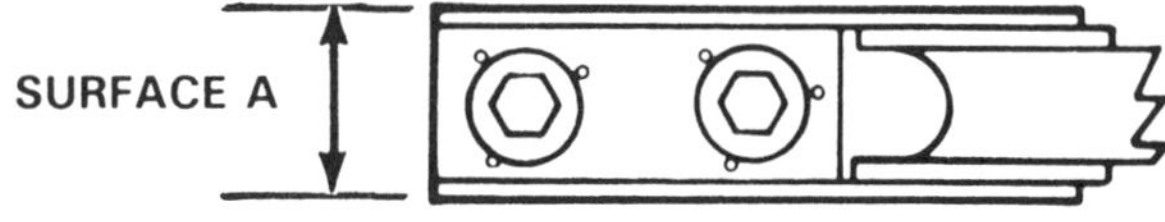

FIELD REPLACEMENT STAKING

5. Reassemble rifle, refer to page 3-77.

CAUTION
If blanks are used, blank firing attachment (BFA) must be attached.

NOTE
If the bolt carrier key is replaced, three to eight rounds of blank or ball ammunition must be fired to ensure a seal is created. Manual operation of the rifle may be required. If blank ammunition is utilized, M16A2 Blank Firing Attachment must be adapted.

3-10. UPPER RECEIVER AND BARREL ASSEMBLY.

This task covers:

a. Disassembly
b. Inspection/Cleaning
c. Repair

d. Reassembly
e. Test

INITIAL SETUP

Test Equipment
   Tool and Gage Set (item 2, app B)

Tools
   (ARMY) Small Arms Repairman Tool Kit
      (item 3, app B)
   Field Maintenance Basic Less Power
      Small Arms Shop Set (item 1, app B)

Materials/Parts
   Brush, cleaning, small (item 3, app D)
   Carbon removing compound (item 8, app
      D)
   Cloth, abrasive (item 13, app D)
   Dichloromethane, technical (item 15, app
      D)
   Dry cleaning solvent (item 16, app D)
   Gloves, chemical and oil protective (item
      18, app D)
   Grease, molybdenum disulfide (item 19,
      app D)
   Lubricant, solid film (item 21, app D)
   Pan, wash (item 24, app D)
   Polyethylene (item 32, app D)
   Sealing compound (item 28, app D)
   Target (item 31, app D)

References
   FM 23-9
   TM 9-1005-319-10

Equipment Conditions
   3-15   Upper receiver and barrel assembly
          removed
   2-43   Handguard assemblies removed

General Safety Instructions
   To avoid injury to your eyes, use care
      when removing and installing spring-
      loaded parts.
   When using solid film lubricant or
      dichloromethane, be sure the area is
      well ventilated.
   When using carbon removing compound,
      avoid skin contact. If carbon removing
      compound comes in contact with the
      skin, wash thoroughly with running
      water. Using a good lanolin base cream
      after exposure to the compound is
      helpful. Using gloves and protective
      equipment is required.

3-10. UPPER RECEIVER AND BARREL ASSEMBLY (CONT).

a. DISASSEMBLY

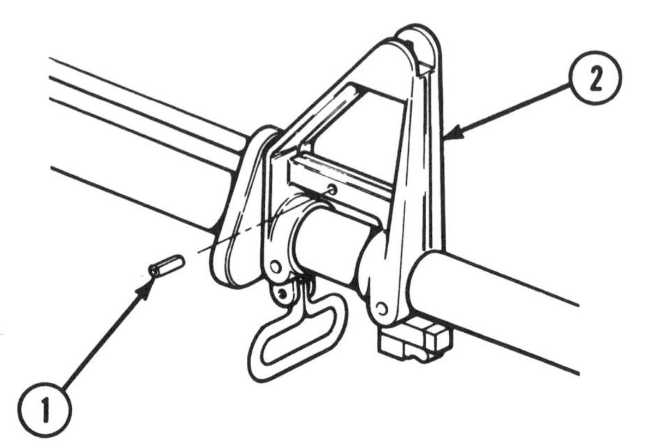

1. Using ball-peen hammer and 5/64 inch diameter drive pin punch, drive spring pin (1) (which retains the gas tube) out of front sight assembly (2).

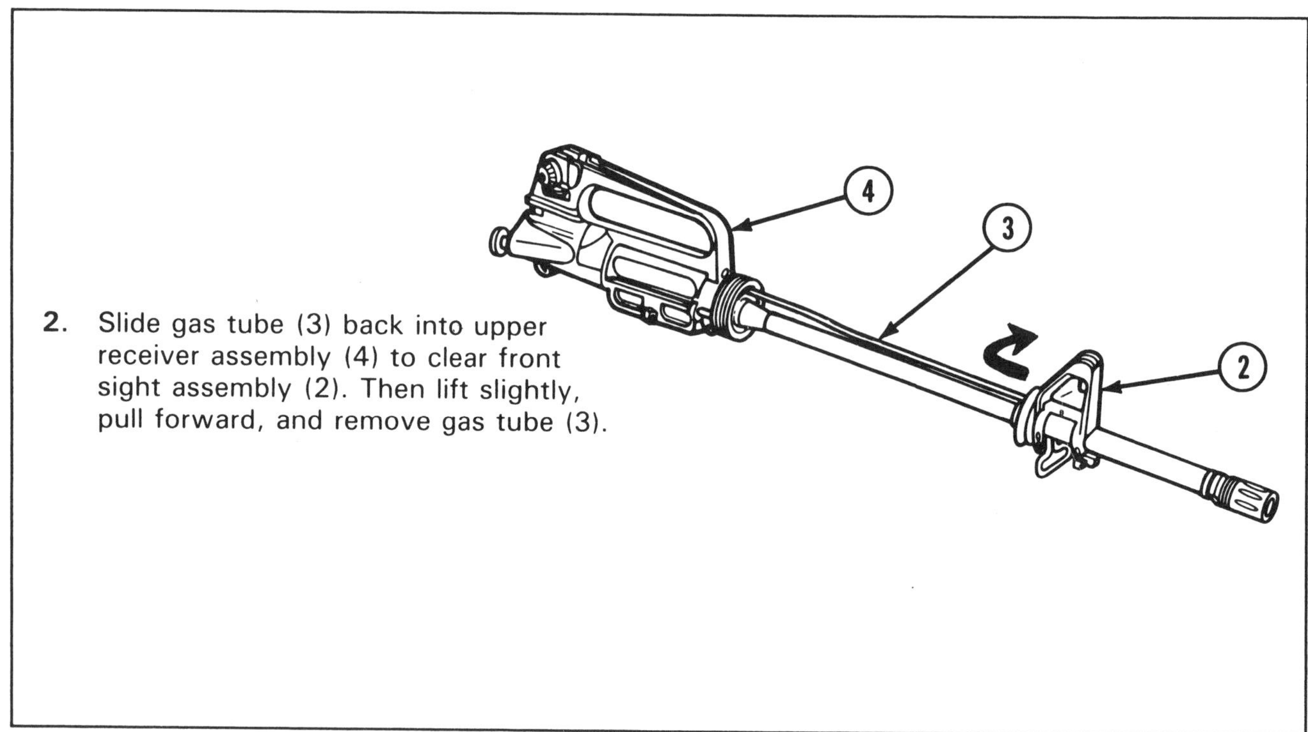

2. Slide gas tube (3) back into upper receiver assembly (4) to clear front sight assembly (2). Then lift slightly, pull forward, and remove gas tube (3).

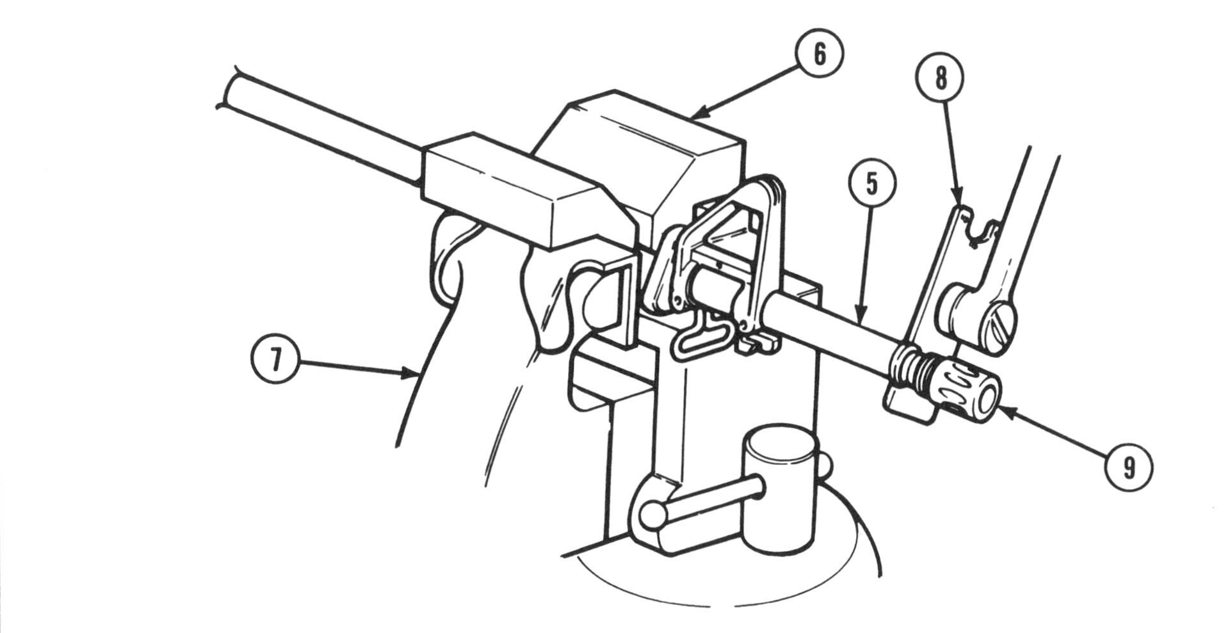

3. Position upper receiver and barrel assembly (5) in barrel removal fixture (6) and secure both in machinist's vise (7).

4. Using combination wrench (8) and ½ inch drive handle, remove compensator (9).

5. Remove peel washer (10) being careful not to lose or bend thin sections.

6. Remove upper receiver and barrel assembly (5) from barrel removal fixture and machinist's vise.

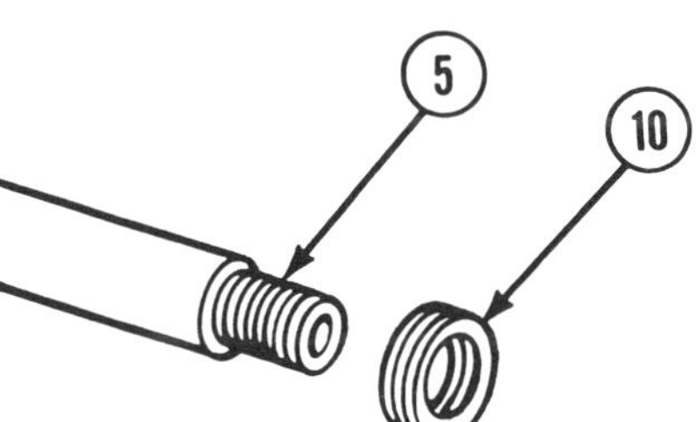

3-10. UPPER RECEIVER AND BARREL ASSEMBLY (CONT).

a. DISASSEMBLY (CONT)

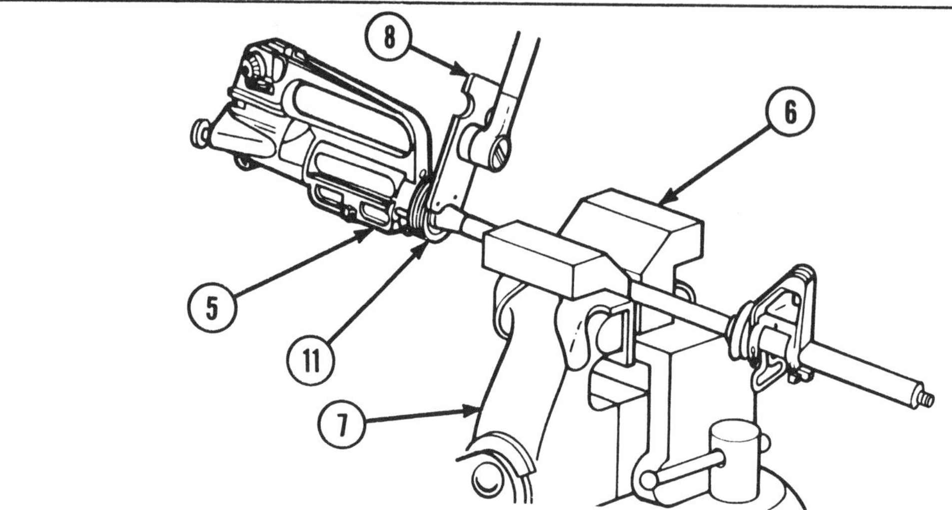

7. Place upper receiver and barrel assembly (5) into barrel removal fixture (6) and clamp into machinist's vise (7).

NOTE

Be sure all three drive pins on combination wrench are fully engaged with barrel nut assembly. Wrench must be pushed toward upper receiver assembly to compress the slip ring spring in barrel nut assembly. Do not use a torque wrench to loosen the barrel nut assembly.

8. Using ½ inch drive handle and combination wrench (8), loosen barrel nut assembly (11).

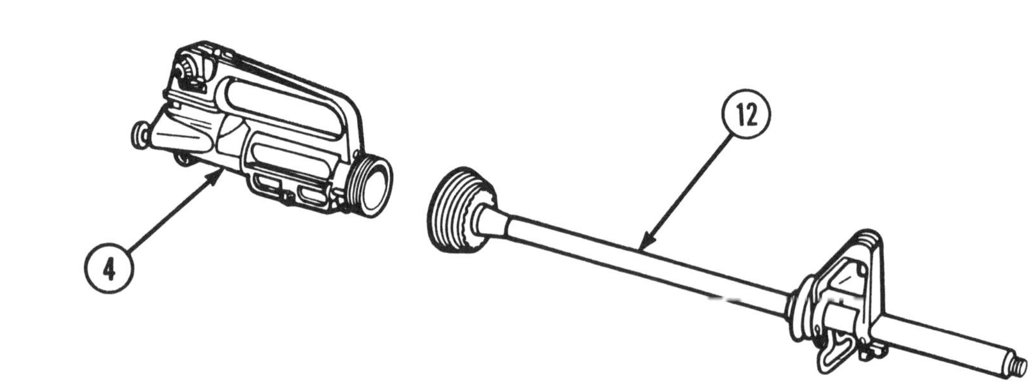

9. Separate upper receiver assembly (4) from barrel assembly (12).

10. Remove barrel assembly (12) from machinist's vise and barrel removal fixture.

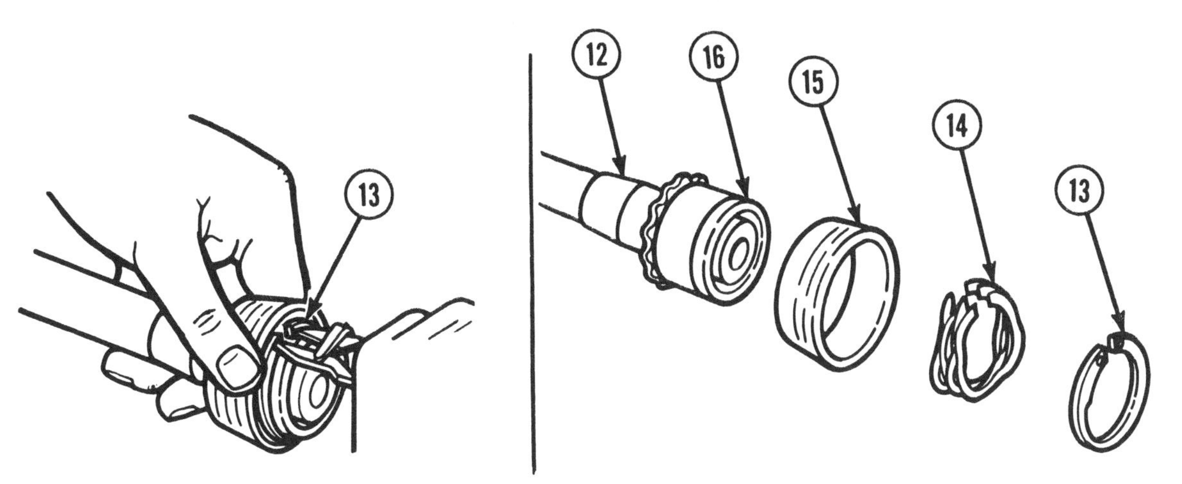

11. Remove retaining ring (13) using retaining ring pliers.

12. Remove slip ring spring (14) and handguard slip ring (15).

13. Do not remove barrel nut (16) from barrel assembly (12).

b. INSPECTION/CLEANING

WARNING

When using carbon removing compound, avoid skin contact. If carbon removing compound comes in contact with the skin, wash thoroughly with running water. Using a good lanolin base cream after exposure to compound is helpful. Using gloves and protective equipment is required.

1. Inspect gas tube for cracks. Replace if defective.

2. Use carbon removing compound to remove carbon deposits from interior and exterior of gas tube. If a large amount of carbon is found and cannot be removed, replace the gas tube.

NOTE

A small arms cleaning brush (bore) (item 4, app D) may be used to clean interior of front sight assembly where gas tube is secured.

3-10. UPPER RECEIVER AND BARREL ASSEMBLY (CONT).

b. INSPECTION/REPAIR (CONT)

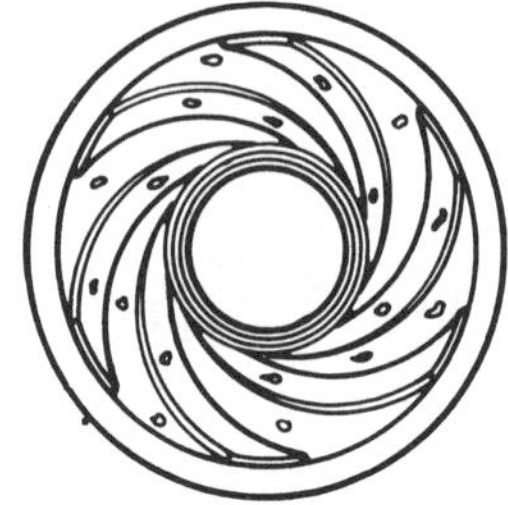

3. Inspect bore for burrs, cracks, rust, bulges, and pits using the following guidelines.

    (a) Pits no wider than a land or groove and no longer than 3/8 inch (0.95 cm) are allowed in the bore.

    (b) Lands that appear dark blue due to coating of gliding metal from projectiles are allowable.

    (c) Definitely ringed bores or bores ringed sufficiently to bulge the outside surface of the barrel are cause for rejection. Replace barrel assembly if defective.

4. If the upper receiver is separated from the barrel assembly, inspect chamber for pits utilizing a flashlight. Pits 1/8 inch (0.32 cm) in length are cause for rejection. Replace rifle barrel assembly if defective.

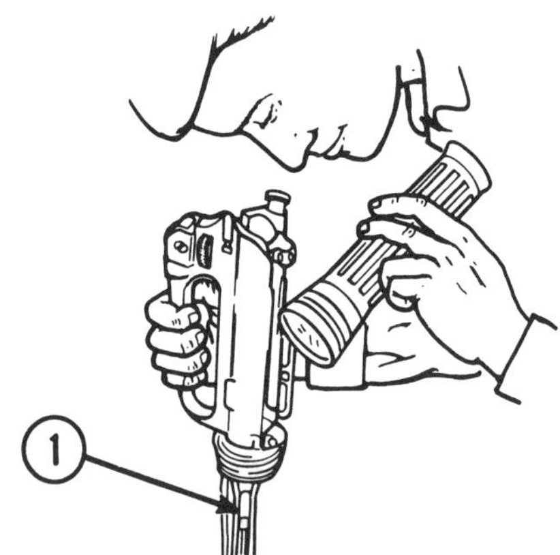

5. If upper receiver and barrel assembly is assembled, inspect chamber using reflector tool (1) and flashlight. Pits 1/8 inch (0.32 cm) in length are cause for rejection. Replace rifle barrel assembly if defective. If rifle barrel assembly is replaced, inspect headspace (p 3-45).

6. Inspect upper receiver assembly for cracks, corrosion, wear, or damage.

(a) Small dents or gouges that do not affect functioning will not be cause for rejection.

(b) If upper receiver assembly contains cracks or holes, the upper receiver assembly will be replaced.

7. Inspect all parts for damage and wear. Replace all defective parts.

NOTE

Damaged or missing teeth of the barrel nut is not cause for rejection provided the proper torque value can be obtained during installation using the tools depicted. If removal of the barrel is not possible with the combination tool, a pipe wrench or other such tool may be used during removal.

8. Inspect front sight guards for bends, if bent see page 3-37 for repair procedures.

c. REPAIR

1. Repair corroded upper receiver assembly surfaces as follows:

(a) Sand corroded area with abrasive cloth and make sure all corrosion has been removed.

WARNING

When using solid film lubricant or dichloromethane, be sure the area is well ventilated.

(b) Wash area with technical dichloromethane (methylenechloride) to remove all dirt, grease, and foreign material.

(c) Apply sealing compound, mixed in accordance with manufacturer's directions, to areas to be filled.

(d) Spread sealing compound as smoothly as possible into defective area using a putty knife or similar tool.

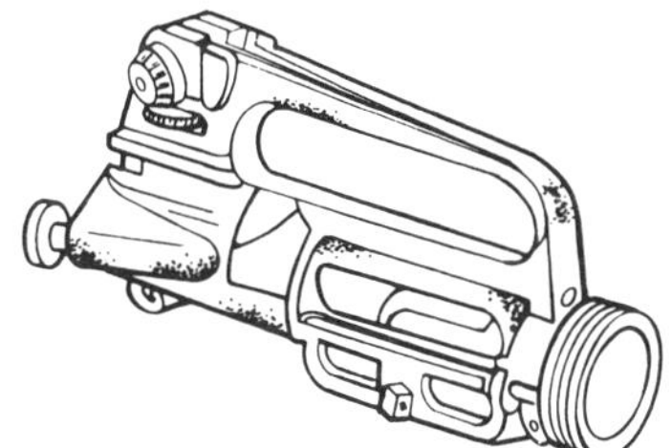

CORRODED
(REPARABLE)

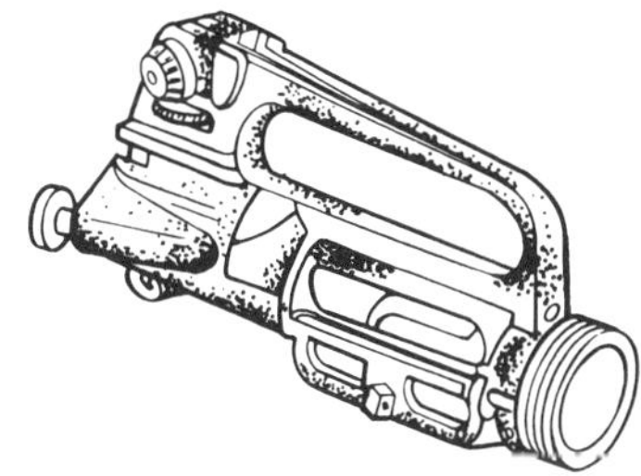

CORRODED
(NONREPARABLE)

3-10. UPPER RECEIVER AND BARREL ASSEMBLY (CONT).

c. REPAIR (CONT)

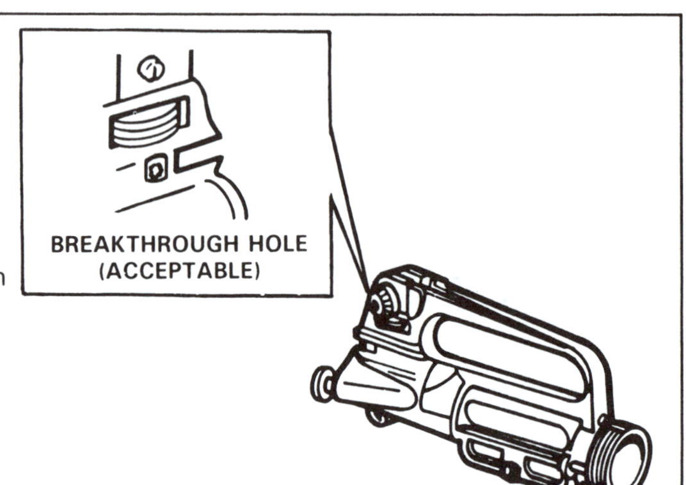

NOTE

Do not feather edges.

(e) Place a sheet of polyethylene (item 32, app D), cut to size, over filled area. Rub by hand to smooth.

2. After curing, remove polyethylene sheet in accordance with instructions by the manufacturer.

WARNING

When using solid film lubricant or dichloromethane, be sure the area is well ventilated.

CAUTION

Solid film lubricant is to be used only as an exterior surface protective finish and touchup. If solid film lubricant comes in contact with recoiling parts or functional surfaces of the rifle, remove immediately by washing with technical dichloromethane.

3. Wash area with technical dichloromethane (methylenechloride) to remove all dirt, grease, and foreign material.

4. Roughen area to be refinished with abrasive cloth and clean surface again. Do not touch the area with fingers.

5. Repair shiny surfaces by spraying a coat of solid film lubricant in accordance with instructions supplied by the manufacturer. Dry 24 hours before handling.

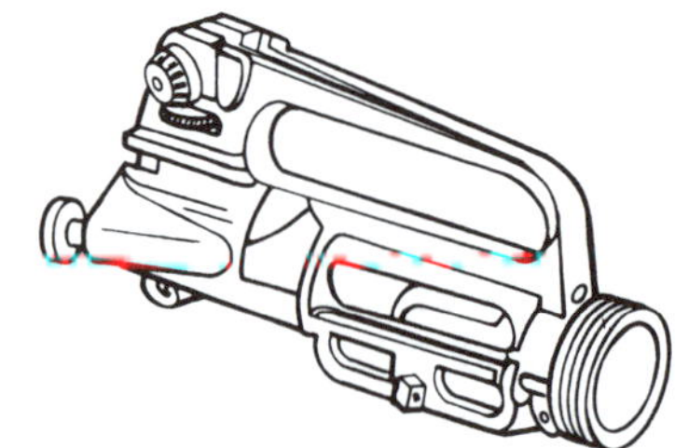

SHINY SURFACES
(REPARABLE)

6. Straighten bent front sight guards (1) as follows:

NOTE

Remove spring before heating. (Heat will damage spring.) The sight post and plunger may be reused unless damaged.

(a) Remove front sight post, detent, and helical spring (see p 2-43).

NOTE

Use copper or brass caps (jaw inserts) on bench vise to prevent damage to front sight base (2) during clamping.

(b) Place front sight base (2) in a bench vise.

(c) Heat front sight guards (1) and bend with pliers. The front sight guards (1) should be put back as nearly as possible to the original position. Allow front sight housing to air cool.

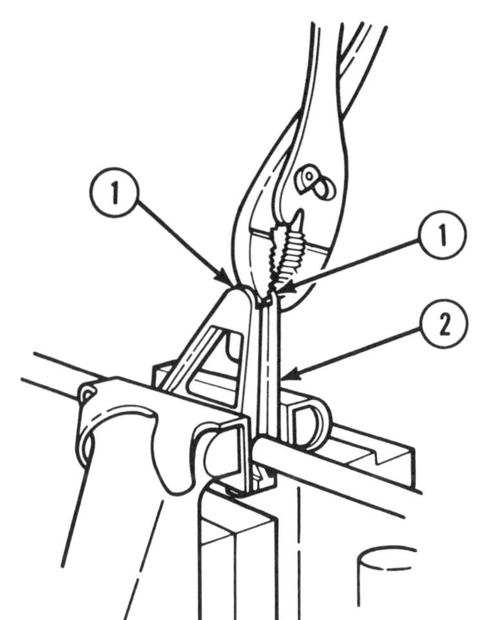

WARNING

Dry cleaning solvent is flammable and toxic and should be used in a well-ventilated area. The use of rubber gloves is necessary to protect skin when washing rifle parts.

(d) Roughen any damaged surface of front sight guards with abrasive cloth and clean with dry cleaning solvent. Always wear rubber gloves when using dry cleaning solvent.

3-10. UPPER RECEIVER AND BARREL ASSEMBLY (CONT).

c. REPAIR (CONT)

CAUTION

Do not allow solid film lubricant to flow into front sight post threaded well.

(e) Apply solid film lubricant to cover the damaged finish.

(f) If front sight guards cannot be straightened utilizing the above procedures, replace the rifle barrel assembly.

7. Slightly bent barrels may be straightened as follows:

(a) Check straightness using straightness gage 8448202 (p 3-45). If the barrel fails the straightness test, and the gage remains in the barrel in the area of the front sight assembly, perform step (b) to determine if it may be straightened.

(b) With the gage remaining in the bore, hold the rifle in a vertical position with the end of the barrel into which the gage was inserted pointing up. Ensure that if/when the gage passes through the barrel it will not be damaged. Using hand pressure ONLY, flex the portion of the barrel between the front sight assembly and the compensator in all four directions (left, right, forward, and back). If the barrel is only slightly bent, the gage will drop through when the barrel is flexed in one of these directions. Note the direction which allowed the gage to drop through the barrel.

CAUTION

Remove the gage from the barrel before continuing.

NOTE

If the gage does not pass through the barrel when it is flexed, replace the rifle barrel assembly.

(c) Place the barrel in a vise using appropriate protective jaws. Clamp the barrel between the front sight assembly and the compensator approximately 1 inch (2.54 cm) from the front sight assembly. The rifle barrel assembly should be in a horizontal position with the side noted in step (b) toward you.

CAUTION
Do not apply pressure to the receiver.

(d) Grasp the BARREL near the receiver so that when force is applied the barrel will flex in the same direction as noted in step (b).

(e) Give the barrel a sharp jerk of approximately 20 to 40 pounds of force.

(f) Remove the barrel from the vise and recheck straightness (step (a)).

(g) If gage still will not pass through the barrel, perform step (b) to determine direction of bend. If the barrel is still bent in the same direction as before, perform steps (c) through (f) using slightly more force. If the barrel is now bent in the opposite direction, replace the rifle barrel assembly.

(h) If the gage passes freely through the barrel, the barrel shall be considered straight and continue in service.

(i) If the barrel has been straightened, the rifle must be targeted (p 3-45).

d. REASSEMBLY

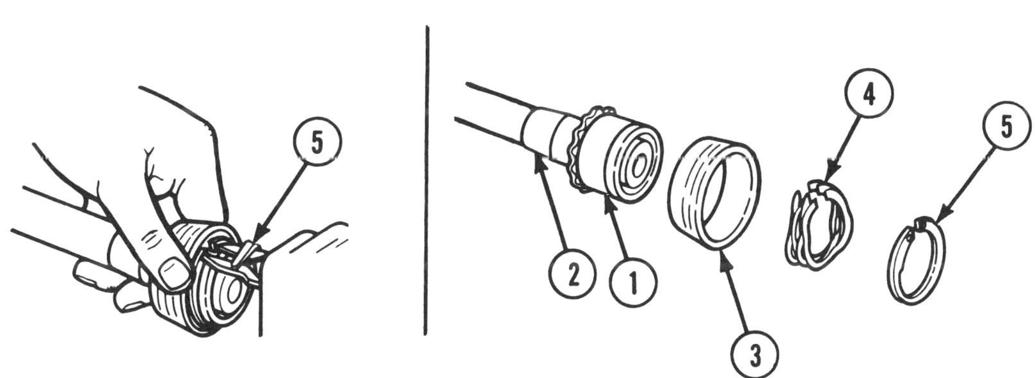

1. Position barrel nut (1) by sliding it to the rear of rifle barrel assembly (2) as far as possible.

2. Slide handguard slip ring (3) over barrel nut (1).

3. Press slip ring spring (4) from both sides and insert it into handguard slip ring (3).

4. Install retaining ring (5) against slip ring spring (4) using retaining ring pliers. Snap retaining ring (5) to barrel nut (1).

NOTE
After cleaning, apply molybdenum disulfide grease to threads of barrel nut assembly before installation.

3-10. UPPER RECEIVER AND BARREL ASSEMBLY (CONT).

d. REASSEMBLY (CONT)

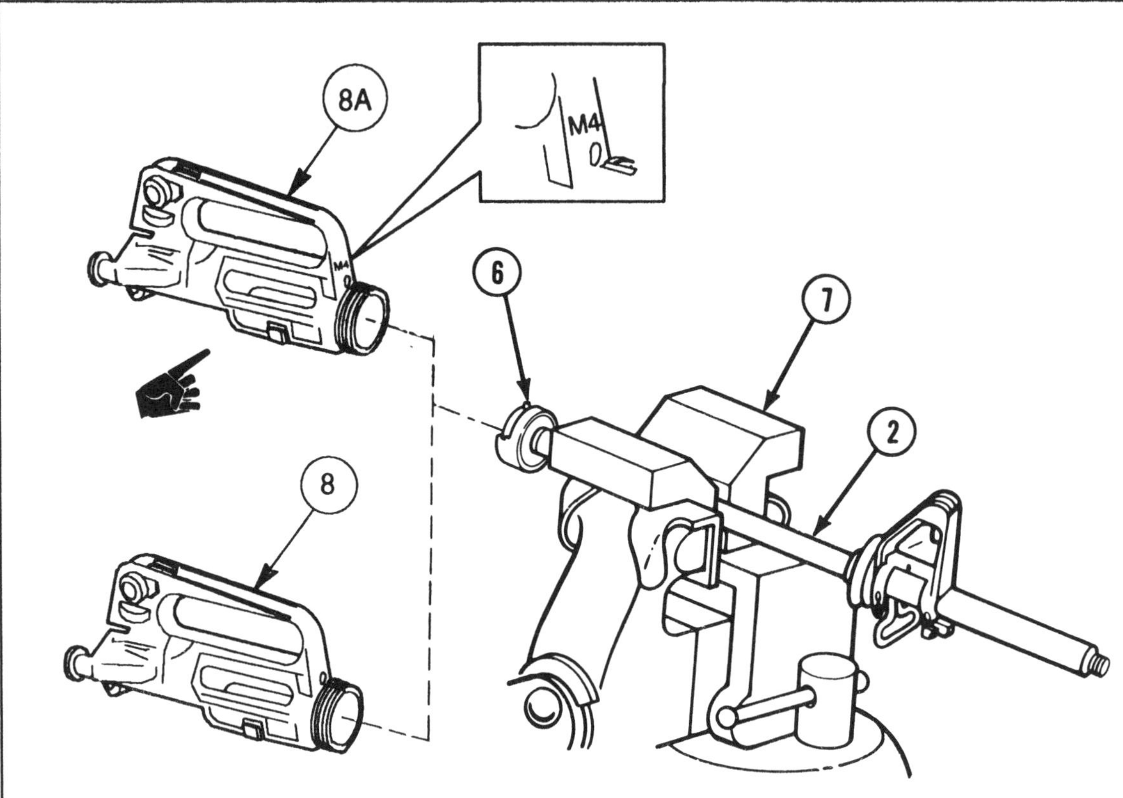

NOTE
The alignment pin (6) must not show any signs of looseness.

5. Position rifle barrel assembly (2) with alignment pin (6) up. Using barrel remover fixture (7), clamp rifle barrel assembly (2) in vise.

NOTE
The slot should fit the alignment pin perfectly with very little or no rotational play present. Note the play in a new rifle barrel assembly and new upper receiver assembly and use this as a guide.

6. Align upper receiver assembly (8) or (8A) using alignment pin (6) and the slot in upper receiver assembly (8) or (8A). Install over end of rifle barrel assembly (2).

7. Wipe upper receiver threads clean and ensure there are no burrs. Apply molybdenum disulfide grease to the threads before installation.

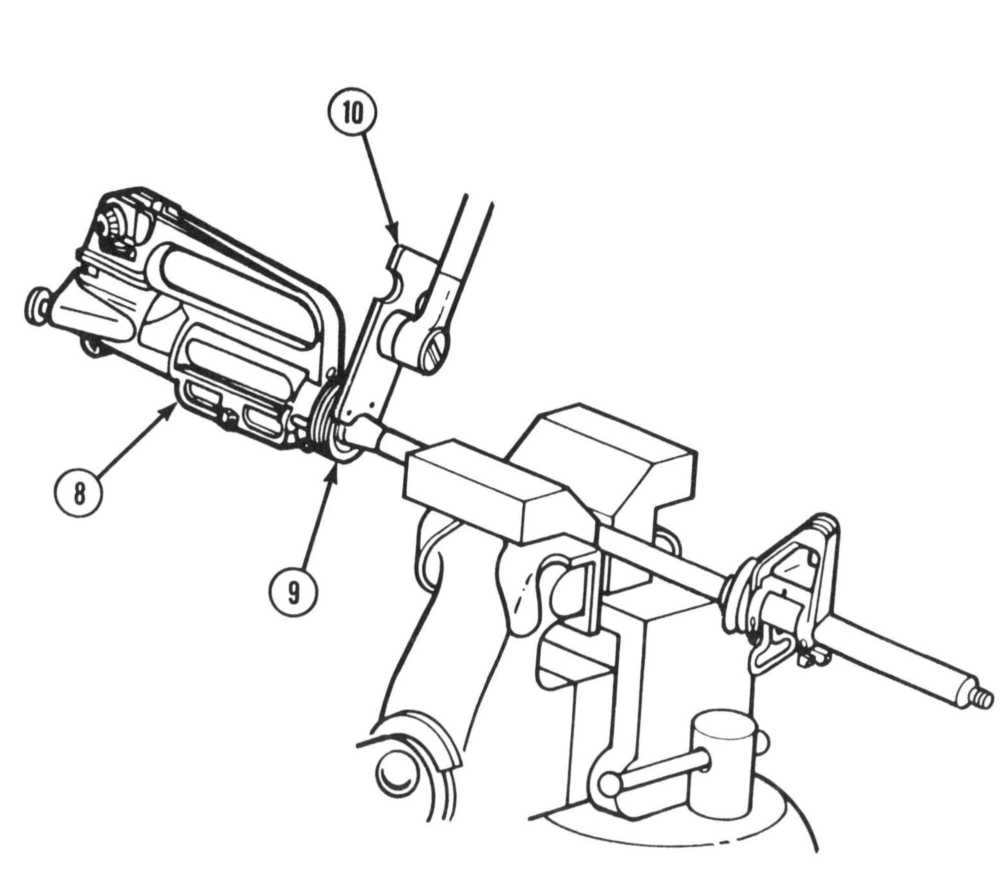

8. Engage threads of barrel nut assembly (9) with upper receiver assembly (8).

9. Using combination wrench (10) and torque wrench, torque barrel nut assembly (9) to 30 ft-lb (40.5 N-m). Torque is measured when both wrenches are used together.

NOTE
Three times torquing procedure provides for a better thread fit and prevents barrel nuts from becoming loose. Do not use the torque wrench for loosening.

10. Make certain all three drive pins on combination wrench are engaged with barrel nut assembly (9). Loosen and repeat torque operation. Then loosen the barrel nut again.

3-10. UPPER RECEIVER AND BARREL ASSEMBLY (CONT).

d. REASSEMBLY (CONT)

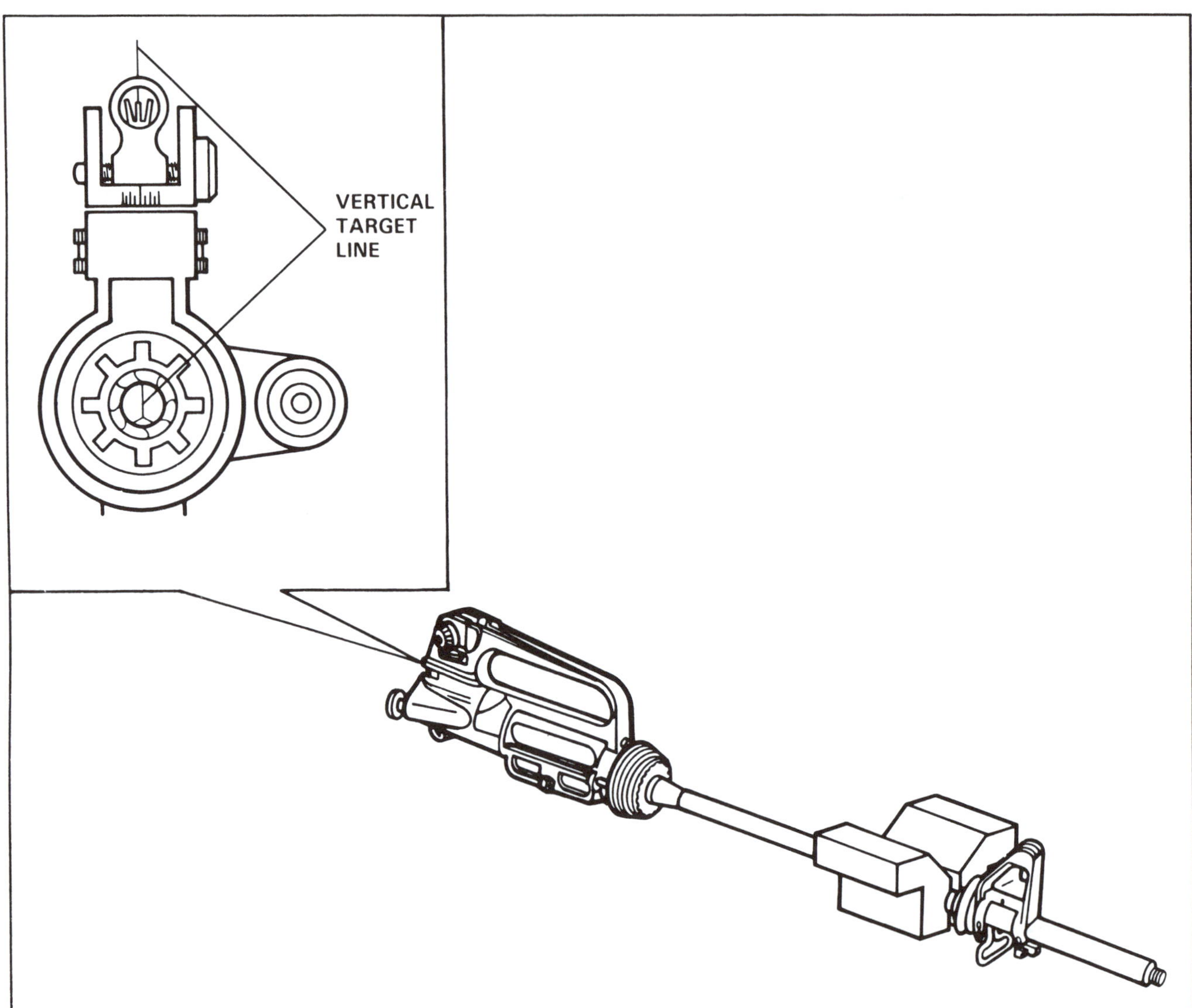

11. Front sight post must be installed. Loosen the vise and align the bore on a distant vertical target. Center the target in the bore from 12 o'clock through 6 o'clock. The front sight post should be on line and vertical with the target. Tighten vise. Adjust the rear sight windage until a proper sight picture is obtained on the vertical target. The rear sight aperture will be approximately in the center of the rear sight base if the rifle barrel assembly is properly aligned in the upper receiver assembly.

NOTE

If rifle barrel assembly (usually new) is not properly aligned in the upper receiver assembly (usually an old part), excessive windage will be present and the upper receiver assembly will require replacement to obtain the proper fit between the alignment pin and slot.

CAUTION

Do not torque over 80 ft-lb (108 N-m) while tightening the barrel nut assembly to the next hole, to allow for proper alignment of gas tube.

NOTE

Do not attempt to hold the upper receiver assembly with a pry bar; however, if the rifle barrel assembly turns in the holding fixture, a pry bar may be used through the front sight assembly base to help prevent the rifle barrel assembly from turning in the holding fixture. Use care not to distort or bend front sight assembly or retaining pins. Use ''buddy system'' to hold pry bar.

12. Torque the barrel nut assembly again to 30 ft-lb (40.5 N-m) while maintaining sight alignment. The barrel nut assembly may be tightened beyond 30 ft-lb (40.5 N-m) to align the barrel nut assembly serrations for proper gas tube clearance. Never loosen the barrel nut assembly to align for gas tube clearance.

13. Check alignment of barrel nut assembly (9) with upper receiver assembly (8). The front 8 inches (20.32 cm) of a gas tube may be used as an alignment tool (see illustration). This is inserted into the bolt carrier key and then inserted into the rear of the receiver. If the parts of the barrel nut assembly are properly aligned, the tool will pass freely and lay top dead center along the top of the barrel. A number 15 twist drill (0.180 inch) may also be needed as an alignment tool. If necessary, tighten barrel nut assembly to next hole to allow proper alignment.

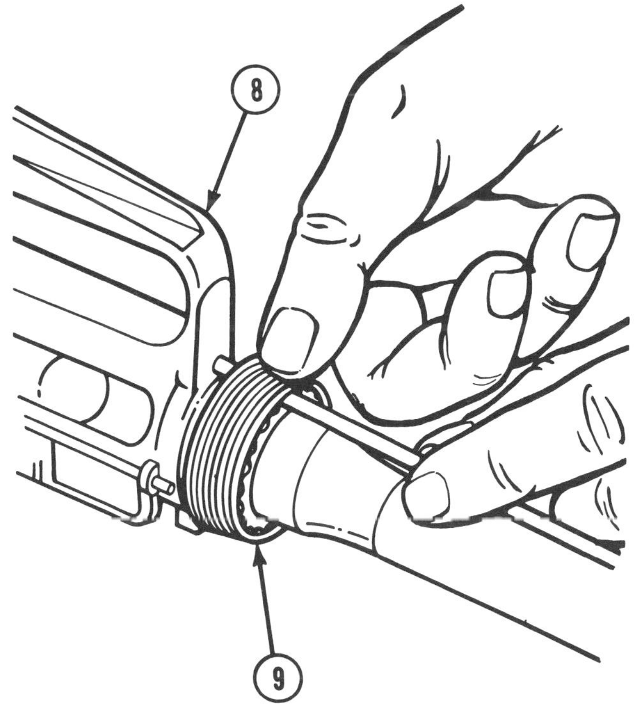

3-10. UPPER RECEIVER AND BARREL ASSEMBLY (CONT).

d. REASSEMBLY (CONT)

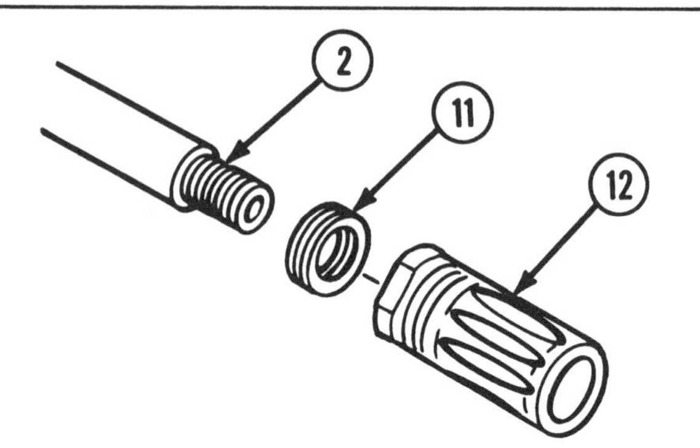

NOTE

The peel washer may be heated to remove thin sections. Always place thin sections to rear.

14. Install peel washer (11) and compensator (12) on rifle barrel assembly (2).

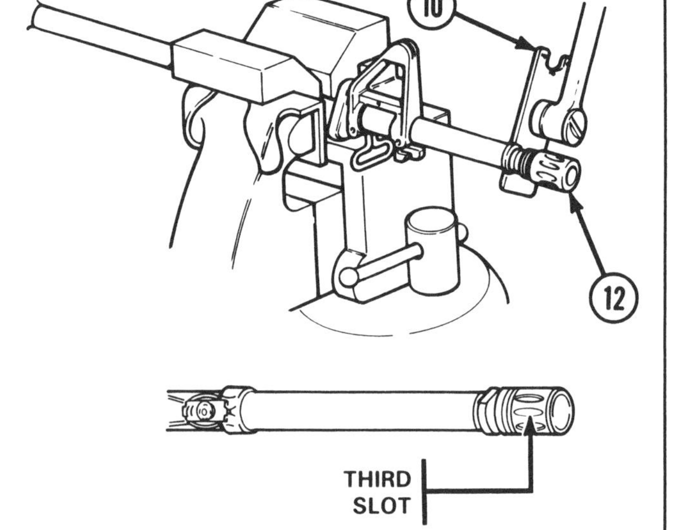

15. Torque compensator (12) to 15-20 ft-lb (20-27 N-m) using combination wrench (10) and torque wrench. Torque is measured when both wrenches are used together. The third or middle slot must be straight up at proper torque level. Thin sections of the peel washer may be removed or added as required. Save unused sections.

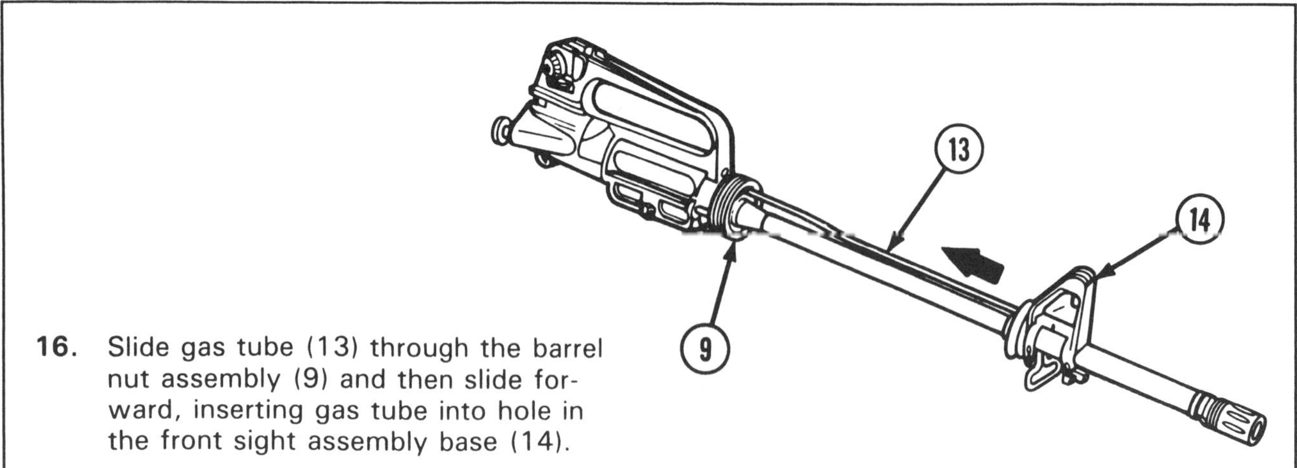

16. Slide gas tube (13) through the barrel nut assembly (9) and then slide forward, inserting gas tube into hole in the front sight assembly base (14).

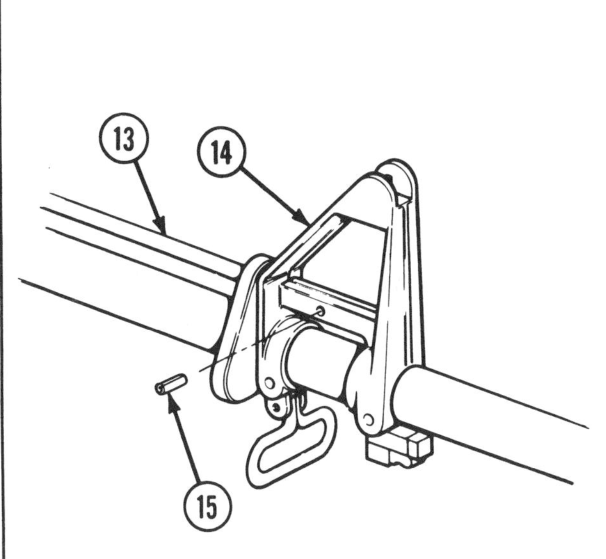

17. Align the holes in gas tube (13) and front sight assembly base (14).

18. Using ball-peen hammer and 5/64-inch diameter drive pin punch, drive spring pin (15) into front sight assembly base (14) to secure gas tube (13).

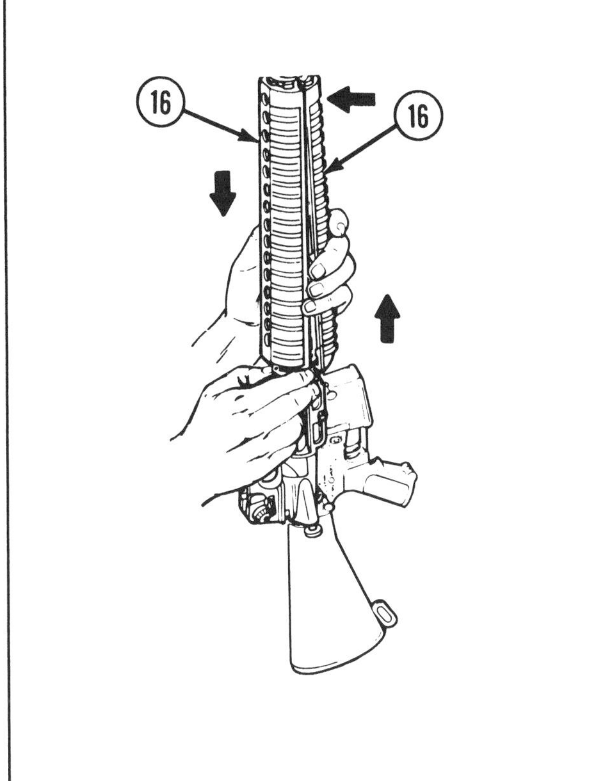

19. Install upper and lower handguard assemblies (16).

---

e. TEST

---

1. The following information pertains to the use of breech, bore, and other gages:

   (a) All M16A2 barrels and chambers are chromed.

   (b) Barrel erosion gage, PN 8448496 (normally used on M16A1 fully-chromed barrels), can be used to gage M16A2 barrels.

   (c) The bore straightness gage, PN 8448202, is required for use on all barrels. The gage must pass through the barrel without being forced.

3-10. UPPER RECEIVER AND BARREL ASSEMBLY (CONT).
---

e. TEST (CONT)

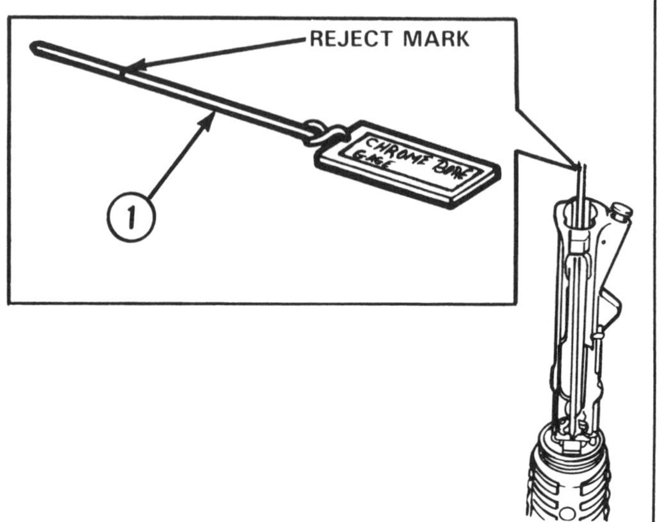

2. Use barrel erosion gage (1) PN 8448496. Install key and bolt carrier assembly with bolt assembly and firing pin removed. Hold rifle vertical with receiver up. Insert gage into rear of key and bolt carrier assembly. The reject line must be read at the rear edge of the key and bolt carrier assembly.

3. If the reject mark passes beyond the rear surface of the key and bolt carrier assembly, the barrel is unserviceable and shall be replaced.

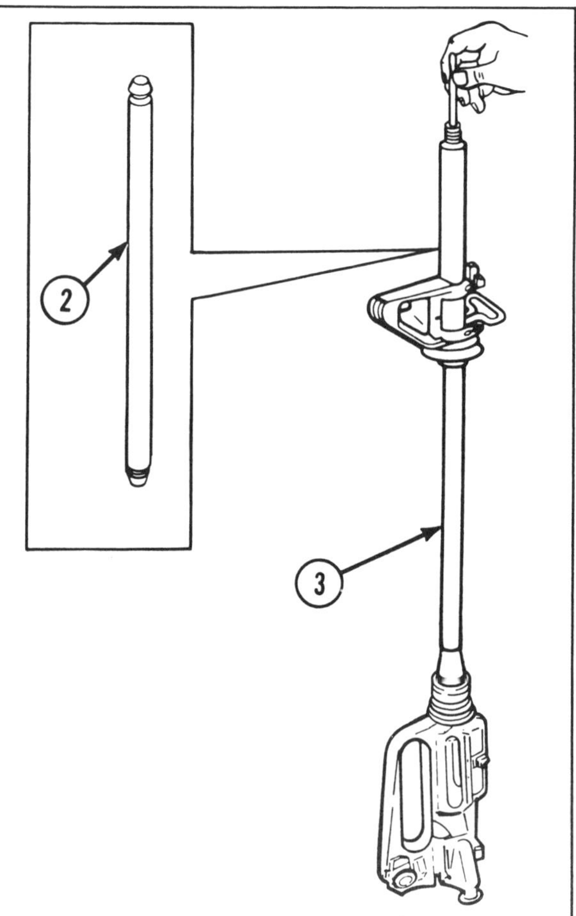

NOTE
Ensure barrel is clean prior to performing the following test.

4. Check straightness of bore using straightness gage (2) PN 8448202. Put gage in barrel. Tilt barrel and allow gage to fall through. Catch gage.

5. Gage must pass freely through barrel. If gage does not pass through barrel, recheck as follows. Hold upper receiver and barrel assembly (3) in vertical position with muzzle pointed down; insert gage into chamber end of barrel. Release gage and catch it as it exits muzzle end. If gage passes freely through the barrel, barrel is acceptable. If it does not, the barrel must be straightened or replaced. (See page 3-38 for straightening instructions/procedures.)

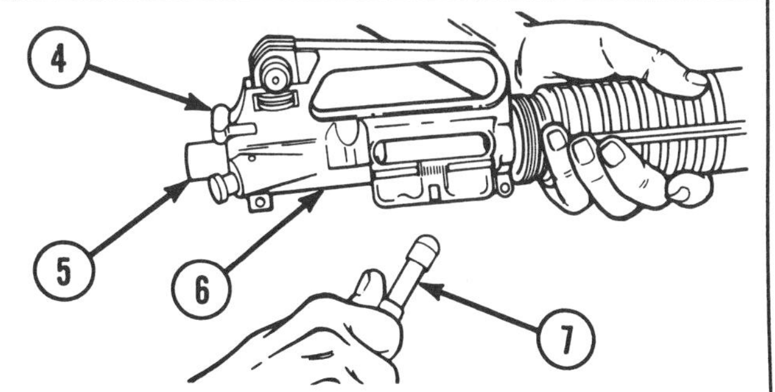

6. Assemble charging handle assembly (4), bolt assembly, and key and bolt carrier assembly (5) into upper receiver assembly (6).

7. Insert headspace gage (7) PN 7799734 in chamber.

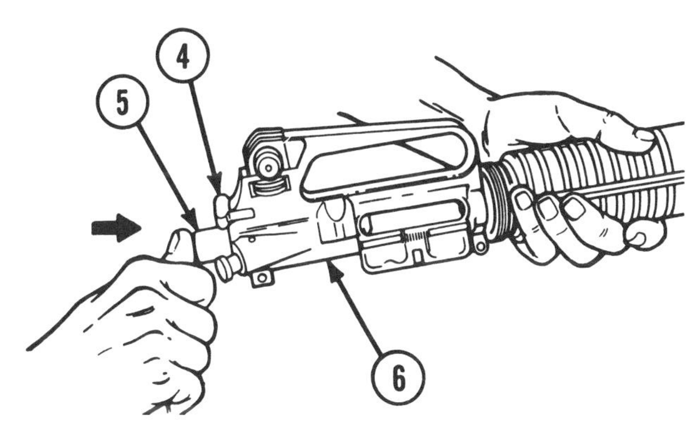

8. Check headspace by pressing key and bolt carrier assembly (5) and charging handle assembly (4) forward using light finger pressure.

9. Bolt should not rotate to locked position. Key and bolt carrier assembly (5) must protrude from rear of upper receiver assembly (6) for proper headspace. If key and bolt carrier assembly is flush with or indented to rear surface, this indicates excessive headspace. If excessive headspace, first replace old bolt assembly with an unused bolt assembly and then recheck. If headspace is not corrected, replace rifle barrel assembly; then recheck with the original bolt assembly to determine if the bolt assembly is still good or if the bolt assembly should be replaced also.

10. Remove key and bolt carrier assembly, bolt assembly, charging handle assembly, and headspace gage.

11. Reassemble rifle, refer to page 3-77.

NOTE

Rifles which have been rebarreled must be function fired with seven rounds of 5.56mm ball ammunition. After rebarreling, the rifle must be targeted with three rounds of 5.56mm ball ammunition at 25 meter range using target. Refer to TM 9-1005-319-10 and FM 23-9.

3-11.   UPPER RECEIVER ASSEMBLY AND REAR SIGHT ASSEMBLY.

This task covers:

a.	Disassembly	**d.**	Lubrication
b.	Inspection	**e.**	Reassembly
c.	Repair	**f.**	Mechanical Zero Procedures (A.F. Only)

INITIAL SETUP

Tools
  (ARMY) Small Arms Repairman Tool Kit
    (item 3, app B)
  Field Maintenance Basic Less Power
    Small Arms Shop Set (item 1, app B)

Materials/Parts
  Cleaner, lubricant, and preservative (CLP)
    (item 9, app D)
  Lubricant, solid film (item 21, app D)
  Index screw (9349065)

Equipment Conditions
  3-29   Upper receiver assembly removed.

General Safety Instructions
  To avoid injury to your eyes, use care
    when removing and installing spring-
    loaded parts.
  When using solid film lubricant or
    dichloromethane, be sure the area is
    well ventilated.

---

a.   DISASSEMBLY

WARNING

To avoid injury to your eyes, use care when removing and installing spring-
loaded parts.

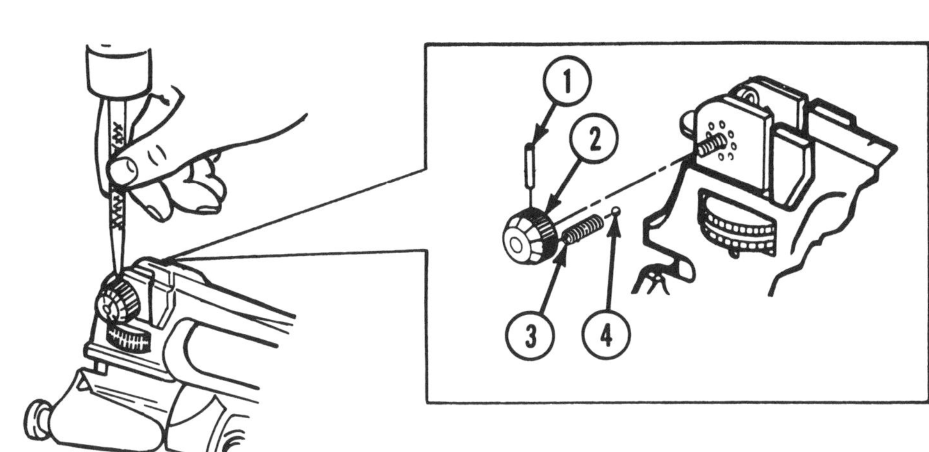

CAUTION

Be sure to catch small parts (1, 2, 3, and 4). **AIR FORCE ONLY**: Use magnet
to keep from losing small parts (1, 3, and 4).

1. Drive out spring pin (1) using a hammer and 1/16 inch punch.

2. Catch rear sight assembly windage knob (2), helical spring (3), and ball bearing (4).

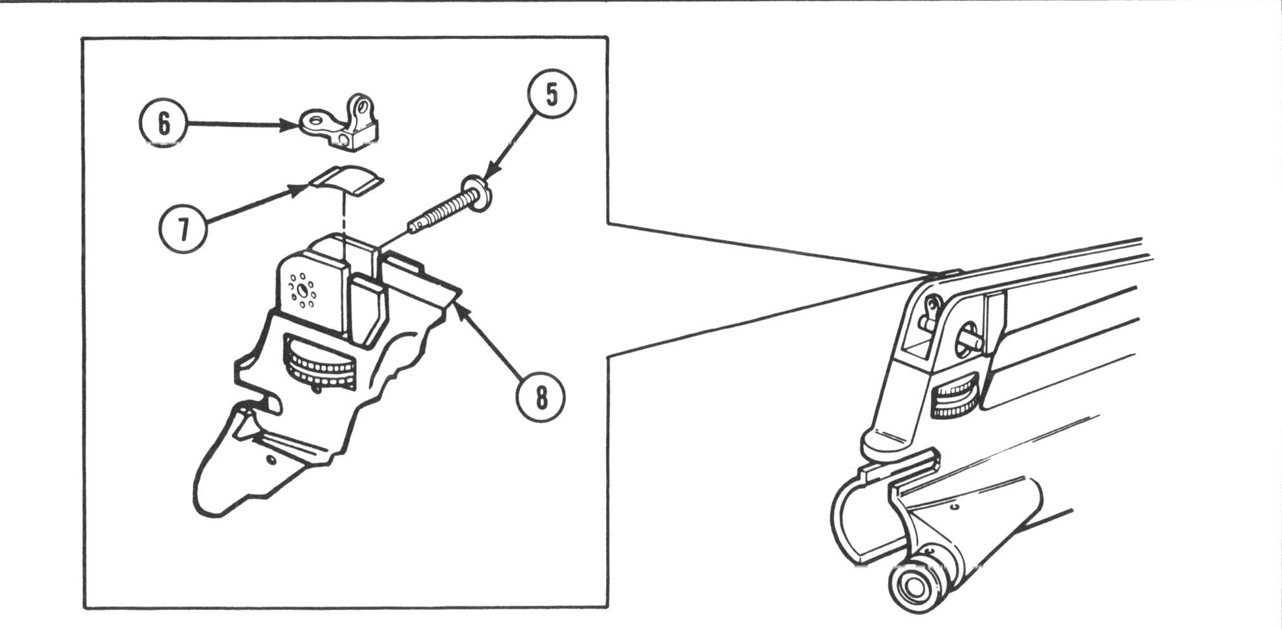

3. Using a flat-bladed screwdriver, remove rear sight assembly windage screw (5).

4. Remove sight aperture (6) and flat spring (7) from upper receiver (8).

3-11. UPPER RECEIVER ASSEMBLY AND REAR SIGHT ASSEMBLY (CONT).

a. DISASSEMBLY (CONT)

WARNING

To avoid injury to your eyes, use care when removing and installing spring-loaded parts.

5. Drive out spring pin (9) using a 3/32 inch punch. Catch helical spring (10) when punch is withdrawn.

6. Rotate elevation index (11) until rear sight assembly base (12) clears upper receiver (8). Catch ball bearing (13) and helical spring (14) as rear sight base clears.

7. Push elevation index (11) out with thumb using slight rotation motion. Catch ball bearing (15) and helical spring (16).

8. Use 1/16 inch allen wrench to remove index screw (17). Discard index screw (17). Separate elevation index (11) from elevation knob (18) by hand.

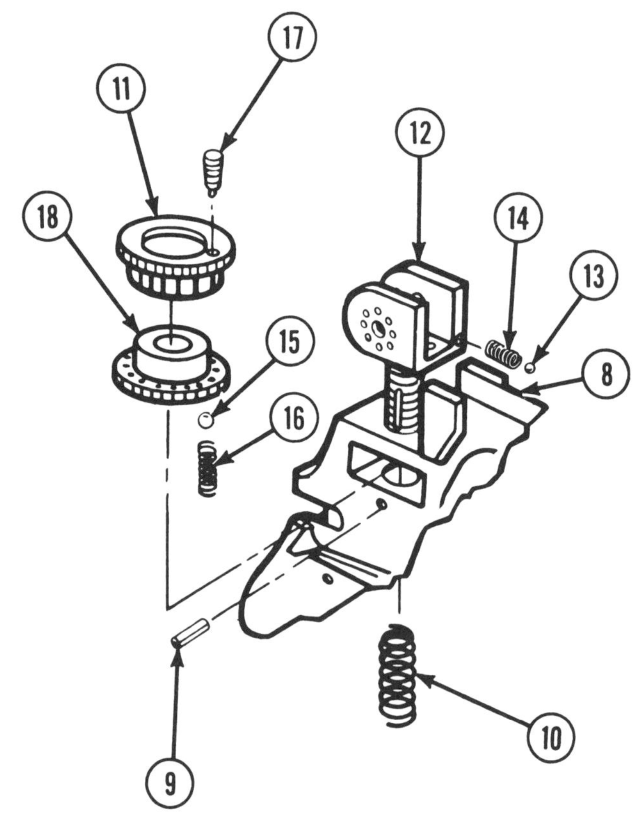

9. Remove pin (19) using 3/32 inch drive pin punch and hand hammer.

WARNING

To avoid injury to your eyes, use care when removing and installing spring-loaded parts.

10. Remove forward assist assembly (20) and helical spring (21) from upper receiver (8). For further disassembly of forward assist assembly see page 3-59.

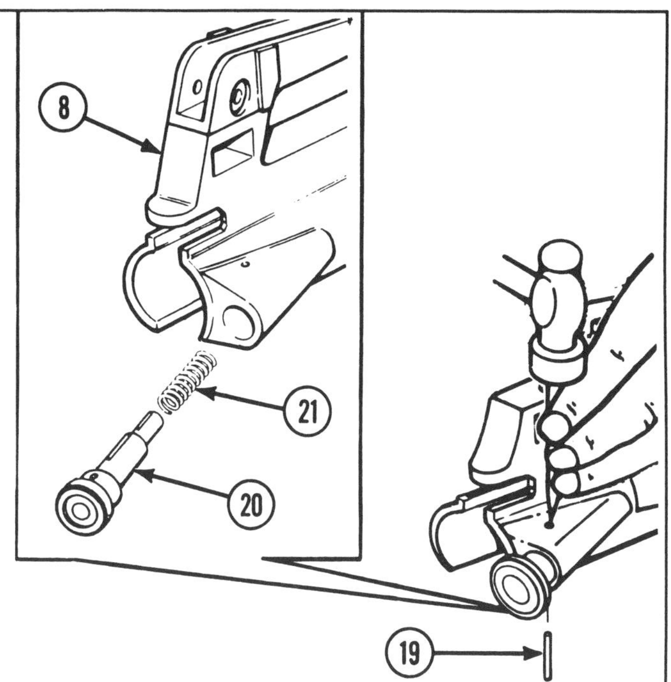

---

b. **INSPECTION**

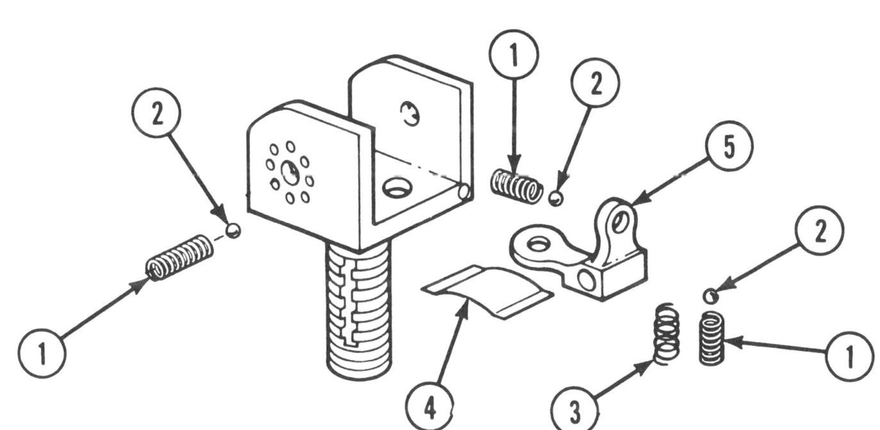

1. Check rear sight parts for serviceability. Inside of apertures should be round and distinct. Replace if defective.

2. Visually inspect rear sight assembly helical springs (1), ball bearings (2), and helical spring (3) for breaks, bends, and missing parts. Ball bearings should be smooth and round. Replace if defective.

3. Check upper receiver for cracks, corrosion, and damage. Clear drain hole with a piece of wire. Repair (p 3-29) or replace if defective.

4. Check that flat spring (4) retains sight aperture (5) firmly in either position. Replace flat spring (4) if sight aperture is not firm.

3-11. UPPER RECEIVER ASSEMBLY AND REAR SIGHT ASSEMBLY (CONT).

b. INSPECTION (CONT)

5. Check elevation index (6) and windage knob (7) for legibility of markings. Check underside of windage knobs for cracks. Detent indexing surfaces should be well formed.

6. Check rear sight assembly base (8) for serviceability. Clear drain holes for springs. Threaded portion of rear sight assembly base and elevation knob should be well formed.

7. Inspect rear sight guards for bends, if bent refer to the following page for repair procedures.

8. Inspect all parts for damage and wear. Replace all defective parts.

c. REPAIR

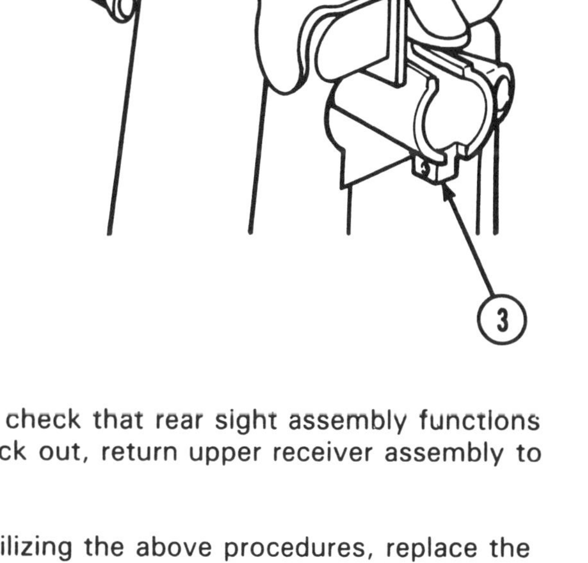

1. To straighten bent rear sight assembly guards (1), remove rear sight assembly components. Place carrying handle (2) in a vise using jaw clamps. Tighten vise to firmly hold upper receiver (3).

2. Using two adjustable wrenches, gradually bend guards (1) to straighten. When bending the guards (1), gradually bend beyond the straight point as the guard will partially return when bending pressure is stopped.

3. After straightening, use a flat file to remove any nicks, kinks, or burrs that remain on the inside of guards (1).

CAUTION
Do not use wire brush on aluminum surfaces.

4. Apply solid film lubricant to brightened area for final protective coating.

5. Replace rear sight assembly components and check that rear sight assembly functions properly. If rear sight assembly functions check out, return upper receiver assembly to service.

6. If rear sight guards cannot be straightened utilizing the above procedures, replace the upper receiver.

d. LUBRICATION

Lubricate upper receiver assembly and rear sight assembly. Apply CLP to helical springs and ball bearings (three each) and threaded portion of screws before installation. Lubricate helical springs and ball bearings through their respective drain holes.

3-11. UPPER RECEIVER ASSEMBLY AND REAR SIGHT ASSEMBLY (CONT).

e. REASSEMBLY

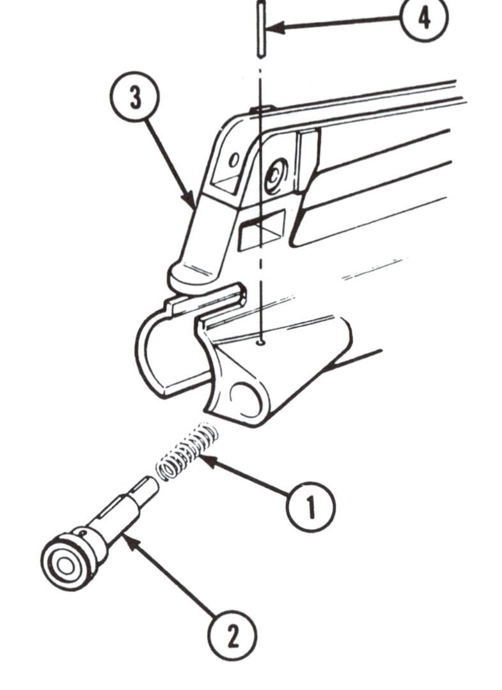

WARNING
To avoid injury to your eyes, use
care when removing and installing
spring-loaded parts.

1. Apply CLP to helical spring (1) and for-
    ward assist assembly (2) and install
    them into upper receiver (3).

2. Install spring pin (4) using 3/32 inch
    drive pin punch and hand hammer.

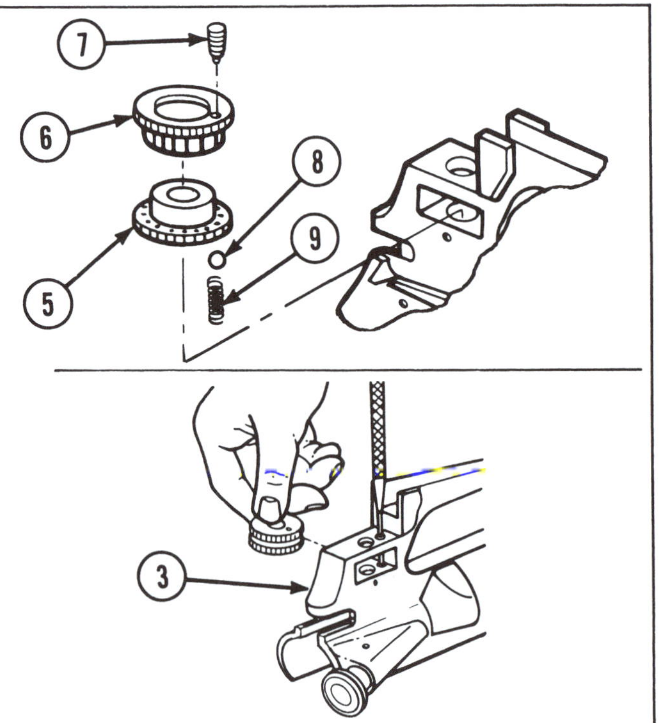

3. Assemble elevation knob (5), elevation
    index (6), and new index screw (7) us-
    ing 1/16 inch allen wrench. Do not
    overtighten index screw as scale will
    require adjustment.

4. Install ball bearing (8) and helical spring
    (9) using needle-nose pliers or
    tweezers.

5. Depress ball bearing (8) with a punch
    inserted through access hole, and slide
    elevation knob assembly in upper
    receiver (3) from the side. Center
    elevation knob assembly.

WARNING

To avoid injury to your eyes, use care when removing and installing spring-loaded parts.

NOTE

All springs are identical when new. Once disassembled from the rifle, their free length may vary due to different amounts of compression when installed. If the length of springs varies, use longest spring as #11, the next longest as #18, and the shortest as #9.

6. Insert threaded portion of rear sight assembly base (10) into upper receiver (3) and rotate elevation knob assembly until threads engage.

7. Insert helical spring (11) and ball bearing (12) in their hole as rear sight assembly base is lowered into upper receiver as elevation knob assembly is further rotated. Rotate elevation knob assembly until rear sight assembly base is all the way down. Then come up 22 clicks before installing spring pin. Check spring action of helical spring (11) on upper receiver.

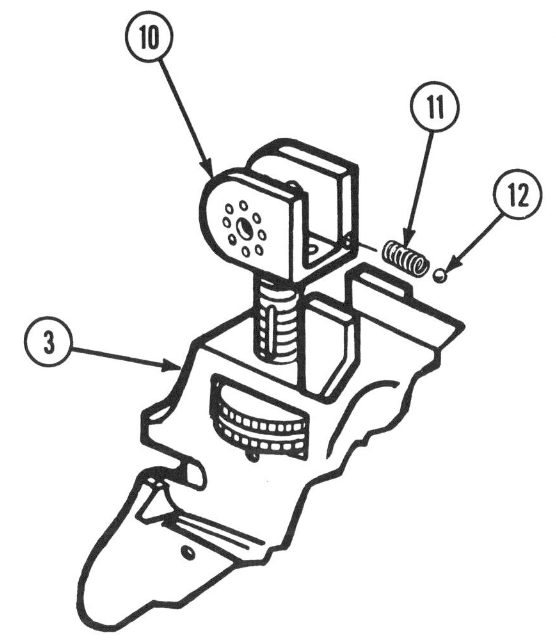

3-11. UPPER RECEIVER ASSEMBLY AND REAR SIGHT ASSEMBLY (CONT).

e. REASSEMBLY (CONT)

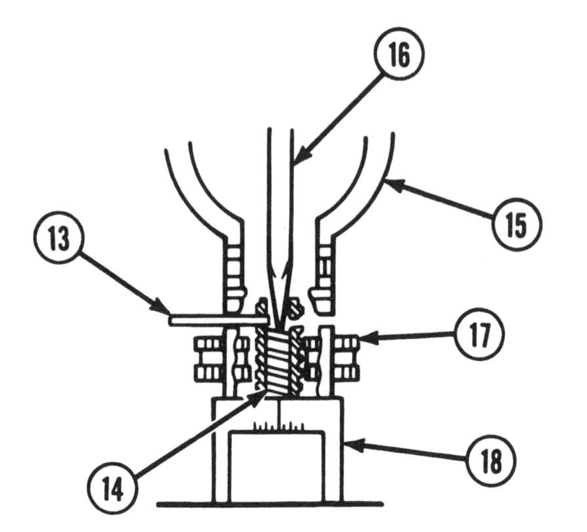

CAUTION

Ensure that spring pin (13) passes over helical spring (14), not through its coils.

8. Insert helical spring (14) through underside of upper receiver (15). Compress helical spring with a small tip screwdriver (16) to install spring pin (13). Spring pin must pass over helical spring, not through its coils. Rotate elevation knob (17) until rear sight base (18) is all the way down.

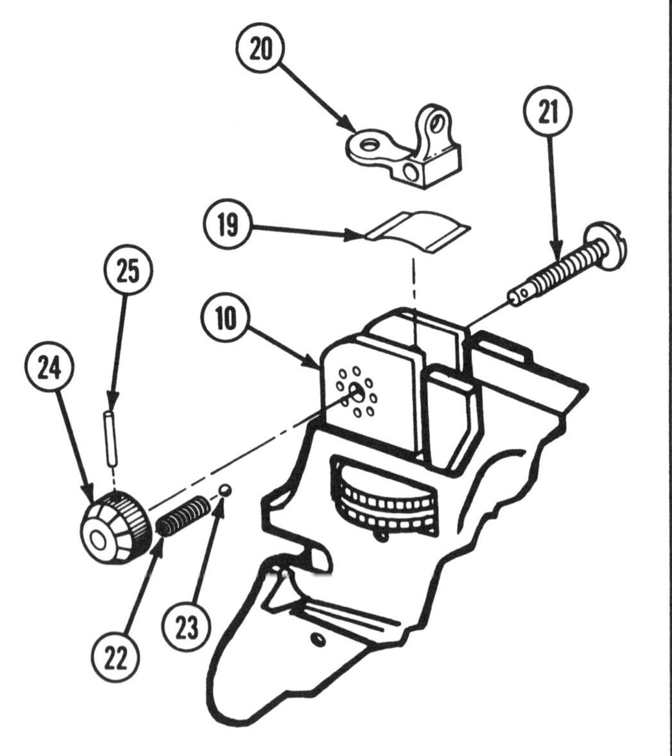

9. Install flat spring (19) and sight aperture (20) in rear sight assembly base (10). Install windage screw (21) with screwdriver.

10. Insert helical spring (22) and ball bearing (23) in windage knob (24).

NOTE

Tilt upper receiver toward windage knob during positioning to prevent loss of ball bearing.

11. Position windage knob on shaft of windage screw (21). Align holes in windage knob with hole of shaft in windage screw and hold in alignment with a small screwdriver or punch. Install spring pin (25).

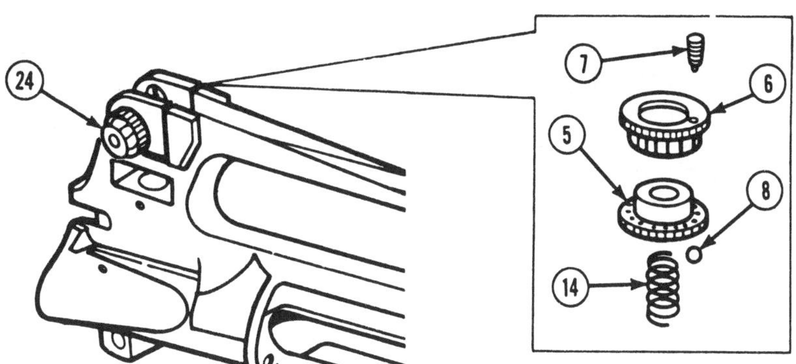

12. Rotate and test elevation index (6) and windage knob (24) for ease of functioning.

13. Inspect elevation knob zero as follows:

(a) Rotate elevation knob (5) counterclockwise until the rear sight assembly is all the way down. If a whole click is not felt as the rear sight assembly stops, the rear sight assembly has bottomed out and will not pivot freely.

(b) Position elevation knob back slightly to its last whole click so the rear sight assembly base is under tension of the ball bearing (8) and helical spring (14). The 300 meter mark should align with the mark on the receiver.

(c) If the 300 meter mark is not aligned with the mark on receiver, slip the range scale in the following manner:

(1) Position the 300 meter mark with the mark on the receiver.

(2) Insert a 1/16 inch allen wrench through the access hole of the rear sight assembly base base and into the index screw (7).

(3) Loosen the index screw three turns and leave the wrench in place.

(4) Rotate lower portion of elevation knob (5) counterclockwise until it stops (range scale should not have moved). Elevation knob should be positioned on its last whole click.

(5) Tighten index screw (7) and remove wrench.

(6) Check for proper setting.

NOTE
After the rifle is assembled, the rear sight is centered, and placed at the 300 meter mark, perform the following check: While looking at a light background, obtain good sight alignment. If the hole in the rear sight aperture appears oval instead of round, the rear sight base or upper receiver should be replaced. To determine which part requires replacement, replace the rear sight base first. If this does not resolve the problem, replace the upper receiver.

14. Assemble rifle, refer to page 3-77.

3-11. UPPER RECEIVER ASSEMBLY AND REAR SIGHT ASSEMBLY (CONT).

f. MECHANICAL ZERO PROCEDURES (A.F. ONLY)

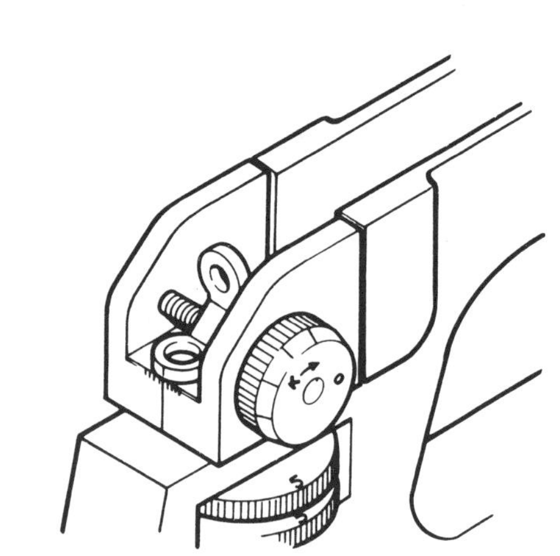

1. Center rear sight by moving windage knob in the appropriate direction.

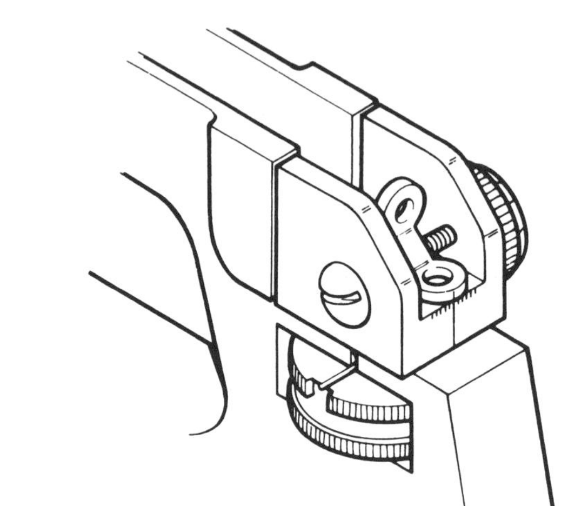

2. Always push in on windage screw head after making rear sight adjustments.

3. Visually check rear sight to ensure it is centered after making adjustments. Also, ensure the rear sight is set in the short-range (unmarked aperture) position.

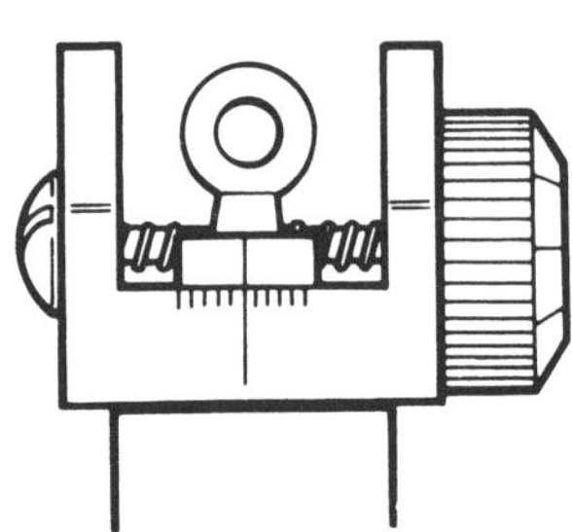

NOTE

This procedure, when used in conjunction with front sight mechanical zero adjustment (p 2-54), will give an approximate battle site zero to most M16A2 rifles. The above steps can also be used before firing a new or newly assigned rifle. Use the procedure to check rifles stored in preferred packaging during routine inspections. This will help ensure people armed with the rifles will stand a better chance of hitting an enemy if the rifles must be used before a live fire zero can be made. Whenever possible, zeroing of the rifle should be accomplished using ball ammunition on a 25 meter zeroing target using the ''L'' aperture.

3-12. FORWARD ASSIST ASSEMBLY.

This task covers:

a. Disassembly
b. Inspection
c. Repair

d. Lubrication
e. Reassembly

INITIAL SETUP

Tools
  (ARMY) Small Arms Repairman Tool Kit
    (item 3, app B)
  Field Maintenance Basic Less Power
    Small Arms Shop Set (item 1, app B)

Materials/Parts
  Cleaner, lubricant, and preservative (CLP)
    (item 9, app D)

Equipment Conditions
  3-48   Forward assist assembly removed

General Safety Instructions
  To avoid injury to your eyes, use care
    when removing and installing spring-
    loaded parts.

a. DISASSEMBLY

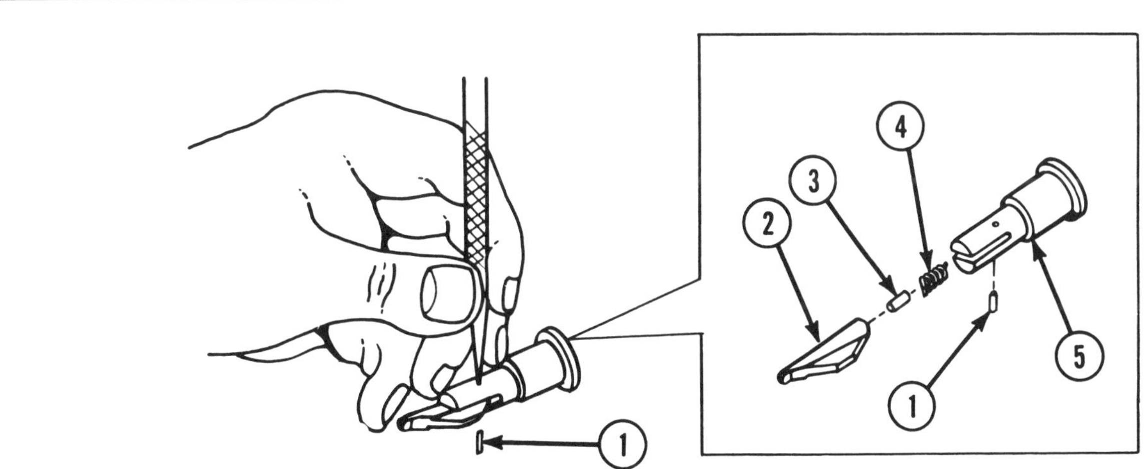

WARNING
To avoid injury to your eyes, use care when removing and installing spring-
loaded parts.

1. Drive out spring pin (1) using 1/16 inch drive pin punch and hand hammer.

2. Remove forward assist pawl (2), pawl detent (3), and helical spring (4) from plunger
assembly (5).

3-12. FORWARD ASSIST ASSEMBLY (CONT).

b. INSPECTION

1. Inspect forward assist pawl (1) for burrs, chips, and cracks. Minor burrs may be removed using fine files or stones, as required. Replace forward assist pawl if defective.

2. Inspect pawl detent (2) for burrs and cracks. Minor burrs may be removed using fine files or stones, as required. Replace pawl detent if defective.

3. Inspect helical spring (3) for kinks, breaks, and wear. Replace helical spring if defective.

4. Inspect plunger assembly (4) for wear, burrs, chips, and breaks. Minor burrs may be removed using fine files or stones, as required. Replace plunger assembly if defective.

5. Inspect spring pin (5) for wear. Replace if defective.

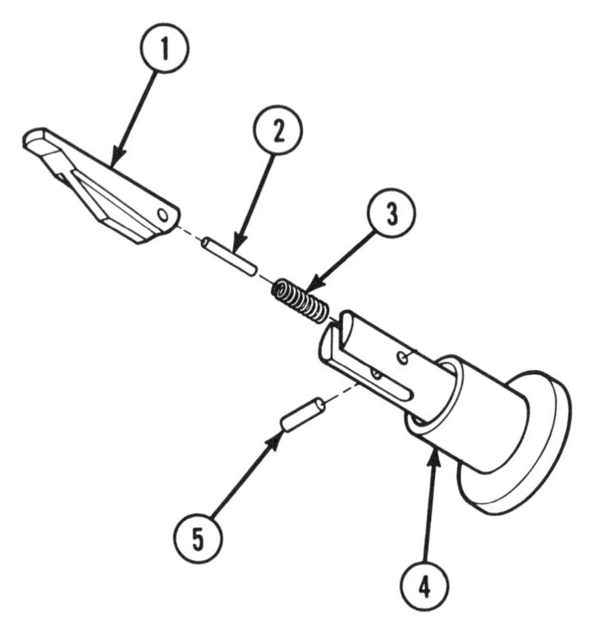

c. REPAIR

1. Repair forward assist pawl using fine files or stones, as required, to smooth burrs. Do not deform forward assist pawl.

2. Repair pawl detent using fine files or stones, as required, to smooth burrs. Do not deform pawl detent.

3. Repair plunger assembly using fine files or stones, as required, to smooth burrs. Do not deform plunger assembly.

d. LUBRICATION

Lubricate helical spring, pawl detent, and forward assist pawl with CLP (p 2-33) before installation.

e. REASSEMBLY

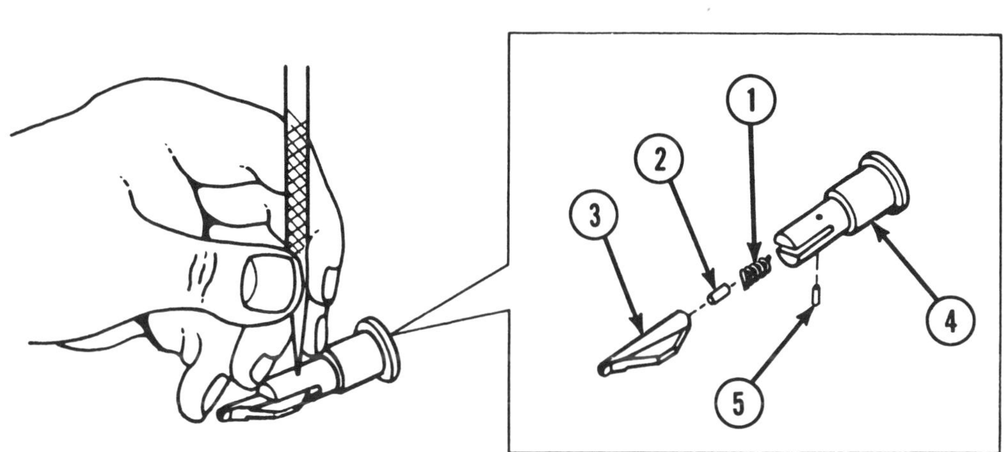

WARNING

To avoid injury to your eyes, use care when removing and installing spring-loaded parts.

1. Install helical spring (1), pawl detent (2), and forward assist pawl (3) into plunger assembly (4).

2. Align holes and install spring pin (5) using 1/16 inch drive pin punch and hand hammer. Spring pin must be flush or slightly below flush after reassembly.

3. Assemble rifle, refer to page 3-77.

3-13. LOWER RECEIVER AND BUTTSTOCK ASSEMBLY.

This task covers:

a. Disassembly
b. Inspection
c. Repair

d. Test
e. Reassembly

INITIAL SETUP

Test Equipment
Tool and Gage Set (item 2, app B)

Tools
(ARMY) Small Arms Repairman Tool Kit
(item 3, app B)
Field Maintenance Basic Less Power
Small Arms Shop Set (item 1, app B)
Pivot pin removal tool (fig. E-3, app E)
Slave pin (fig. E-6, app E)

Materials/Parts
Cleaner, lubricant, and preservative (CLP)
(item 9, app D)
Dichloromethane, technical (item 15,
app D)

Lubricant, solid film (item 21, app D)
Screw, self-locking (item 6, p C-11)

Equipment Conditions
3-15   Lower receiver and buttstock
assembly removed
2-57   Buttstock assembly and pistol grip
removed

General Safety Instructions
To avoid injury to your eyes, use care
when removing and installing spring-
loaded parts.
When using solid film lubricant or
dichloromethane, be sure the area is
well ventilated.

a. DISASSEMBLY

WARNING

To avoid injury to your eyes, use care when removing and installing spring-
loaded parts.

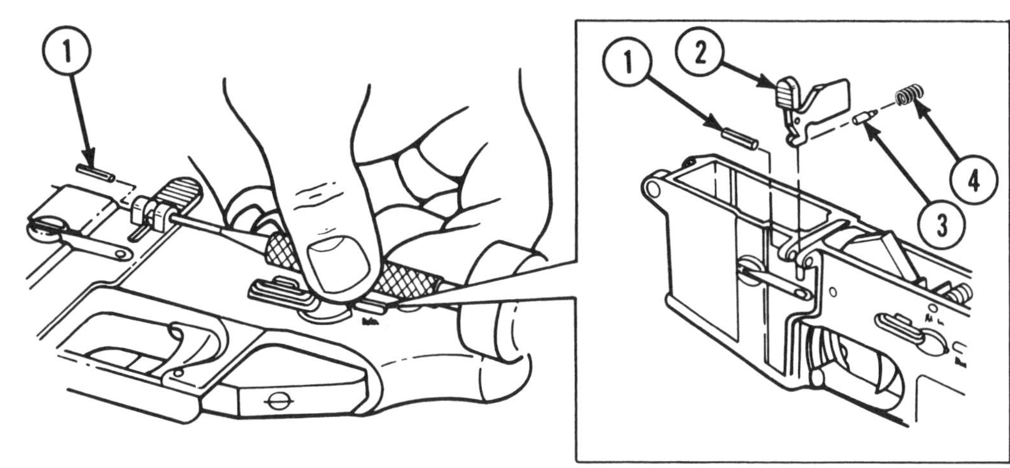

1. Remove spring pin (1) using 3/32 inch drive pin punch and hand hammer.

2. Remove bolt catch (2), bolt catch plunger (3), and bolt catch spring (4).

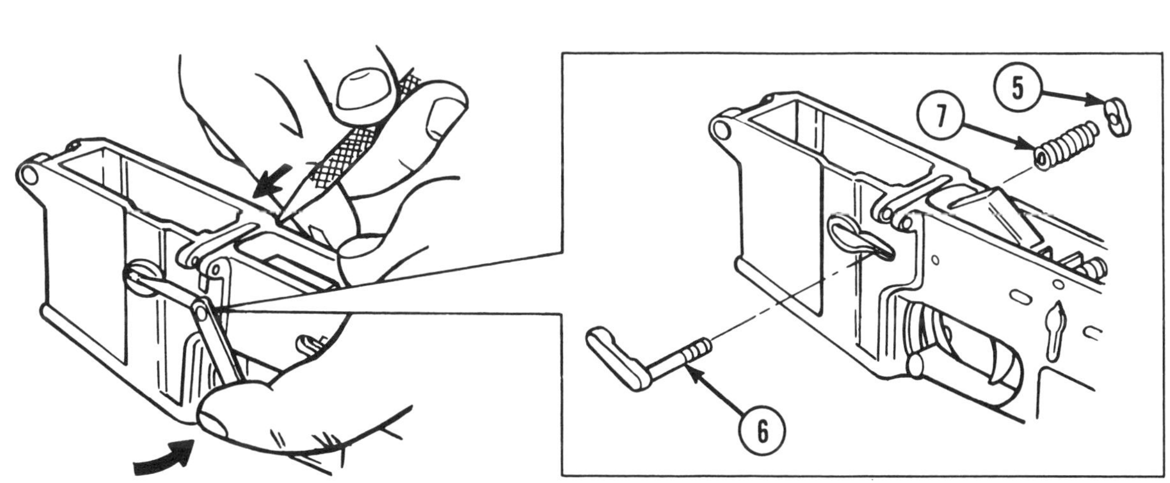

WARNING

To avoid injury to your eyes, use care when removing and installing spring-loaded parts.

3. Using drive pin punch, press in on magazine button (5) and turn magazine catch (6) counterclockwise to unscrew and remove.

4. Remove magazine button (5) and magazine catch spring (7).

3-13. LOWER RECEIVER AND BUTTSTOCK ASSEMBLY (CONT).

a. DISASSEMBLY (CONT)

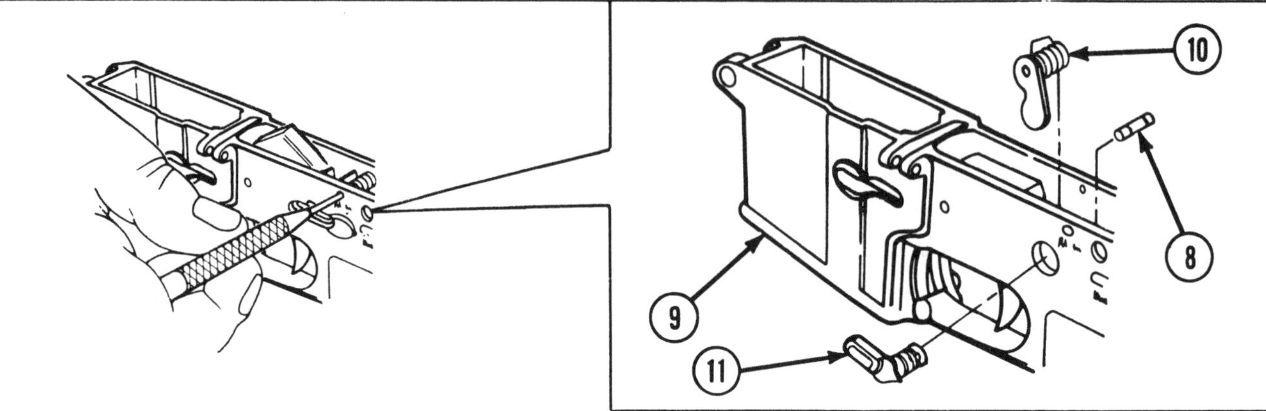

5. Use a rounded tip drive pin punch to push automatic sear pin (8) out of lower receiver (9).

NOTE

To remove automatic sear, selector lever must be positioned to BURST (if installed).

6. Remove automatic sear (10) and selector lever (11).

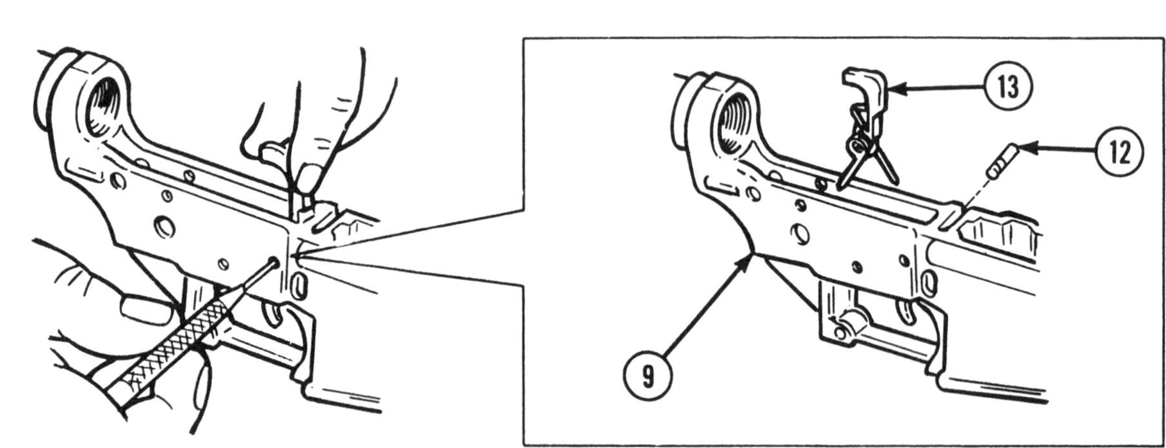

NOTE

To remove (hammer assembly should be forward), place selector lever (if installed) to SEMI position.

7. Use drive pin punch to push hammer pin (12) from lower receiver (9).

8. Remove hammer assembly (13). If further disassembly is required, see page 3-73.

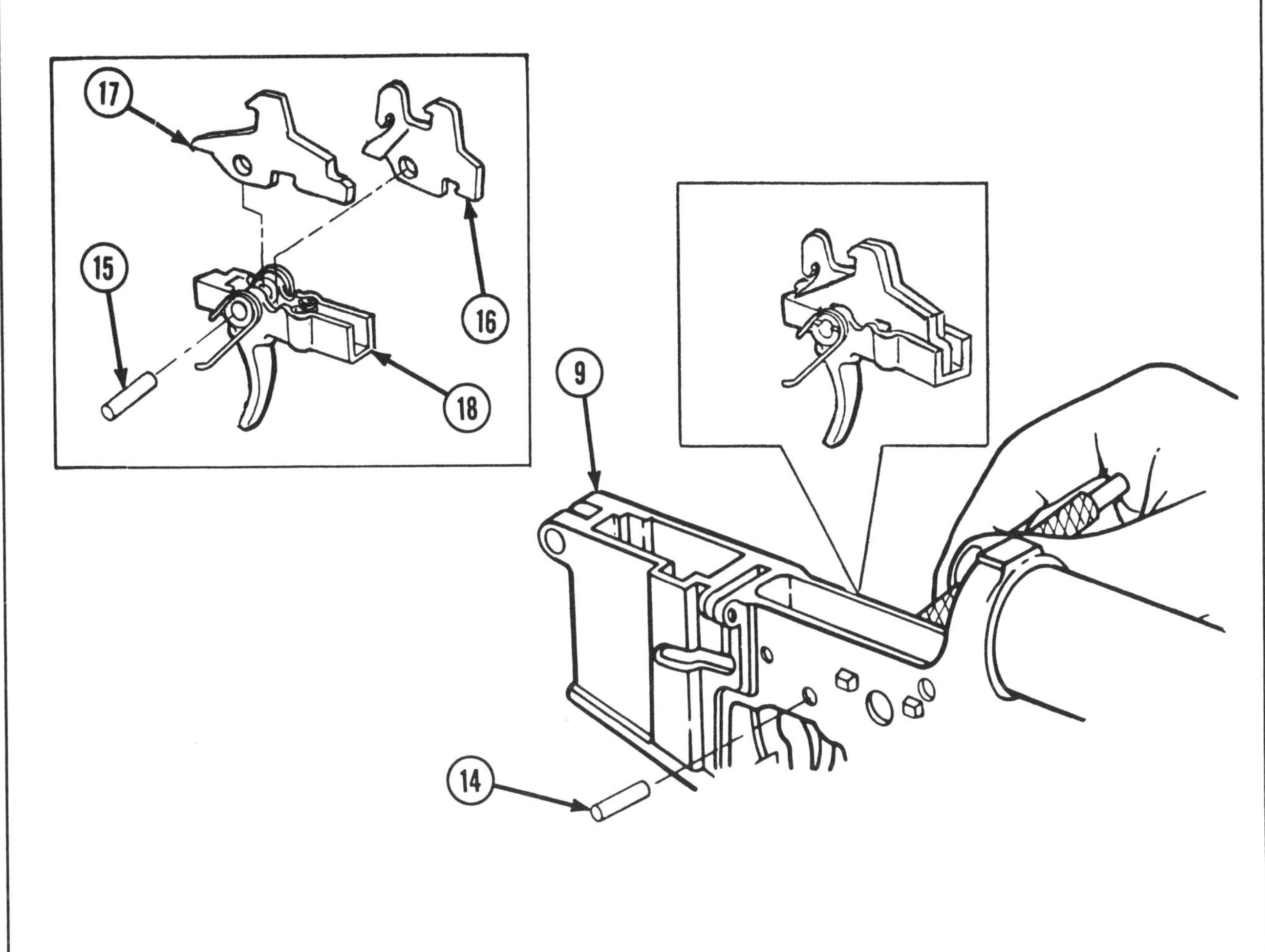

NOTE
Use of fabricated slave pin will allow removal of the following parts as a unit.

9. Remove trigger pin (14) by pushing from the left side of lower receiver (9) with fabricated slave pin (15) and a drive pin punch.

10. Remove semiautomatic disconnector (16), burst disconnector (17), and trigger assembly (18). If further disassembly of trigger assembly is required, see page 3-75.

3-13. LOWER RECEIVER AND BUTTSTOCK ASSEMBLY (CONT).

b. INSPECTION

NOTE
Refer to page 3-35 for repair of corroded surfaces.

Inspect lower receiver and buttstock assembly for corrosion in the lower receiver lobes of the pivot area or hinge pin area. If extensive corrosion appears in these areas, the receiver will not be repaired; turn rifle in for replacement.

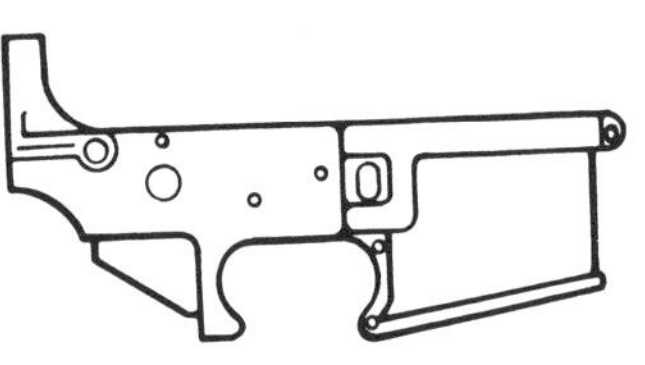

SHINY SURFACES
(REPARABLE)

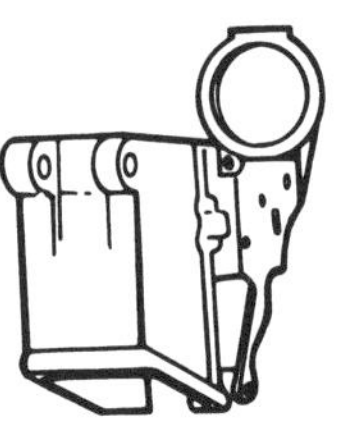

SHINY SURFACES
(REPARABLE)

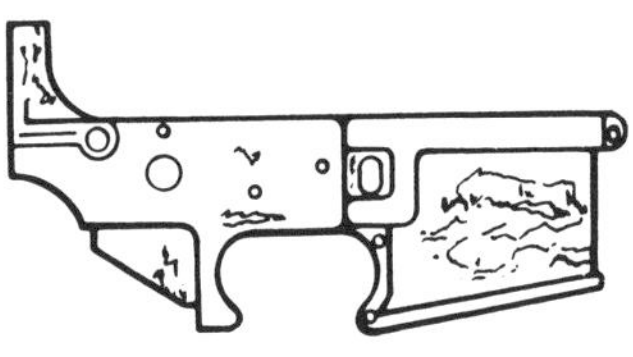

CORRODED AND NO HOLES
(REPARABLE)

CORRODED
(REPARABLE)

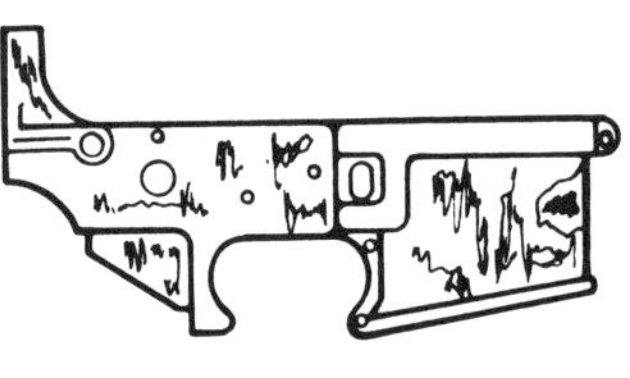

CORRODED WITH HOLE
(NONREPARABLE)

CORRODED LOBES-WEAKENING
PIVOT PIN AREA
(NONREPARABLE)

SHINY SURFACES
(REPARABLE)

CORRODED
(REPARABLE)

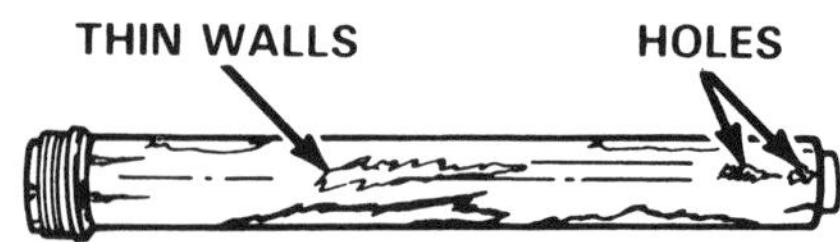

HOLES AND THIN WALLS

WARNING

When using solid film lubricant or dichloromethane, be sure the area is well ventilated.

CAUTION

Do not use a wire brush on aluminum surfaces.

NOTE

If a M16A2 rifle lower receiver is missing one third or more of its exterior protective finish, resulting in an unprotected, light reflecting surface, it is candidate for overhaul. This missing finish will be considered a shortcoming. This shortcoming requires action to obtain a replacement rifle. Once a replacement has been received, evacuate the original rifle to depot for overhaul.

Solid Film Lubricant (SFL) is the only authorized touchup for the M16A2 rifle and may be used on up to one third of the exterior finish of the rifle. **FOR ARMY CONUS USE ONLY AND AIR FORCE TRAINING RIFLES ONLY**: SFL may be used as a touchup without limitation on the upper receiver and barrel assembly. This is to say that units which **DO NOT** fall under the category of Divisional Combat Units or rapid deployment type units may have up to 100 percent of the exterior surface of the upper receiver and barrel assembly protected with SFL if necessary.

---

c. REPAIR

Repair or replace all parts of lower receiver and buttstock assembly if defective.

3-13. LOWER RECEIVER AND BUTTSTOCK ASSEMBLY (CONT).

d. TEST

1. With the upper receiver attached to the lower receiver, and the pivot pin and takedown pins in place, perform the following test:

    (a) Apply hand pressure to push the upper receiver as far to one side as possible.

    (b) Attempt to insert a 0.020 inch thickness gage between the pivot pin lugs of the upper and lower receivers.

    (c) If the thickness gage penetrates to the pivot pin at all accessible locations, repair by replacement of the upper receiver (see (b) below) or replacement of rifle is required.

2. If the rifle fails the above test, remove the upper receiver and install a ''NEW'' upper receiver and perform the test again.

3. If the rifle now passes the above test, it shall be considered serviceable and continue in use.

4. If the rifle fails the test with a new upper receiver, this failure shall be considered a shortcoming. This shortcoming requires action to obtain a replacement rifle. Once a replacement has been received, evacuate the original rifle to depot for overhaul.

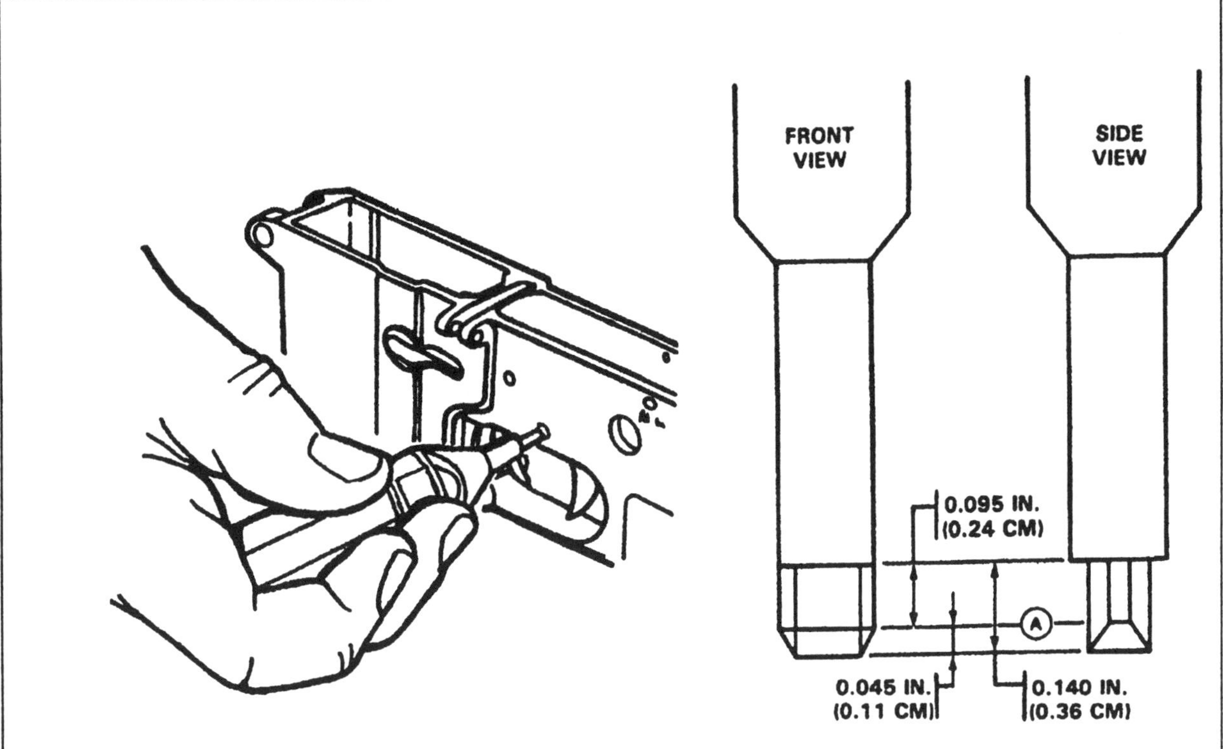

NOTE

If the lower receiver is not disassembled, visually inspect for broken or damaged parts, and to ensure that the hammer and trigger springs are correctly installed before beginning this test. It is not necessary to disassemble the lower receiver for the sole purpose of this visual inspection. If broken or damaged parts are found, disassemble (p 3-62) and repair as authorized.

5. Test two hammer pin holes and two trigger pin holes using not-go plug gage PN 12006472. This test may be conducted by disassembly of the lower receiver (p 3-62) or by pushing the pin far enough to disengage the end of the pin from the side of the receiver which is being tested. If the lower receiver is not disassembled and the not-go plug gage enters any hole to first shoulder (A), the lower receiver must be disassembled and all four holes must be tested again.

6. Gently insert the not-go plug gage and rotate it 180 degrees. If the not-go plug gage passes through any one of the four pin holes, the rifle is unserviceable and will be turned in for replacement. The gage must extend through the wall thickness to be unserviceable.

7. After completion of gaging operation, visually inspect hammer and trigger springs to ensure proper location of spring legs.

3-13. LOWER RECEIVER AND BUTTSTOCK ASSEMBLY (CONT).

e. REASSEMBLY

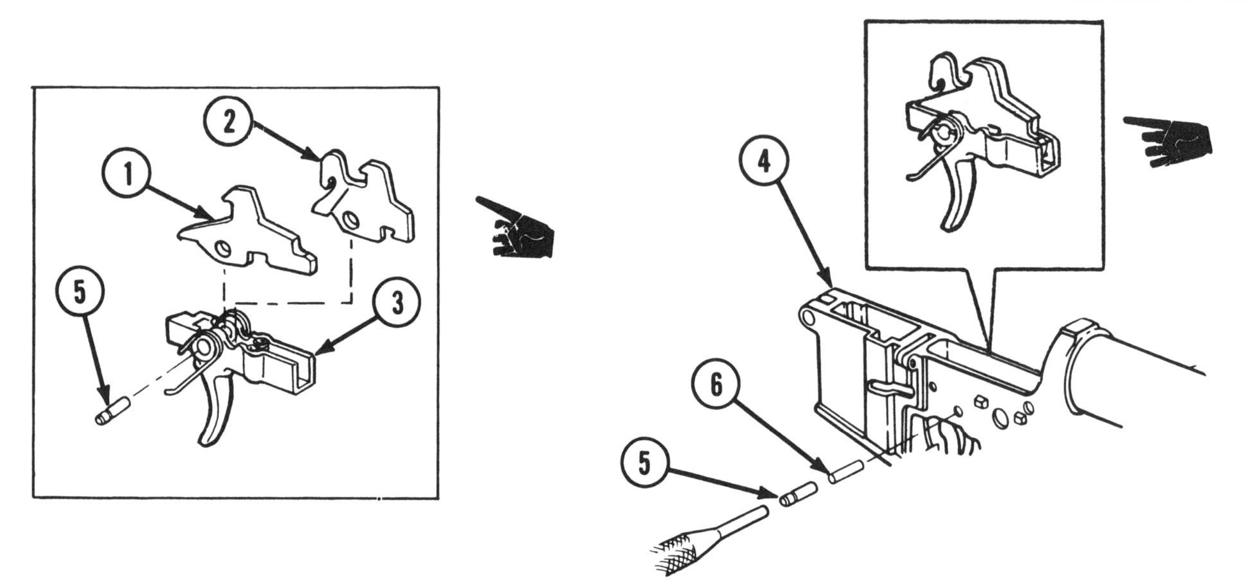

1. Assemble semiautomatic disconnector (1), burst disconnector (2), and trigger assembly (3). Install as a unit in lower receiver (4) using slave pin (5).

2. Install trigger pin (6) using drive pin punch. Push in until flush. Push out slave pin (5).

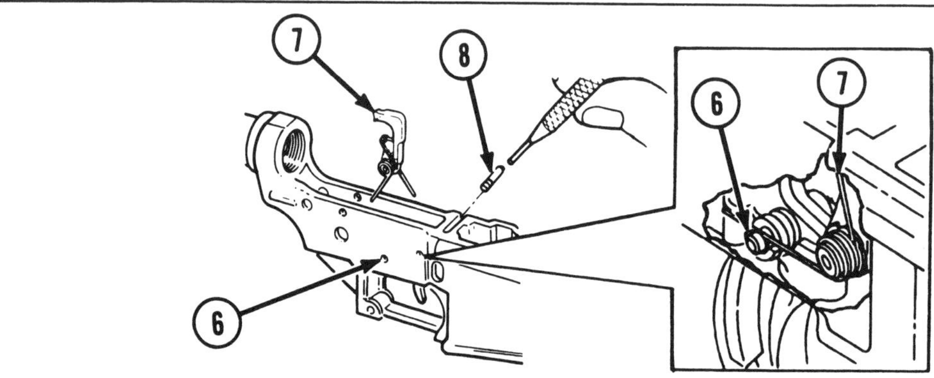

NOTE
Ends of hammer spring are installed to rear of trigger pin (6), resting in the annular groove on upper surface of trigger pin (6).

3. Install hammer assembly (7).

4. Install hammer pin (8) using drive pin punch. Push in until flush.

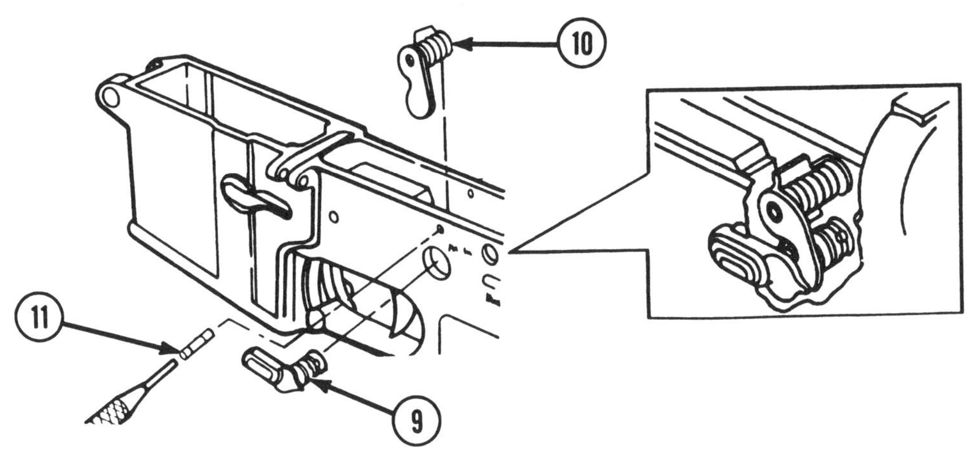

NOTE

Hammer assembly should be cocked prior to installing the selector lever.

Selector lever (9), if installed, must be positioned to BURST. Long leg of automatic sear (10) spring must rest on top of selector lever.

5. Install selector lever (9) and automatic sear (10).

6. Install automatic sear pin (11) (install on the right side) into receiver using drive pin punch. Push in until flush.

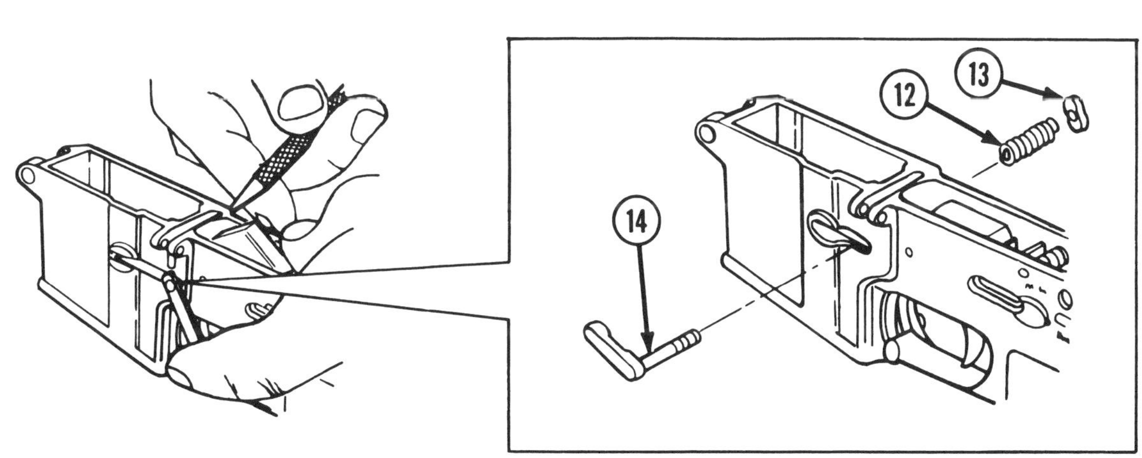

7. Install magazine catch spring (12) and magazine button (13).

NOTE

Drive pin punch should be larger than hole in magazine button.

8. Install magazine catch (14). Push in on magazine button (13) using a drive pin punch and turn magazine catch (14) clockwise until threaded end of magazine catch is flush with magazine button head.

3-13. LOWER RECEIVER AND BUTTSTOCK ASSEMBLY (CONT).

e. REASSEMBLY (CONT)

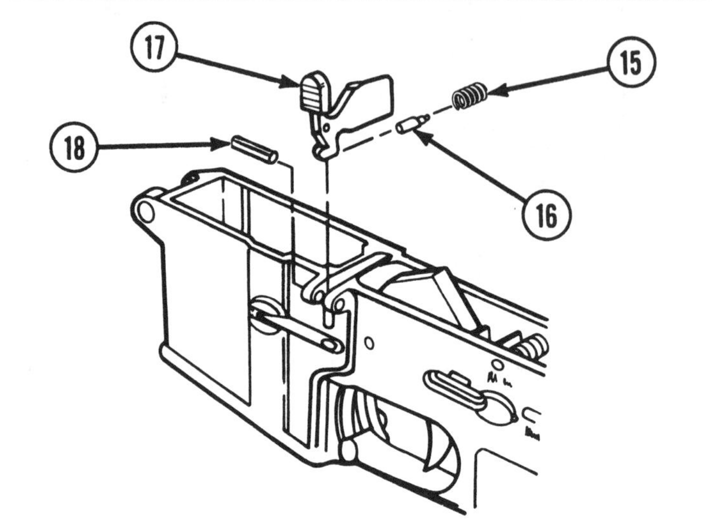

9. Install bolt catch spring (15), bolt catch plunger (16), and bolt catch (17).

10. Secure by installing spring pin (18) using 3/32 inch drive pin punch and hand hammer.

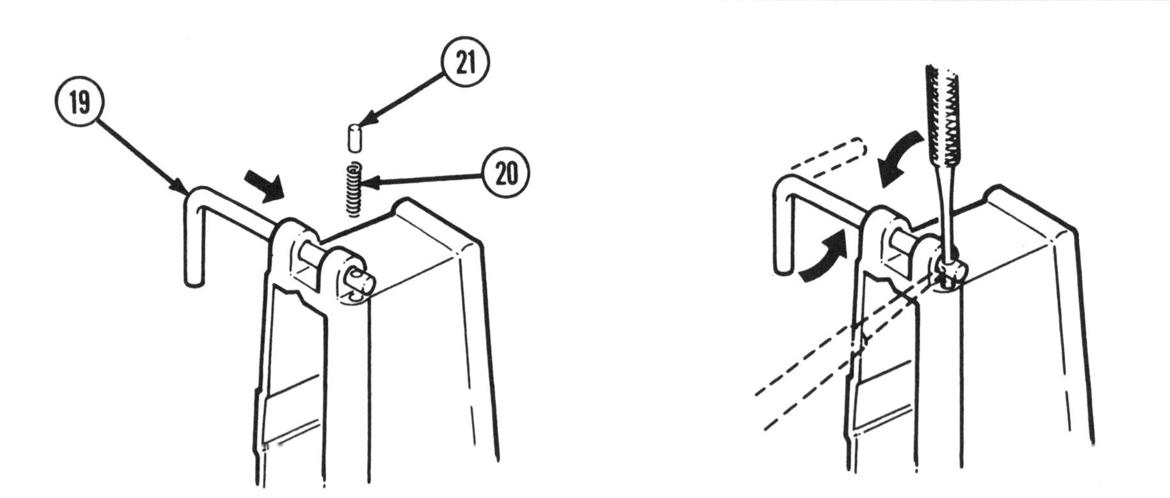

11. Install fabricated pivot pin installation tool (19). Insert helical spring (20) and pivot pin detent (21). Compress pivot pin detent in recess with punch and rotate tool. Remove punch.

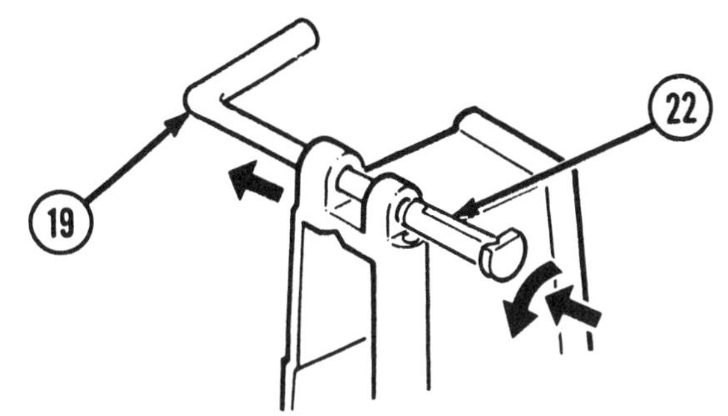

WARNING

To avoid injury to your eyes, use care when removing or installing spring-loaded parts.

NOTE

Rounded end of pivot pin detent must be in the groove of the pivot pin (22) when assembly is complete.

12. Position pivot pin (22) and removing fabricated pivot pin installation tool (19) while maintaining pressure, slide pivot pin (22) into hole. Rotate pivot pin to receive pivot pin detent.

13. Assemble rifle, refer to page 3-82.

3-14. HAMMER ASSEMBLY.

This task covers:

a. Disassembly
b. Inspection/Repair

c. Reassembly

INITIAL SETUP

Equipment Conditions
3-62 Hammer assembly removed

a. DISASSEMBLY

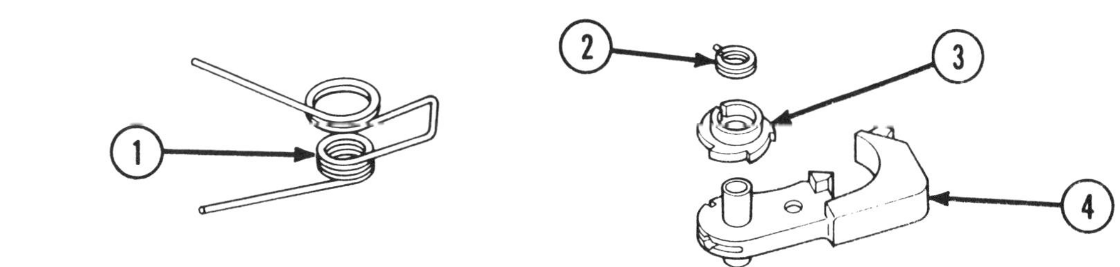

Remove hammer spring (1), cam clutch spring (2), and burst cam (3) from hammer and hammer pin retainer assembly (4).

3-14. HAMMER ASSEMBLY (CONT).

b. INSPECTION/REPAIR

1. Inspect hammer spring for deformities, breaks, and bends. Pay special attention to the large coil. Replace hammer spring if defective.

2. Inspect cam clutch spring and burst cam for deformities, breaks, and bends; replace if defective.

3. Inspect hammer and hammer pin retainer assembly for chips and breaks. Hammer pin should click home under strong finger pressure. Install hammer pin into hole in hammer to check spring retention of the hammer pin. Replace hammer and hammer pin retainer assembly if defective.

c. REASSEMBLY

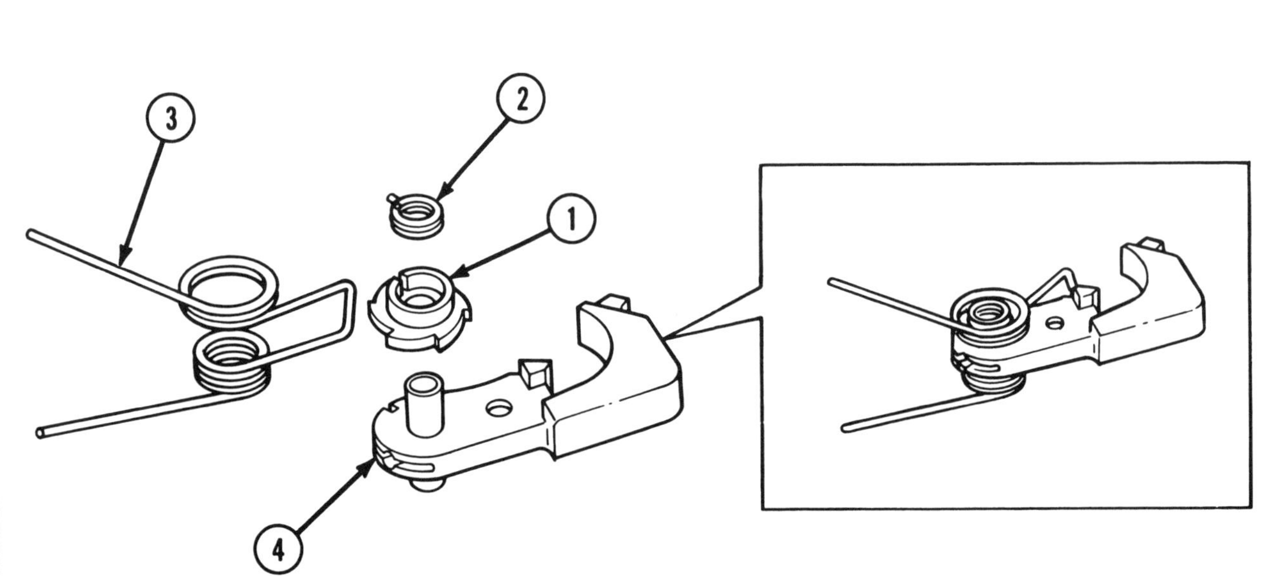

NOTE

Cam clutch spring should be assembled with bend to the inside. The large loop of the hammer spring should be assembled over the burst cam.

Install burst cam (1), cam clutch spring (2), and hammer spring (3) on hammer and hammer pin retainer assembly (4).

3-15. TRIGGER ASSEMBLY AND TRIGGER SUBASSEMBLY.

This task covers:

a. Disassembly   **c.** Reassembly
b. Inspection/Repair

INITIAL SETUP

Tools
  (ARMY) Small Arms Repairman Tool Kit
    (item 3, app B)
  Field Maintenance Basic Less Power
    Small Arms Shop Set (item 1, app B)

Equipment Conditions
  3-62   Trigger assembly removed

General Safety Instructions
  To avoid injury to your eyes, use care
    when removing and installing spring-
    loaded parts.

a. DISASSEMBLY

NOTE
Do not remove disconnector
springs (2) unless required for
repair.

Remove trigger spring (1) and two disconnector springs (2) from trigger (3).

3-15. TRIGGER ASSEMBLY AND TRIGGER SUBASSEMBLY (CONT).

b. INSPECTION/REPAIR

1. Inspect trigger spring for kinks, deformities, and weakness. Replace if defective.

2. Inspect disconnector springs for deformities, bends, breaks, and weakness. Replace if defective.

3. Inspect trigger for chips, wear, and cracks. Inspect for damaged searing surface on the trigger nose (1). Replace if defective.

c. REASSEMBLY

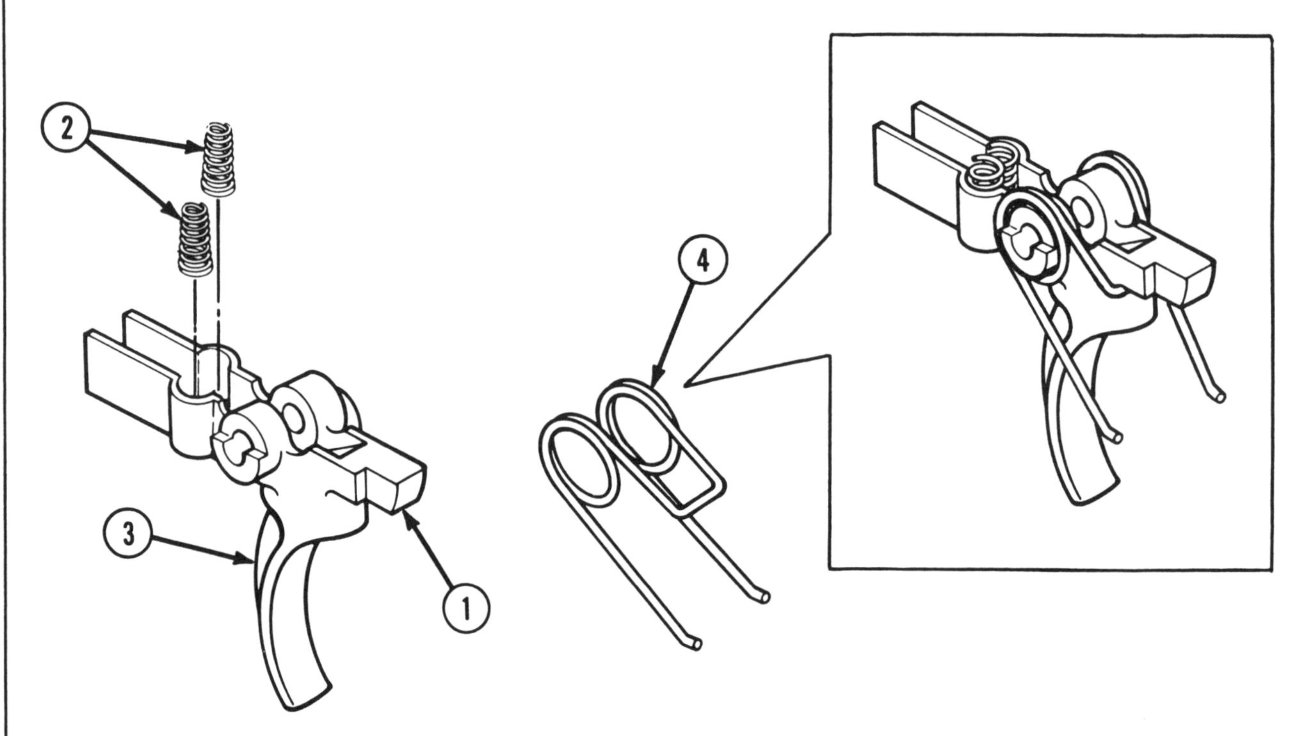

1. Install two disconnector springs (2) in trigger (3). Large end of disconnector springs should snap into trigger recesses.

2. Install trigger spring (4).

3-16. LOWER RECEIVER AND RECEIVER EXTENSION ASSEMBLY.

This task covers:

a. Disassembly
b. Inspection

c. Repair/Modify
d. Reassembly

INITIAL SETUP

Tool Equipment
Tool and Gage Set (item 2, app B)

Tools
(ARMY) Small Arms Repairman Tool Kit (item 3, app B)
Field Maintenance Basic Less Power Small Arms Shop Set (item 1, app B)

Materials/Parts
Cloth, abrasive (item 13, app D)
Grease, molybdenum disulfide (item 19, app D)
Lubricant, solid film (item 21, app D)

Equipment Conditions
3-62 Lower receiver and receiver extension assembly removed

General Safety Instructions
To avoid injury to your eyes, use care when removing and installing spring-loaded parts.
When using solid film lubricant or dichloromethane, be sure the area is well ventilated.

a. DISASSEMBLY

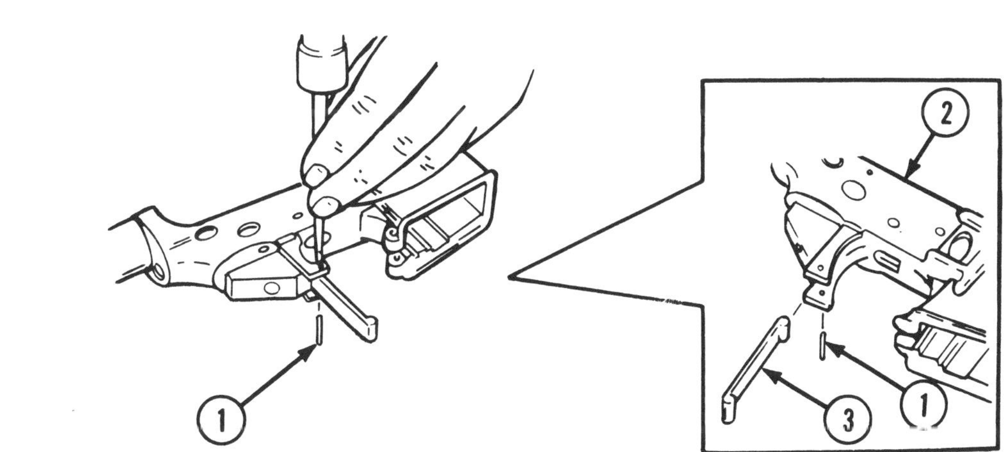

1. Remove spring pin (1) from lower receiver (2) using 1/8 inch drive pin punch and hand hammer.

2. Remove trigger guard (3).

3-16. LOWER RECEIVER AND RECEIVER EXTENSION ASSEMBLY (CONT).

a. DISASSEMBLY (CONT)

3. Use padding between lower receiver and brass vise jaws. Grip the solid portion of the lower receiver with brass vise jaws which conform to the shape of the lower receiver in this area.

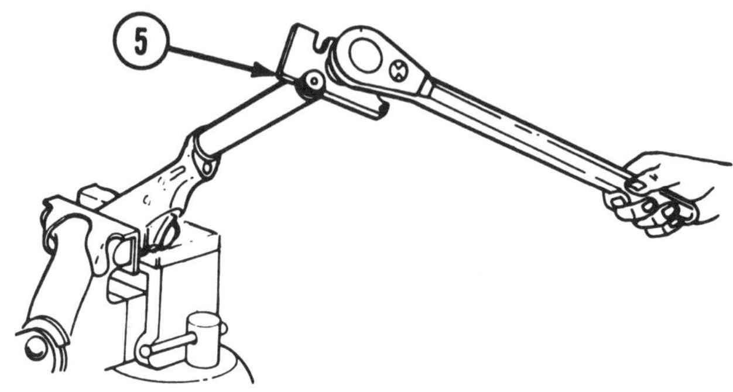

RIFLE ONLY

4. Clamp lower receiver (2) in a machinist's vise using vise jaw caps and tighten on solid portion just tight enough to hold.

WARNING

To avoid injury to your eyes, use care when removing and installing spring-loaded parts.

NOTE

As lower receiver extension is removed, catch buffer retainer and helical spring. Lower receiver is a serial number controlled item.

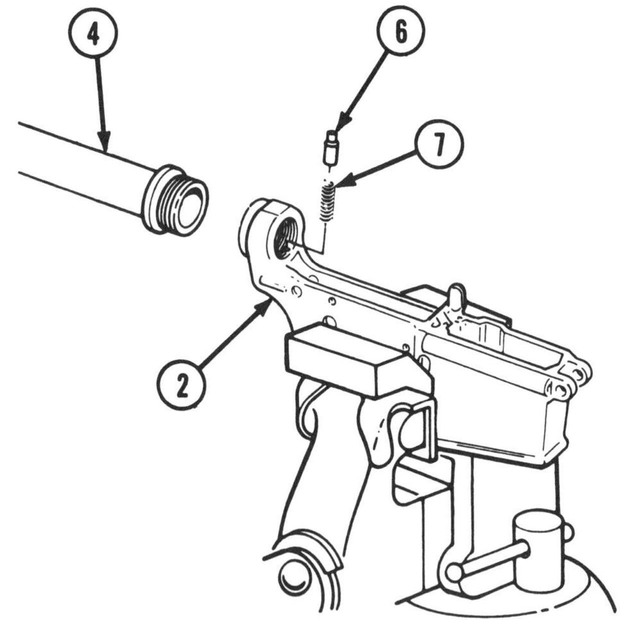

5. Remove lower receiver extension (4) from lower receiver (2) using combination wrench (5) and socket wrench handle. Catch buffer retainer (6) and helical spring (7).

CARBINE ONLY

6. Clamp lower receiver (8) in vise and tighten on solid portion just tight enough to hold.

NOTE
Use wooden vise jaws in place of brass vise jaw caps.

7. Remove the lower receiver extension (9), by loosening the extension locking nut (10) using the special tool (item 12, app C). Catch buffer retainer and spring (11).

CAUTION
While performing the following step, care should be taken to restrain the pivot pin spring and detent.

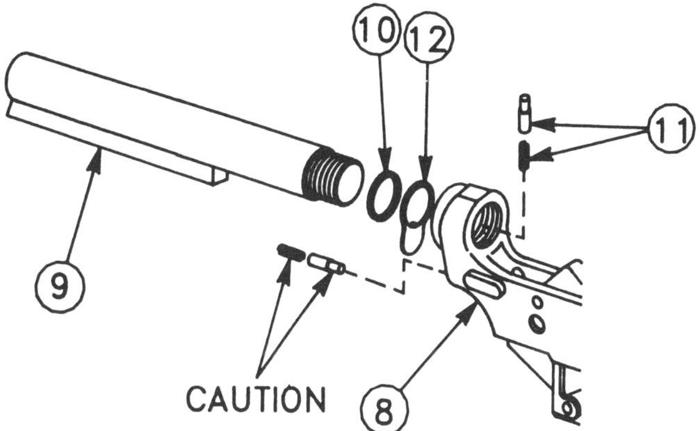

8. Loosen the locking nut (10) to allow the receiver end plate (12) to disengage from the lower receiver. Hold the buffer retainer and spring (11) in position with your index finger and unscrew the lower receiver extension (9) from the lower receiver (8).

b. INSPECTION

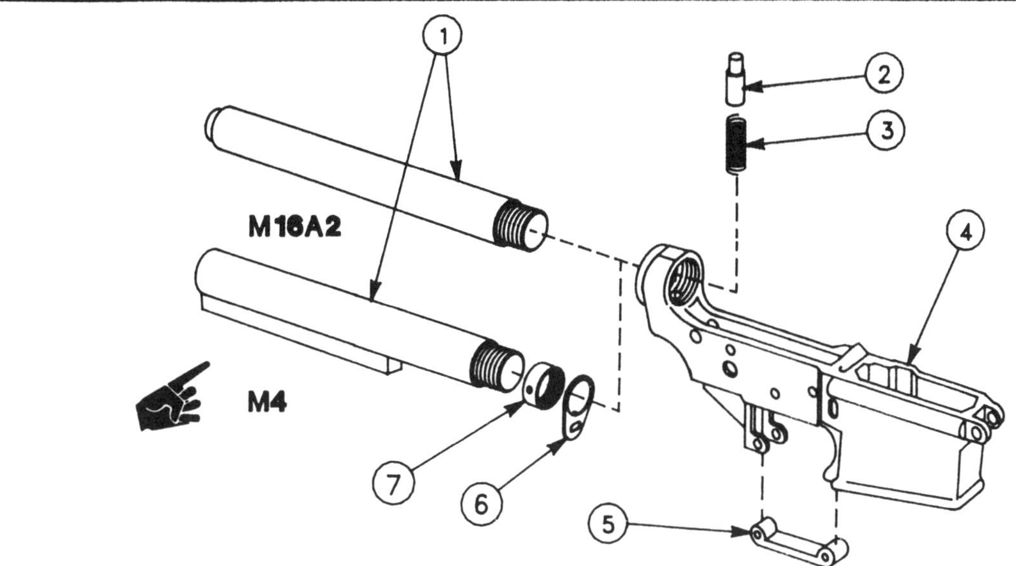

1. Inspect lower receiver extension (1) for corrosion, dents, and wear. Repair or replace if defective.

2. Inspect buffer retainer (2) for wear. Replace if defective.

3. Inspect helical springs (3) for deformities and breaks. Replace if defective.

4. Inspect lower receiver (4). See page 3-66, INSPECTION.

5. Inspect trigger guard (5) for deformities and check operation of plunger and spring. Replace trigger guard if defective.

6. **CARBINE ONLY:** Inspect receiver end plate (6) and locking nut (7) for damage. Replace if damaged.

c. REPAIR/MODIFY

1. Repair lower receiver extension by using abrasive cloth to remove light corrosion. Retouch using solid film lubricant.

NOTE

AIR FORCE ONLY: Only depot maintenance is authorized to restamp the serial number on rifle.

ARMY ONLY: Only direct support level is authorized to restamp serial number.

3-16. LOWER RECEIVER AND RECEIVER EXTENSION ASSEMBLY (CONT).

c. REPAIR/MODIFY (CONT)

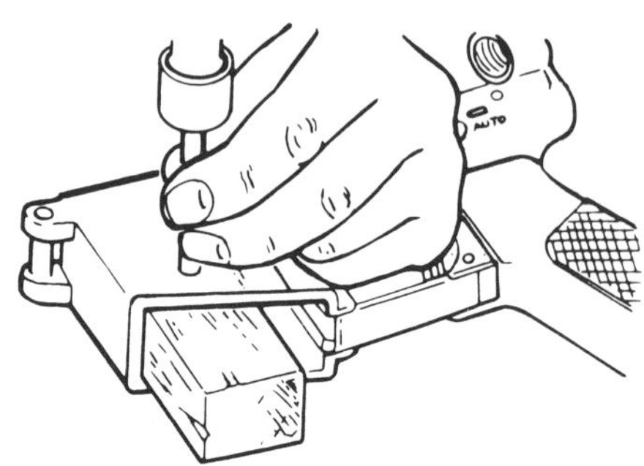

2. **ARMY ONLY:** If serial number is hard to read on rifle, restamp as follows:

   (a) Support the receiver in the stamping area to prevent bending and distortion of the receiver.

   (b) Exercise extreme care to restamp the same serial number as the original.

   (c) Restamp the serial number the same size as the original serial number.

NOTE

Most rifle serial numbers are 1/8 inch (0.31 cm) in height, or close enough that this size is acceptable for such restamping. In the event that a rifle has a serial number that cannot be reproduced by the use of the die sets contained in the Set D Field Maintenance Post, Camp, and Station Small Arms Shop Set, local purchase of an appropriate size die set is authorized.

d. REASSEMBLY

RIFLE ONLY

WARNING

To avoid injury to your eyes, use care when removing and installing spring-loaded parts.

1. Install helical spring (1) and buffer retainer (2) into lower receiver (3).

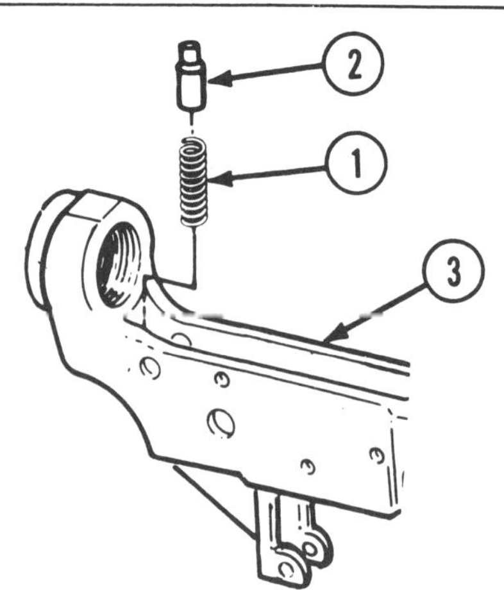

RIFLE ONLY

2. Lubricate threads of lower receiver (3) and lower receiver extension (4) with molybdenum disulfide grease (item 19, app D) before reassembly.

3. Install lower receiver extension (4) into lower receiver (3) while depressing buffer retainer.

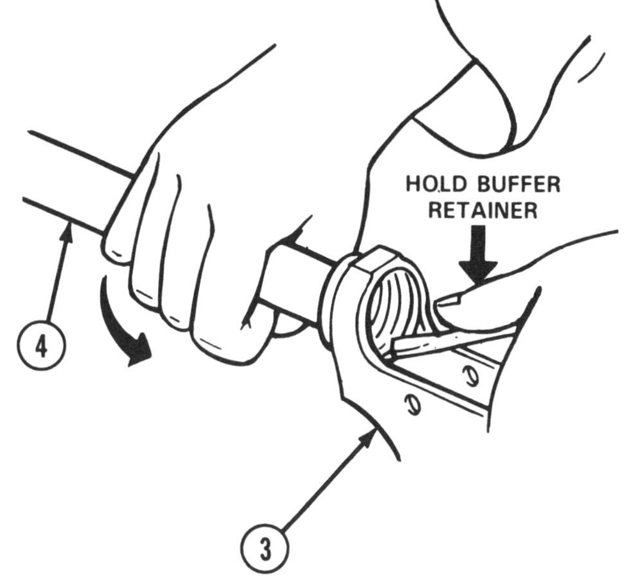

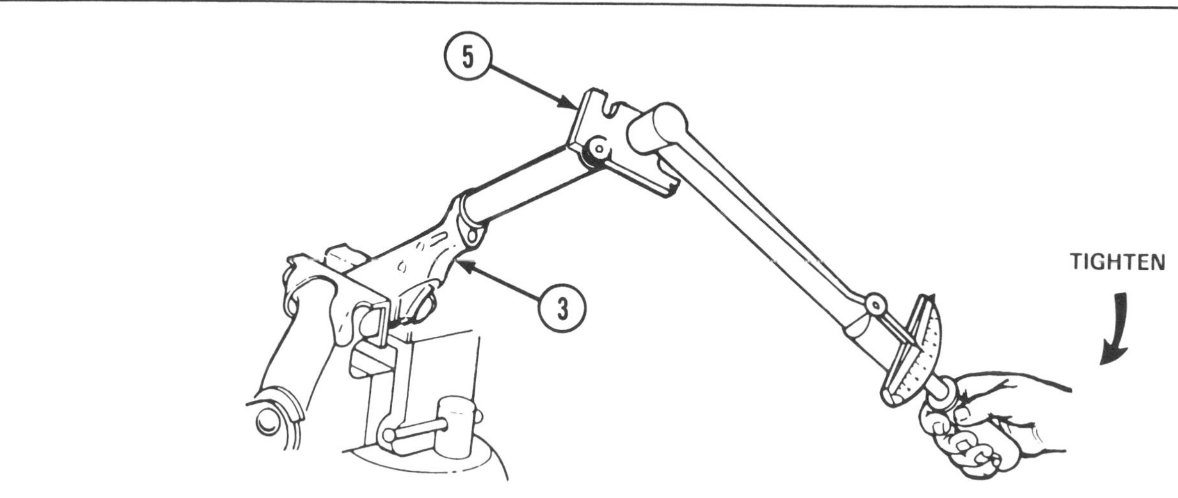

RIFLE ONLY

4. Use padding between lower receiver and brass vise jaws. Use vise jaws in vise and brass vise jaw caps, if available.

5. Clamp solid portion of lower receiver (3) in a machinist's vise using vise jaws. Grip the solid portion of the lower receiver with vise jaws which conform to the shape of the lower receiver in this area.

6. Using combination wrench (5) and torque wrench, torque lower receiver extension to 35-39 ft-lb (47.25 — 52.65 N-m). Torque is read when both wrenches are used together.

3-16. LOWER RECEIVER AND RECEIVER EXTENSION ASSEMBLY (CONT).

d. REASSEMBLY (CONT)

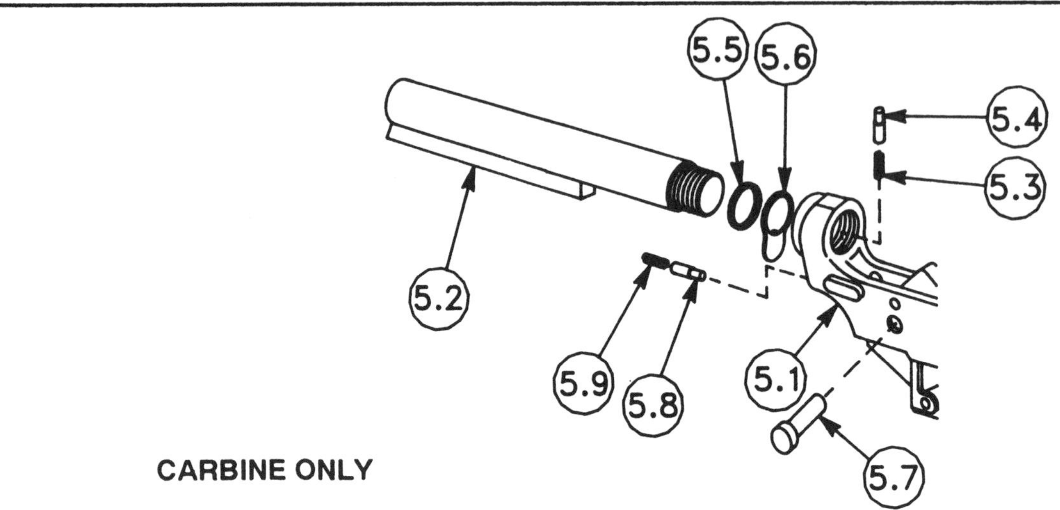

6A. Lubricate threads of lower receiver (5.1) and lower receiver extension (5.2) with molybenum disulfide grease (item 19, app D) before reassembly.

6B. Pre-position the spring (5.3) and buffer retainer (5.4) into the retaining hole of the lower receiver (5.1). Screw the locking nut (5.5) onto the lower receiver extension (5.1) with the three notches on the locking ring (5.5) facing forward.

6C. Align the receiver end plate (5.6) onto the lower receiver extension (5.2) with the lug of the receiver end plate (5.6) facing forward.

6D. Pre-position the takedown pin (5.7), detent (5.8), and spring (5.9) in lower receiver assembly (5.1).

6E. Push down on the buffer retainer (5.4) and spring (5.3) and at the same time, screw the lower receiver extension (5.2) in until it retains the buffer retainer (5.4) in position.

6F. Align the lug of the receiver end plate (5.6) into the rear of the lower receiver (5.1). Screw the locking nut (5.5) forward until it contacts the receiver end plate (5.6).

6G. Using the special tool (item 12, app C), tighten the locking nut (5.5) until snug.

6H. Using the special tool (item 12, app C), and torque wrench, torque locking nut (5.5) to 40 ± 2 inch pounds.

6J. Stake the receiver end plate (5.6) in 2 places across from the notches in the locking nut (5.5).

3-16. LOWER RECEIVER AND RECEIVER EXTENSION ASSEMBLY (CONT).

d. REASSEMBLY (CONT)

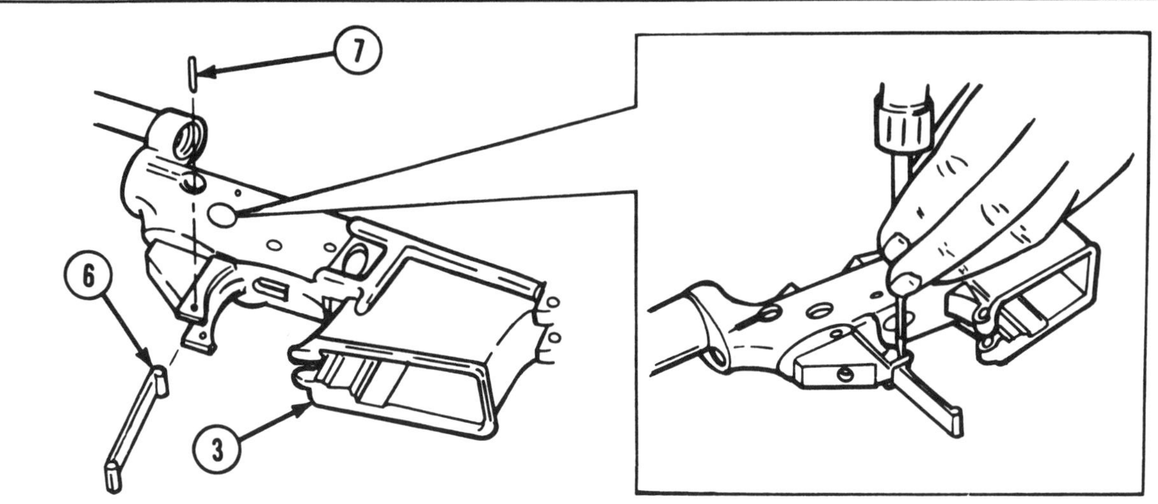

7. Install trigger guard (6) into lower receiver (3).

8. Install spring pin (7) using 1/8 inch drive pin punch and hand hammer.

9. Deleted.

3-17. MAJOR COMPONENTS OF M16A2 RIFLE.

This task covers:

a. Reassembly   b. Test   c. Inspection

INITIAL SETUP

Test Equipment
    Tool and Gage Set (item 2, app B)

Tools
    (ARMY) Small Arms Repairman Tool Kit
    (item 3, app B)

Reference
    TM 9-1005-319-10

General Safety Instructions
    To avoid injury to your eyes, use care
    when removing and installing spring-
    loaded parts.
    Live ammunition must not be near the
    work area.

a. REASSEMBLY

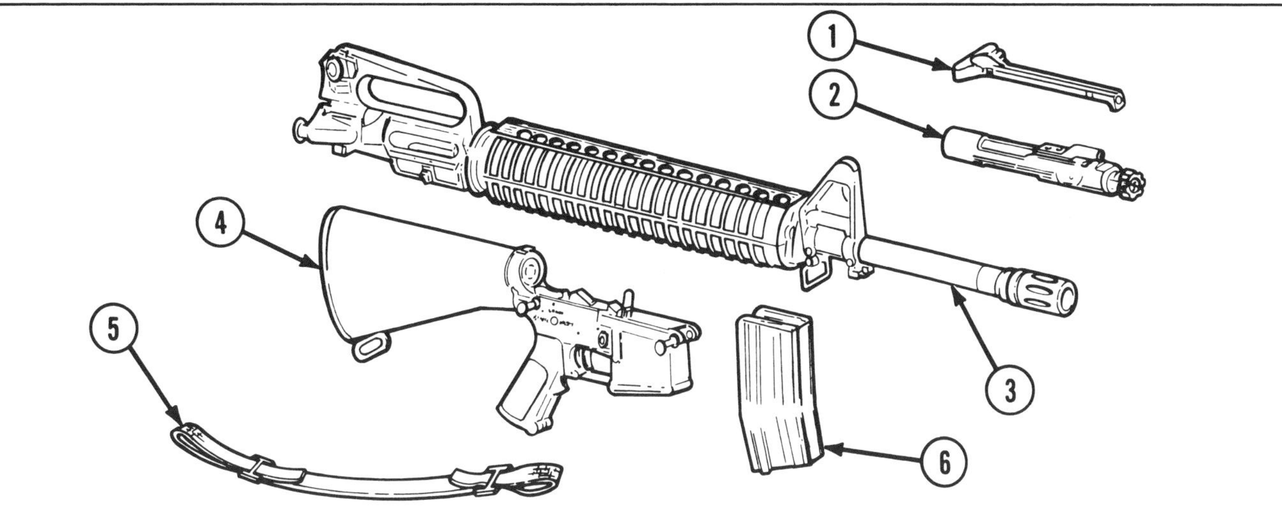

1. Refer to TM 9-1005-319-10.

2. Install charging handle assembly (1) and bolt carrier assembly (2) into upper receiver and barrel assembly (3). Join upper receiver and barrel assembly (3) and lower receiver and buttstock assembly (4).

3. Snap on sling (5) and install magazine (6).

b. TEST

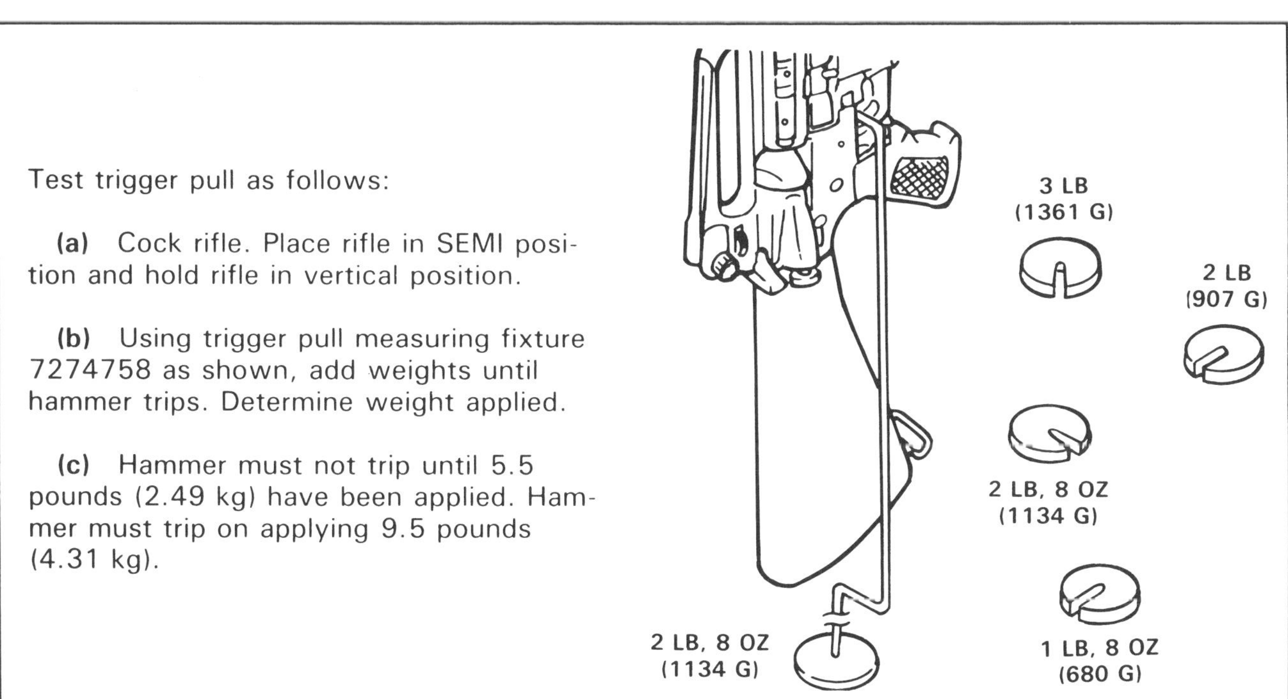

Test trigger pull as follows:

   (a)   Cock rifle. Place rifle in SEMI position and hold rifle in vertical position.

   (b)   Using trigger pull measuring fixture 7274758 as shown, add weights until hammer trips. Determine weight applied.

   (c)   Hammer must not trip until 5.5 pounds (2.49 kg) have been applied. Hammer must trip on applying 9.5 pounds (4.31 kg).

3-17. MAJOR COMPONENTS OF M16A2 RIFLE (CONT).

b. TEST (CONT)

NOTE

Perform the above test three times. These tests must be on consecutive trigger pulls. The rifle must pass the test all three times.

(d) If rifle fails trigger pull test or excessive creep is present, replace trigger and/or hammer.

NOTE

Always gage trigger and hammer pin holes with not go plug gage 12006472 before replacing parts.

c. INSPECTION

Perform final inspection procedures below.

3-18. M16A2 RIFLE FINAL INSPECTION.

This task covers:

a. Final inspection
b. Test

c. Functional theory of three-round burst control

INITIAL SETUP

Test Equipment
  Tool and Gage Set (item 2, app B)

References
  TM 9-1005-319-10

General Safety Instructions
  Live ammunition must not be near the work area.
  Air Force rifles will be inspected in accordance with AFR 50-36, Vol. 1.

a. **FINAL INSPECTION**

1. Visually inspect general appearance of rifle. Rifle should look almost new. All metal surfaces are to have a dull, rust- or corrosion-resistant finish with no burrs or deep scratches.

2. Visually inspect barrel for serviceability. Barrels must be straight, clean, free of rust, powder fouling, and free of bulges and rings. Fine pitting is allowable.

3. Visually inspect rifle for missing parts. All parts must be attached and all modifications must be applied. Steel parts must be rust free. Spring pins must be secure and screws must be tight.

4. Functionally inspect key and bolt carrier assembly and gas tube alignment. Refer to TM 9-1005-319-10 and use the following procedures:

   (a) Disengage the takedown pin and open the receiver.

   (b) Remove bolt carrier assembly.

   (c) Remove bolt assembly from bolt carrier assembly.

   (d) Insert key and bolt carrier assembly into upper receiver and barrel assembly. The bolt assembly must not be installed while performing test.

   (e) Slide key and bolt carrier assembly forward to detect binding between key and bolt carrier assembly and gas tube by feel. Badly bent gas tube could cause damage to both the key and bolt carrier assembly and the gas tube. A slightly bent gas tube will cause unnecessary wear of the key and bolt carrier assembly and gas tube.

   (f) Correct slight binding by removing handguard assemblies and slightly bending gas tube in the handguard area while repeating step (e) above until no binding is detected. Badly bent gas tubes will be replaced and realigned.

   (g) Remove key and bolt carrier assembly from upper receiver and barrel assembly.

   (h) Reassemble bolt assembly into key and bolt carrier assembly.

   (i) Reinstall key and bolt carrier assembly into upper receiver and barrel assembly.

5. Make a functional check on an assembled rifle with selector lever in SAFE, SEMI, and BURST positions. Any portion of this check may be used alone to determine the operational condition of any specific firing position selected.

6. Check rear sight assembly as follows:

   (a) Rotate elevation knob counterclockwise until the rear sight assembly is all the way down. If a whole click is not felt as the rear sight assembly stops, the rear sight assembly has bottomed out and will not pivot freely.

3-18.   M16A2 RIFLE FINAL INSPECTION (CONT).

a.   FINAL INSPECTION (CONT)

(b) Position elevation knob back slightly to its last whole click as the rear sight assembly base is under tension of the ball bearing and helical spring. The 300 meter mark should align with the mark on the receiver.

(c) If the 300 meter mark is not aligned with the mark on receiver, slip the range scale in the following manner:

(1) Position the 300 meter mark with the mark on the receiver.

(2) Insert a 1/16 inch allen wrench through the access hole of the rear sight assembly base and into the index screw.

(3) Loosen the index screw three turns and leave the wrench in place.

(4) Rotate lower portion of elevation knob counterclockwise until it stops (range scale should not have moved). Elevation knob should be positioned on its last whole click.

(5) Tighten index screw and remove wrench.

(6) Check for proper setting.

7. Pull charging handle assembly to rear. Check that chamber is clear. Leave hammer in cocked position.

8. Place selector lever in SAFE position and pull trigger. Hammer should not fall.

WARNING
If rifle fails any of the following inspections, continued use of the rifle could result in injury to, or death of, personnel.

9. Place selector lever in SEMI position.

10. Pull trigger. Hammer should fall.

11. Hold trigger to the rear, charge rifle and release the trigger with a slow, smooth motion without hesitations or stops, until the trigger is fully forward; an audible click should be heard. Hammer should not fall.

12. Repeat the above selector lever SEMI position test (steps 10 and 11) five times. The rifle must not malfunction during any of these five repetitions. If the rifle malfunctions during any of these five tests, refer to page 3-7.

13. Place selector lever in BURST position. Charge rifle and pull trigger. Hammer should fall.

14. While holding the trigger to the rear, pull charging handle assembly to the rear and release it three times. Hammer should not fall. The burst disconnector should have held the hammer to the rear while the trigger was in the pulled position.

15. Pull trigger. Hammer should fall. This would be the first round of a three-round burst.

NOTE
A detailed explanation of the three-round burst control can be found in the following pages.

16. **With the hammer in the forward position, using moderate finger/thumb pressure, attempt to place the selector lever in the SAFE position. Selector lever should not go in the SAFE position.**

17. Perform the following additional functional checks and adjustments on assembled rifle:

    (a) Press magazine catch button. Make sure it functions properly.

    (b) Press bolt catch. Make certain it operates smoothly and holds bolt in open position.

    (c) Inspect front sight and rear sight assembly. Make certain they can be adjusted properly.

    (d) Actuate forward assist assembly. It must work freely.

    (e) Inspect upper receiver and barrel assembly. Barrel assembly should not rotate within upper receiver assembly.

    (f) Check that third or middle slot of compensator is straight up (TDC).

3-18. M16A2 RIFLE FINAL INSPECTION (CONT).

b. TEST

1. Check headspace using headspace gage PN 7799734. See page 3-45, TEST.

2. Check firing pin protrusion using firing pin protrusion gage PN 7799735. See page 3-45, TEST.

3. Check extent of barrel erosion using barrel erosion gage PN 8448496. See page 3-45, TEST.

4. Check barrel straightness using barrel straightness gage PN 8448202. See page 3-45.

c. FUNCTIONAL THEORY OF THREE-ROUND BURST CONTROL

NOTE

First become familiar with the functioning of the firing mechanism especially when in the SAFE and SEMI positions. You should also understand the role that the automatic sear plays when firing in the BURST position. Functioning of the mechanism is explained below in a step by step manner. This actually will seem to complicate something that is really very simple and happens in less than 1 second. The diagrams below and on the following pages do not show the associated springs for the sake of simplicity. The positioning of the burst cam is shown in detail.

Functional check of three-round burst is as follows:

NOTE

Assume the rifle is fully loaded with a live round in the chamber and the selector lever on BURST.

(a) Hammer is cocked.

(b) Front hook of burst lever is in stop notch.

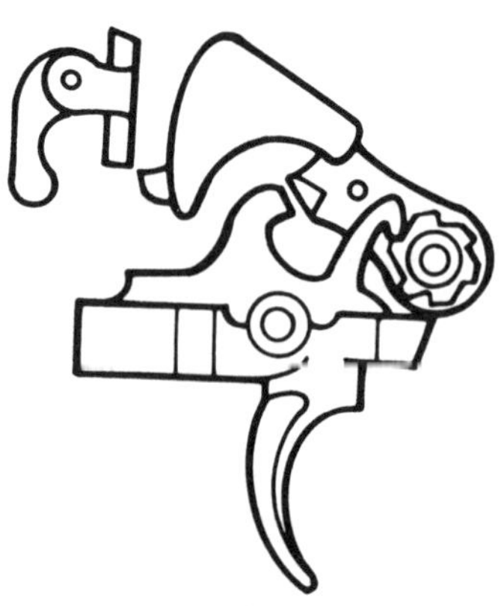

(c)   Trigger is pulled.

(d)   Trigger nose drops and hammer falls firing the FIRST ROUND.

(e)   Front hook of burst disconnector holds burst cam in place as hammer falls.

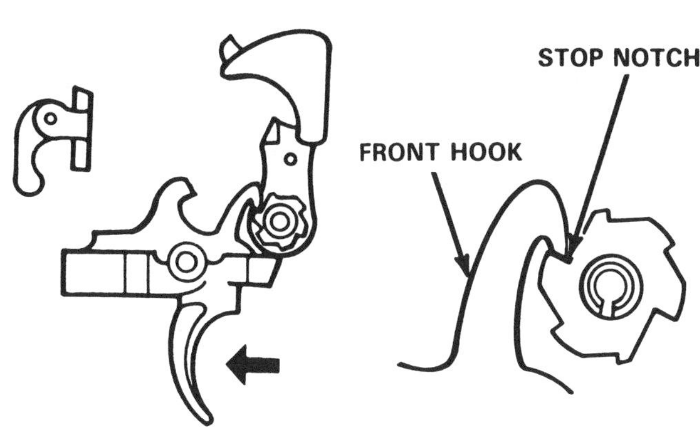

NOTE
Anytime the hammer falls forward, the clutch spring releases the burst cam and allows the front hook of the burst disconnector to keep it in place.

(f)   As the key and bolt carrier assembly moves to the rear, the hammer is forced to the rear.

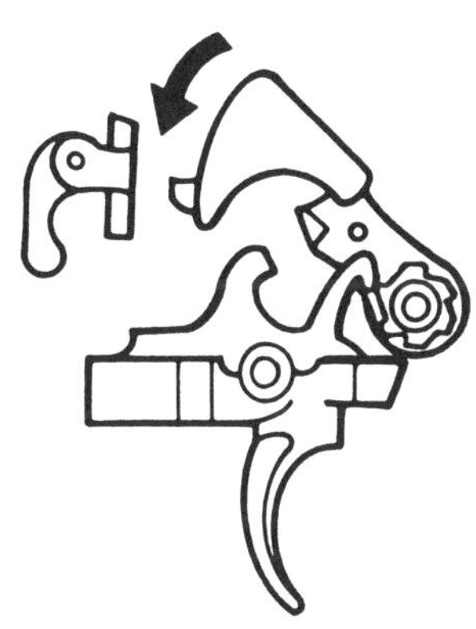

(g)   The clutch spring of the burst cam clutches the burst cam and causes it to rotate one notch as the hammer is forced back.

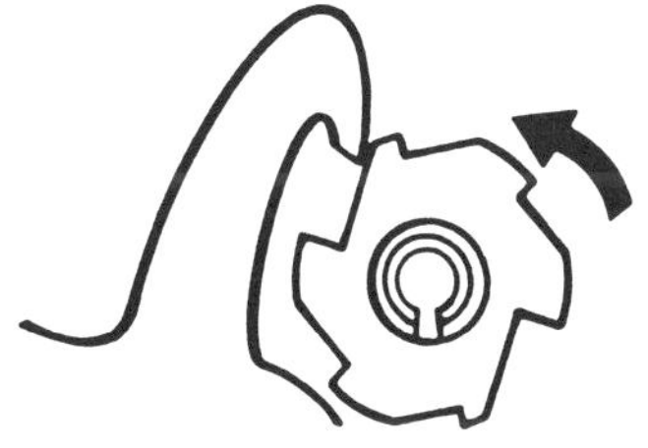

(h)   When hammer is fully to the rear, the automatic sear catches it.

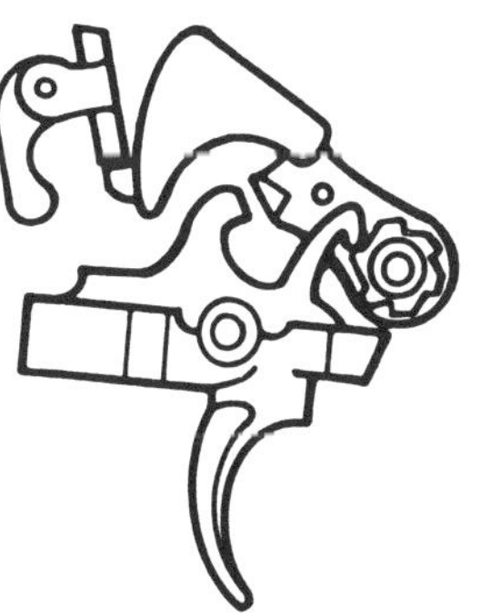

3-18. M16A2 RIFLE FINAL INSPECTION (CONT).

c. FUNCTIONAL THEORY OF THREE-ROUND BURST CONTROL (CONT)

(i) The front hook of the burst disconnector is now fully in the second notch.

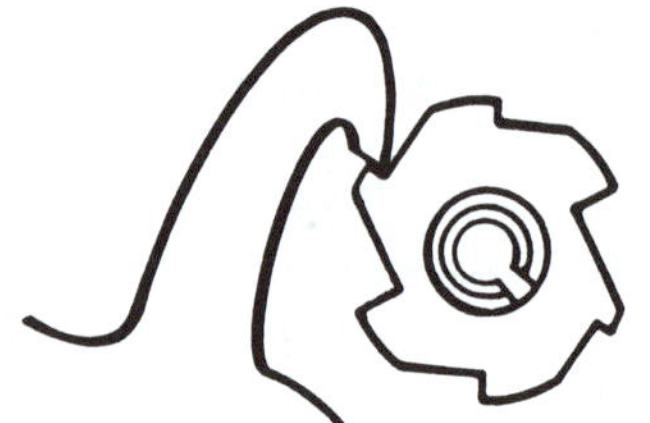

(j) As the key and bolt carrier assembly
travels forward, the automatic sear releases
the hammer and the hammer falls.

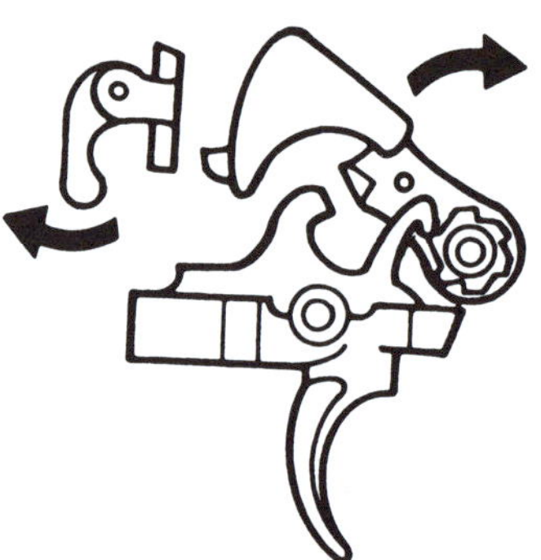

(k) When the hammer falls, the SEC-
OND ROUND is fired.

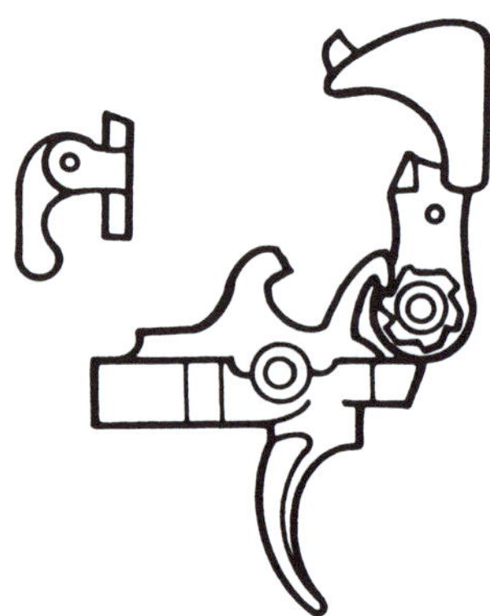

(l) As the key and bolt carrier assembly
moves to the rear, the hammer is forced
back to the rear.

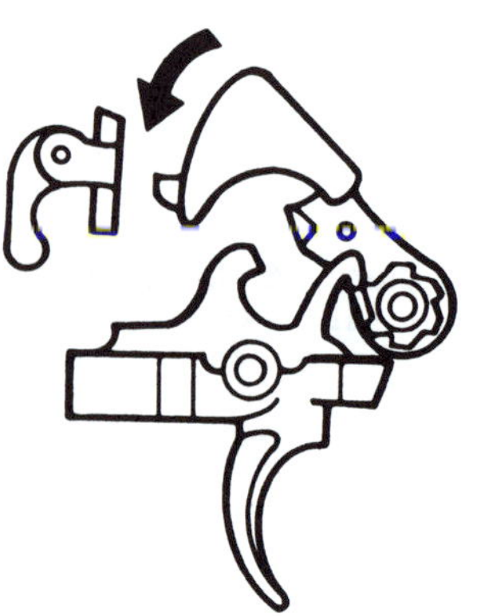

(m)   The clutch spring of the burst cam clutches against the burst cam and causes it to rotate one notch as the hammer is forced back.

(n)   When the hammer is fully to the rear, the automatic sear catches it.

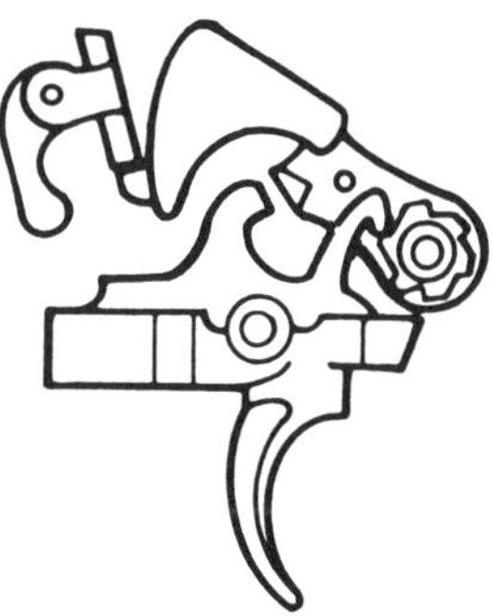

(o)   The front hook of the burst disconnector is now fully in the third notch.

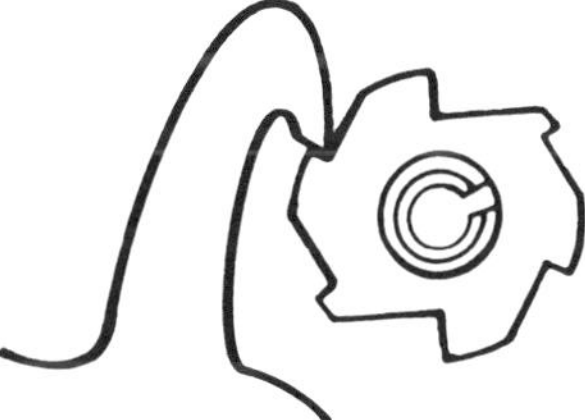

(p)   As the key and bolt carrier assembly travels forward, the automatic sear releases the hammer and the hammer falls.

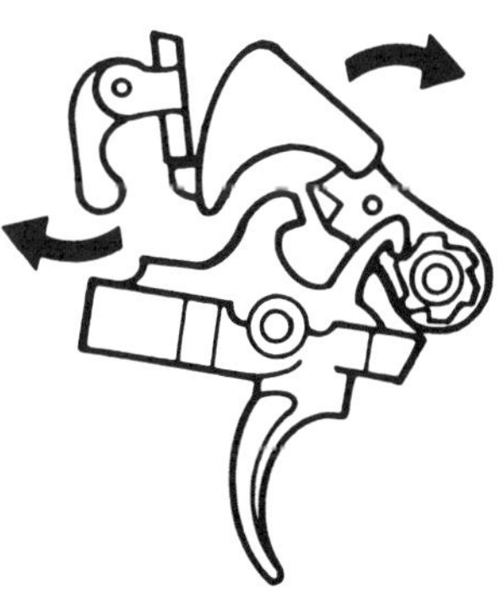

3-18. M16A2 RIFLE FINAL INSPECTION (CONT).

c. FUNCTIONAL THEORY OF THREE-ROUND BURST CONTROL (CONT)

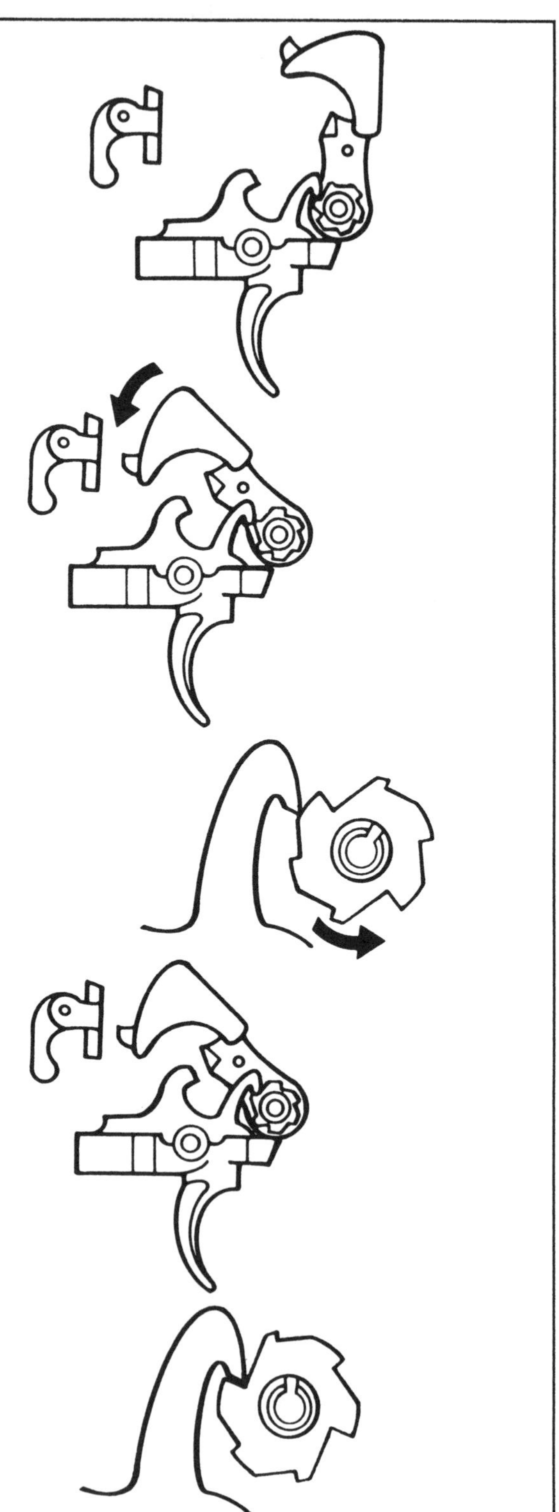

(q) When the hammer falls, the THIRD ROUND is fired.

(r) As the key and bolt carrier assembly moves to the rear, the hammer is forced back to the rear.

(s) The clutch spring of the burst cam clutches against the burst cam and causes it to rotate one notch as the hammer is forced back.

(t) When the hammer is fully to the rear, it is initially caught by the automatic sear. However, the front hook of the burst disconnector is now fully in the next stop notch which is deeper than the others.

(u) Because a stop notch is deeper than the others, it allows the front hook of the burst disconnector further forward than before. This allows the rear hook of the burst disconnector to latch on the rear hammer notch. This holds the hammer fully to the rear even though the trigger is still to the rear. This happens when the burst is over and the firing is stopped.

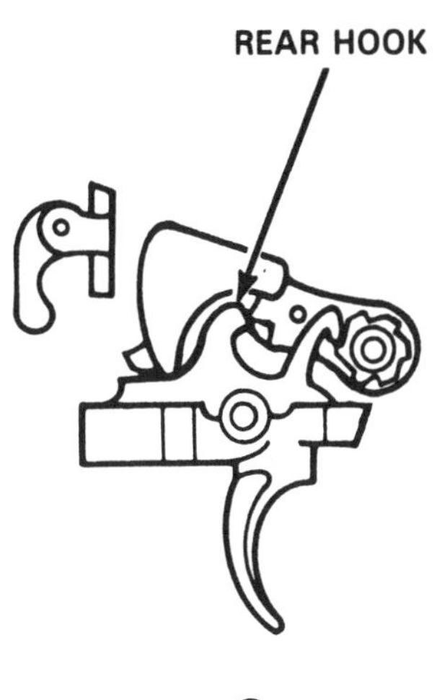

(v) Once the trigger is released, the trigger nose comes up and holds the hammer back.

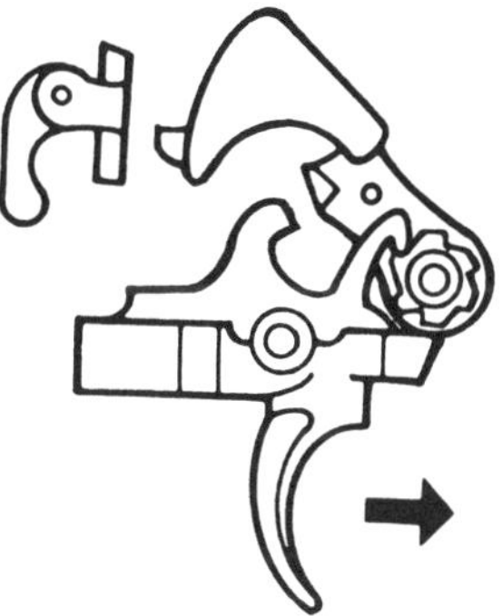

NOTE

Pulling the trigger to the rear and holding it back again will fire another three-round burst. This will continue until the magazine is empty. However, the trigger must be released between each burst.

3-19. M16A2 RIFLE ANNUAL GAGING REQUIREMENTS.

This task covers:

a. Inspection

b. Gaging

INITIAL SETUP

Test Equipment
  Tool and Gage Set (item 2, app B)

Tools
  (ARMY) Small Arms Repairman Tool Kit
    (item 3, app B)
  Field Maintenance Basic Less Power
    Small Arms Shop Equipment (item 1,
    app B)

References
  DA PAM 738-750
  TM 9-1005-319-10
  AFR 50-36
  AFTO Form 105

General Safety Instructions
  • To avoid injury to your eyes, use care
    when removing and installing spring-
    loaded parts.
  • All M16A2 rifles must be gaged at
    least once annually for safety.
  • All Army Reserve and Army National
    Guard M16A2 rifles must be in-
    spected and gaged at least once

every 2 years, after the initial in-
spection/gaging procedures have
been accomplished. This 2 year interval may be maintained unless
preventive maintenance checks and
services (PMCS), or other physical
evidence, indicates that an individual
unit's rifles require inspection/gaging
at a more frequent interval. If it is
determined that a yearly inspection
is necessary for an individual unit,
only that unit will be affected. This
will not affect other units in regard
to the interval of inspection.
  • It is recommended that training units
    inspect/gage all rifles at the end of
    each training cycle. Training units
    will inspect/gage all rifles at least
    once annually.
  • Air Force M16 rifles will be inspected
    in accordance with AFR 50-36, Vol.
    1, Chapter 5.

a. Inspection

1. Visually inspect general appearance of rifle. Overall appearance will be approximately that of a new weapon. For inspection criteria refer to final inspection, page
3-84. All visual and functional inspection requirements must be met.

NOTE
To perform the following tests, disassemble weapon only as far as allowed
in chapter 2 (Unit Maintenance Instructions), unless a deficiency is un-
covered.

2. Perform a general inspection of rifle per section III of this chapter. Repair as required and authorized.

b. Gaging

1. Gage bolt carrier assembly for firing pin protrusion using firing pin protrusion gage PN 7799735. See page 3-24, TEST.

2. Gage bolt carrier assembly for firing pin hole wear using not-go plug gage PN 12620101. See page 3-24, TEST.

3. Inspect chamber in upper receiver and barrel assembly using chamber reflector tool PN 8448201. See page 3-33, INSPECTION/CLEANING.

4. Gage barrel in upper receiver and barrel assembly using barrel erosion gage PN 8448496 and bore straightness gage PN 8448202. See page 3-45, TEST.

5. Check headspace in upper receiver and barrel assembly by inserting headspace gage PN 7799734 in chamber. See page 3-45, TEST.

6. Gage pivot pin lug area clearance in lower receiver assembly using 0.020 thickness gage. See page 3-68.

7. Gage hammer and trigger pin holes in lower receiver assembly using taper plug gage PN 12006472. See page 3-68, TEST.

8. Gage trigger pull using trigger pull measuring fixture PN 7274758. See page 3-83, TEST.

9. Document inspection with DA PAM 738-750, AFTO Form 105, or NAVMC 11003 when completed.

Section IV. PREEMBARKATION INSPECTION OF MATERIEL IN UNITS ALERTED FOR OVERSEAS MOVEMENT

3-20. PURPOSE. This section establishes standards for overseas shipment (preembarkation inspection criteria) of M16A2 rifle. These standards are provided to ensure that the user is furnished equipment which will perform its mission without early failure or major maintenance problems.

3-21. SCOPE.

a. The standards prescribed provide for a high percentage of remaining life in affected rifles; therefore, rifles designated for overseas shipment must qualify under the standards contained in the following paragraph, table, and in referenced DA publications, before they can be approved for shipping action.

b. Provisions of this standard apply to all US Army agencies/activities selecting or preparing rifles for shipment to US troops overseas. It also applies to CONUS troops preparing rifles for shipment overseas. Provisions do not apply either to rifles being prepared for shipment to MAP/MAS recipients unless specifically prescribed by MAP/MAS transaction for the materiel or to rifle being returned to CONUS from overseas. The maintenance instructions and standards contained herein do not apply to rifles once the materiel has arrived at the overseas destination. At that time, maintenance instructions contained in the applicable TM's will be used.

c. This applies to rifles which are the logistic responsibility of the US Army Armament, Munitions and Chemical Command.

3-22. GENERAL.

a. Only rifles which have been classified as serviceable condition code A, B, or C under AR 725-50 will be considered for overseas shipment. All items of equipment for which equipment serviceability criteria have been published must, as a minimum, be rated green under the ESC as a prerequisite to overseas shipment. In addition to the condition code standard, as enumerated above, and the required ESC rating prescribed herein, the rifle being considered for overseas shipment must meet the requirements of this section. The ESC will be discontinued as new operator manuals are revised which will be used to determine serviceability condition of rifle.

b. Waivers to provisions can only be granted by the gaining command of any particular end item being considered for issue, deployment, or shipment. The issuing services may recommend issue or shipment of rifles not meeting the provisions when all the following conditions exist:

(1) Repair parts in required quantities cannot be obtained from the supply system prior to delivery of the end item.

(2) The gaining command concurs in the receipt of the end item for storage until required repair parts become available. The gaining command must also state that capability, facilities, and funds are available to perform the necessary work when parts become available.

(3) Department of the Army approval is obtained on a case-by-case basis.

(4) Required repair parts are requisitioned by the issuing command for delivery to the gaining command.

c. All Department of the Army MWO's applicable to the specific rifle being considered for shipment overseas must have been applied.

d. Refer to SB 746-1 for pertinent publications relating to equipment processing and marking information.

e. Refer to AMC-P 310-9 for publications containing applicable overhaul standards.

3-23. SHIPMENT OR ISSUE.

a. Organizational Repair Parts, Tools, and Equipment. Rifles must be complete with all items required by applicable Department of the Army publications, including those in the basic issue items list of the appropriate operator's manual.

b. Publications. Operator publications applicable to the equipment log book must accompany the equipment. All log book entries must be complete and up-to-date including those covering any repairs, replacements, or adjustments made to the rifle in complying with this section.

c. Documentation. Prepare DA Form 2408-9 (Equipment Control Record) at time of overseas shipment or issue to another stock record or property book account, in accordance with the provisions of DA PAM 738-750.

d. Preparation. Process rifles for shipment as required by shipping documents and pertinent regulations.

3-24. DISPOSITION. Disqualified rifles which do not qualify for shipment will either be redistributed within the camp, post, or station, be repaired, or become candidates for overhaul, cannibalization, or other disposition as required by existing regulations.

3-25. GENERAL INSPECTION CRITERIA.

WARNING
Before starting an inspection, be sure to clear the rifle. Do not actuate the trigger until the rifle has been cleared. Inspect the bore and chamber to ensure that it is empty and free from obstructions, and check to see that no ammunition is in position to be introduced.

a. Before inspection, the materiel must be thoroughly cleaned of all grease, dirt, or other foreign matter that might interfere with its proper function or the use of gages and tools during inspection.

b. Materiel must be free of burrs, rust, or corrosion on functional surfaces.

3-25. GENERAL INSPECTION CRITERIA (CONT).

c. Parts must not be cracked, bent, distorted, or damaged and must be free of detrimental wear or looseness.

d. Minor defects in metal components do not normally affect their acceptability. For example, scratches and tool marks are ordinarily of no importance.

e. Inspect finish of metal surface.

(1) General. Satisfactory metal surfaces for rifles range from black to light gray. A worn shiny metal surface is objectionable only when it is capable of reflecting light. No rifle will be rejected unless exterior parts have a shine. All rear sights must have a dull gray or black finish on all surfaces that would cause a glare.

(2) M16A2 Rifle. Minor loss of finish (shiny spots, nicks, scratches) on exterior surfaces of the barrel and flash suppressor shall not be cause for rejection of M16A2 rifles located in hands of troops at training centers. Large shiny surfaces, nicks, scratches, etc., can be restored by the use of solid film lubricant (item 21, app D). Rifles (small arms) missing in excess of one-third or more of the exterior finish resulting in an unprotected, light-reflecting surface, are considered candidates for overhaul. The only authorized level of maintenance to phosphate finish small arms is depot.

f. Plastic components must not be cracked or damaged in such a way as to interfere with their structural strength. Surface cracks, bruises, or dents that do not affect their strength will not be cause for rejection. Cracks will be cause for rejection. Criteria for determining which cracks are repairable are on page 2-14.

g. Barrels must be clean and free of corrosion such as that caused by moisture and powder fouling. Standards of serviceability are indicated in (1) through (9).

(1) Pits in the chamber are allowable if they do not cause extraction difficulties.

(2) Pits as wide as a land and 3/8 inch (0.95 cm) or less in length are allowable for 5.56-mm barrels. Pits not greater than the width of a land and less than 3/8-inch (0.95 cm) long are permissible.

(3) Scattered or uniformly fine pits, or fine pits in a densely pitted area are allowable.

(4) Tool marks are acceptable regardless of length. They will appear as lines running laterally in the grooves, or may run spirally across the top of lands.

(5)   Ringed bores or bores ringed sufficiently to bulge the outside surface of the barrel are cause for rejection. However, faint rings or shadowy depressions do not indicate an unserviceable barrel and will not be cause for rejection. Gap in lined barrels will not be classified as a ringed bore.

(6)   Lands that appear dark due to coating of gilding metal from projectiles will not be cause for rejection.

(7)   Breech bore diameter will be checked on unlined barrels using the appropriate breech bore gage.

(8)   Barrel erosion gages are provided for lined barrels. Bore wear will be checked using barrel erosion gage for the M16A2. For detailed instructions in the use of the above gage and for serviceability limits, refer to page 3-46.

(9)   Flaking or checking (fine cracks) of chromium plate in barrels or chambers will not be cause for rejection, unless accompanied by pitting to the degree that extraction difficulties are encountered or accuracy is unacceptable.

h.   Springs must be free of distortion and broken coils. Springs must have sufficient tension to perform their intended function.

i.   Screw heads must be in serviceable condition and threads must not be stripped. Internal threads must not be stripped.

j.   The sear, hammer, and/or cocking notches must be in good condition. Chipped engaging corners will be cause for rejection. Slight wear on functional surfaces, including engaging corners, shall be acceptable, providing the minimum trigger pull requirements and selector lever checks are met in accordance with instructions on page 3-83.

k.   Chips, flat spots, or bent striker points on firing pins will be cause for rejection.

l.   The cartridge engaging surfaces on extractors must not be chipped or deformed.

m.   Evidence of any damage to sights will be cause for a sight alignment check. Rear sight bases should have no movement.

n.   Rear sight elevating and windage mechanisms must operate with distinct clicks, without binding. Sights must have sufficient tension to retain their setting during firing. Graduations and numerals must be legible. Graduation filler is not required.

o.   Safeties must positively position in both the ON and OFF position. When in the ON or safe position, the rifle must not fire when the trigger is squeezed; when in the OFF or fire position, the rifle must fire when the trigger is squeezed.

3-25. GENERAL INSPECTION CRITERIA (CONT).

p. All locking devices such as latches, magazine latches, or detents must be positive in action and must not become disengaged due to normal handling and firing. Retaining pins and similar devices must not be subject to accidental loss during use or transportation.

q. Each rifle must be hand functioned to check for unusual binding, positive cocking action, and general operation. Dummy ammunition must be used to assure positive feeding, chambering, extraction, and ejection action.

Table 3-1. 5.56MM Rifle M16A2

Item	Standard
RIFLE: General	Clear rifle of any ammunition and inspect in accordance with paragraph 3-25.
Barrel and barrel extension	Check barrel erosion. Use barrel erosion gage 8448496 for chrome lined barrels. Stripping of lands and grooves shall not be cause for rejection unless so determined by barrel erosion gage. Visually inspect, using chamber reflector tool 8448201. Pits 1/8 inch (0.31 cm) in length and those pits large enough to extend from the body of the chamber into the shoulder stop area and forcing cone area are cause for rejection. Large pits are defined as those 1/8 inch (0.31 cm) or more in diameter as determined by visual inspection. Only closed flash suppressors are acceptable. Check barrel for straightness using bore straightness gage 8448202. Gage must pass freely through the bore to be acceptable, either dropped from the muzzle or chamber end.
Front sight and gas tube	Inspect gas tube for proper alignment with carrier key. Gas tube must not bind when mating with the key. Evidence of gas leaks around the front sight connection of the gas tube shall be cause for rejection until rifle has been function fired to determine if the loss of gas is sufficient to cause malfunction. If function firing malfunctions occur, repairs are necessary. Inspect front sight for damage.

Table 3-1. 5.56MM Rifle M16A2 - Cont

Item	Standard
Bolt carrier group	Inspect bolt for elongated or oversized firing pin hole using plain cylinder gage 12620101.
	Firing pin holes which permit the plain cylinder plug gage to fully penetrate at any position on the circumference will be rejected.
	Bolt face with a cluster of pits which are touching or tightly grouped, covering an area measuring approximately 1/8 inch (0.31 cm) across will be rejected.
	Bolts which contain pits extending into the firing pin hole will not be rejected unless firing pin hole gaging check determines rejection.
	Bolts which contain individual pits or scattered pits will not be cause for rejection.
	Only phosphated bolt carriers are acceptable. Both phosphated and chrome plated bolts are acceptable.
Bolt locking lugs and bolt cam pin hole	Inspect for cracks in the locking lugs and cam pin hole area. Use a black light, if available; otherwise, use a glass of no more than 3X magnification or use inspection penetrant (item 25, app D). Use instructions contained in kit for application. If cracks are detected, the bolts will be replaced.

NOTE

Particular attention must be given to the area where the lugs meet the bolt body and around the side walls of the cam pin hole.

Item	Standard
	Bolt rings must not be broken. Ring gaps must be properly spaced approximately ⅓ turn apart and not in line.
	Firing pin protrusion must be not less than 0.028 inch (0.071 cm) or more than 0.036 inch (0.091 cm). (Use firing pin protrusion gage 7799735.)
	Socket head capscrews must be staked.
	Carrier key must not be dented where end mates with gas tube.
	Repair or replace damaged carrier keys.
Headspace	Inspect headspace using headspace gage 7799734. Excessive headspace will be cause for rejection.
Trigger pull	Inspect trigger pull using trigger measuring fixture 7274758. Trigger pull must be minimum 5.5 pounds (2.49 kg), maximum 9.5 pounds (4.31 kg). Test trigger pull, refer to page 3-83, steps a, b, c, and d.

3-25. GENERAL INSPECTION CRITERIA (CONT).

Table 3-1. 5.56MM Rifle M16A2 - Cont

Item	Standard
Lower receiver group	Inspect hammer and trigger pin holes using plain cylinder plug gage 12006472. Penetration of the gage in any one or more of the four holes will be cause for rejection. Inspect for cracks, corrosion, or mutilation which would affect functioning. Small dents or gouges will not be cause for rejection. Inspect receiver for corrosion in the lobes of the pivot or hinge pin area. Width between lobes shall not exceed 0.515 inch (1.30 cm). Inspect receiver for breakthrough of metal. Inspect receiver and receiver extension for initial loss of protective coating.
Action spring	Free length of spring shall be between 11-3/4 and 13-1/2 inches (29.84 and 34.29 cm).
Handguard	Inspect handguard assembly for breaks, separations, and cracks. Breaks and separations of material which prevent proper retention or interfere with functioning of the rifle will be cause for handguard rejection and replacement. Cracks up to 1 inch (2.54 cm) in length are acceptable provided they do not extend into the handguard retaining flange (1) (critical area). Each handguard assembly may have up to two of the three front retaining tabs (2) missing. If all three front tabs are missing, the handguard assembly must be replaced. Replace severely cracked handguards. Handguards that have a heat shield loose enough to rattle must be replaced.
Stock assembly	Inspect buttstock assembly for dents, cracks, and chips. Check for breaks and separation of material which could prevent proper functioning of rifle.

Table 3-1. 5.56MM Rifle M16A2 - Cont

Item	Standard

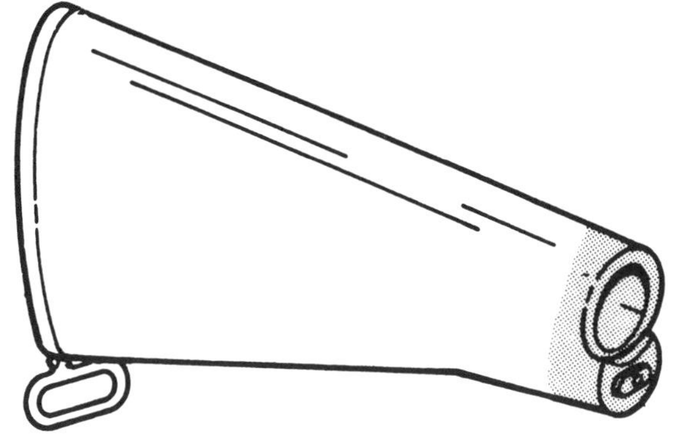 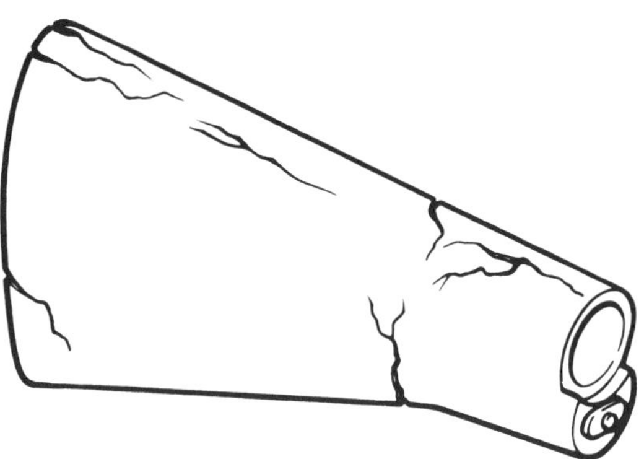

Under the following conditions, hairline cracks originating from buttplate end of buttstock are acceptable. No chipped away material is allowed.

a. One hairline crack, not to exceed 1 inch (2.54 cm) in length, per side of buttstock.

b. Two additional hairline cracks up to 0.22 inch (0.55 cm) in length, per side of buttstock.

c. A total of three cracks per side of buttstock, originating from buttplate end, are allowable.

Buttstocks with unauthorized markings stamped into their surfaces will be replaced. Unauthorized markings, scratched, etched, carved, etc., are acceptable if they do not extend into the fiber of the buttstock which may weaken it. These marks may lie at any location on the buttstock.

Cracks in the critical area at the front end of the buttstock are not acceptable and these buttstocks must be replaced.

CHAPTER 4
MAINTENANCE OF AUXILIARY EQUIPMENT

CHAPTER OVERVIEW

This chapter contains information and instructions to keep auxiliary equipment used with your rifle in good repair.

Section I. AUXILIARY EQUIPMENT REPAIR

4-1. GENERAL.

    a. The following items of auxiliary equipment are used in conjunction with the M16A2 Rifle:

        (1) 40mm Grenade Launcher M203, NSN 1010-00-179-6447. **(RIFLE ONLY)**

        (2) Lock Plate, NSN 1005-00-233-9031.

        (3) Top Sling Adapter, NSN 1005-00-406-1570.

        (4) Blank Firing Attachment M15A2, NSN 1005-00-118-6192. **(M16A2 ONLY)**
        Blank Firing Attachment M23, NSN 1005-01-361-8208. **(M4 ONLY)**

        (5) ARMY ONLY: Conversion Kit, M261 (caliber .22 rimfire adapter), NSN 1005-01-010-1561.

        (6) Bayonet-Knife M7, NSN 1005-00-073-9238

        (7) Bayonet-Knife Scabbard M10, NSN 1095-00-223-7164.

        (8) M9 Multi-Purpose Bayonet System, NSN 1005-01-227-1739.

        (9) Night Vision Sight, Individual Served Weapon, AN/PVS-4, NSN 5855-00-629-5334.

    b. Refer to TM 9-1010-221-24&P for unit maintenance for the Grenade Launcher M203.

    c. ARMY ONLY: Refer to TM 9-6920-363-12&P for unit maintenance of the M261 Conversion Kit (caliber .22 rimfire adapter) M16, M16A1, and M16A2 Rifles.

    d. Refer to TM 9-1005-237-23&P for repair instructions and repair parts for Bayonet-Knife M7 and Bayonet-Knife Scabbard M10 and M9 Multi-Purpose Bayonet System.

4-2. LOCK PLATE.

This task covers:

a. Installation
b. Removal

c. Inspection

INITIAL SETUP

Tools
  (ARMY) Small Arms Repairman Tool Kit
    (item 3, app B)

WARNING

The lock plate prevents the selector lever from being placed in BURST and will be installed at the discretion of the unit commander. It is mandatory for use in civil disturbance (riot control). Do not keep live ammunition near work area.

a. INSTALLATION

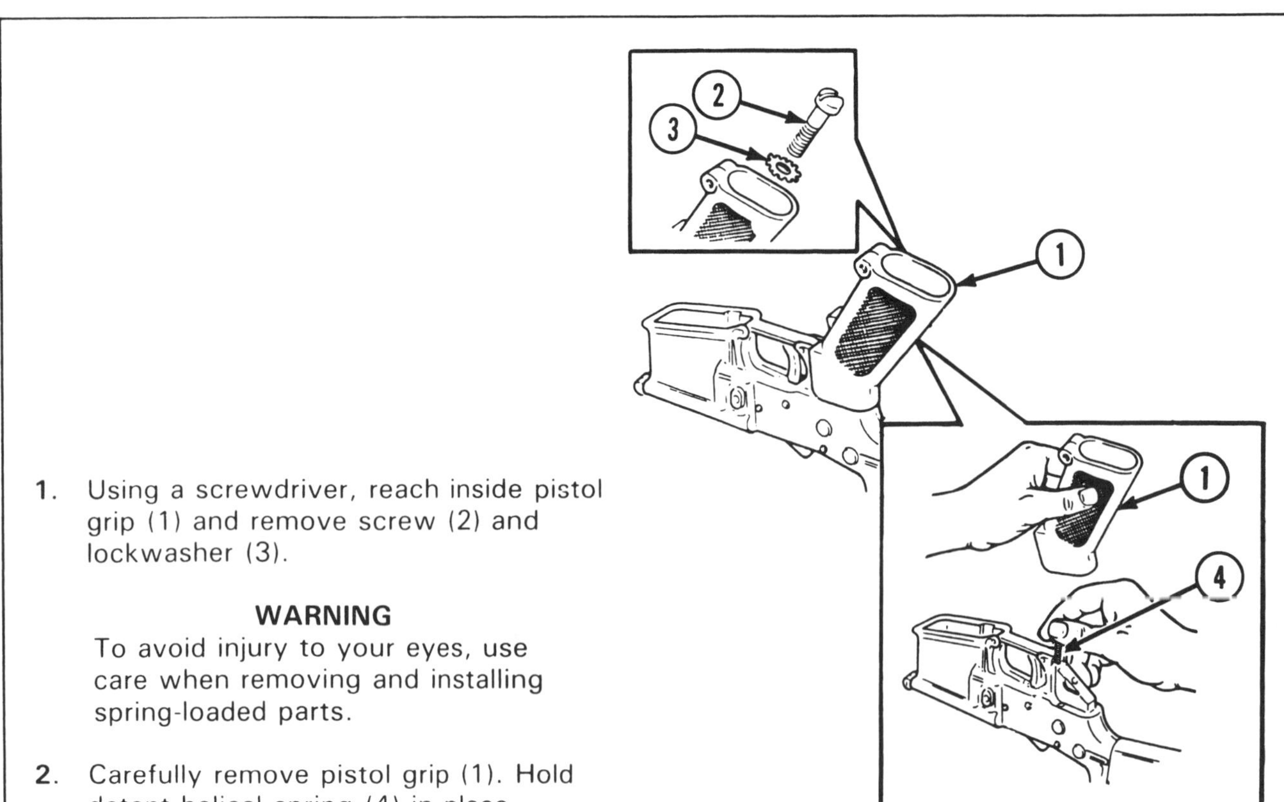

1. Using a screwdriver, reach inside pistol grip (1) and remove screw (2) and lockwasher (3).

WARNING
To avoid injury to your eyes, use care when removing and installing spring-loaded parts.

2. Carefully remove pistol grip (1). Hold detent helical spring (4) in place.

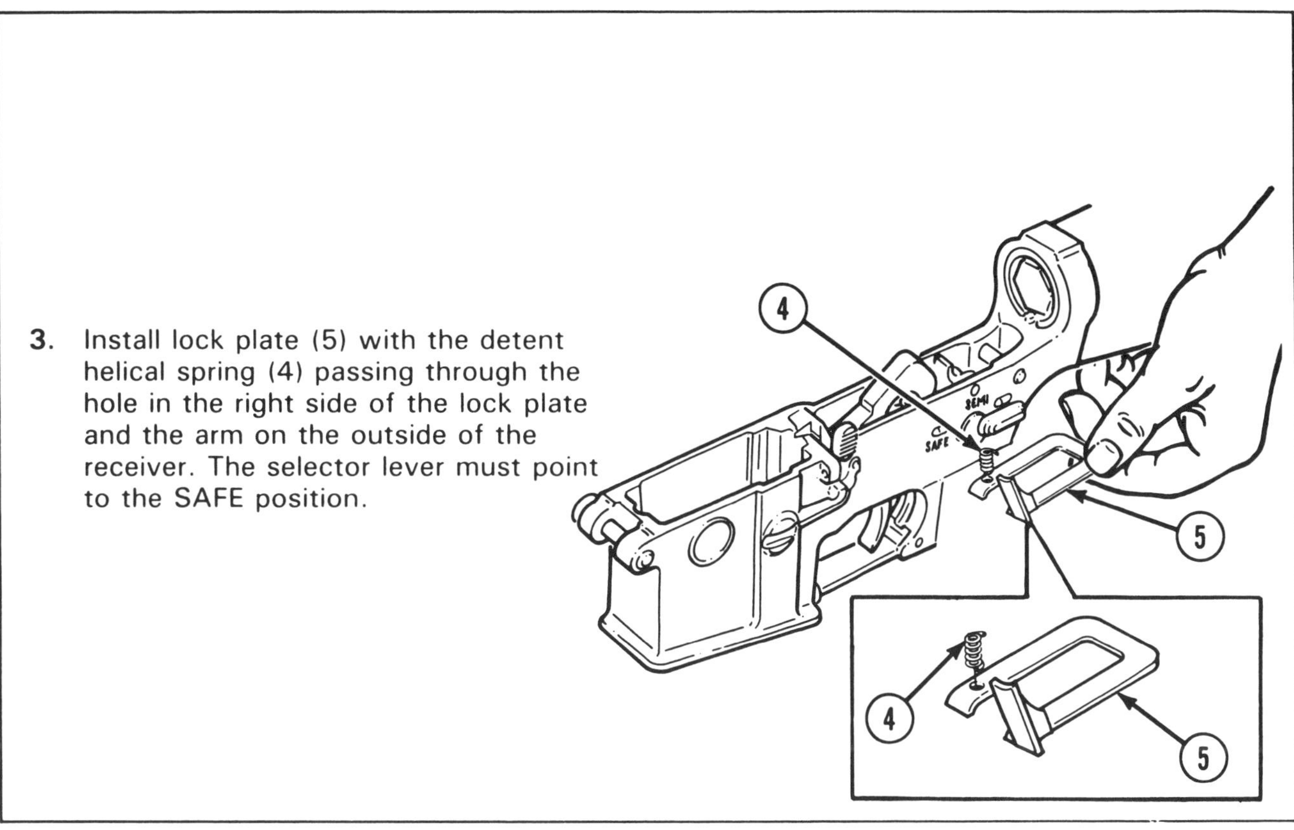

3. Install lock plate (5) with the detent helical spring (4) passing through the hole in the right side of the lock plate and the arm on the outside of the receiver. The selector lever must point to the SAFE position.

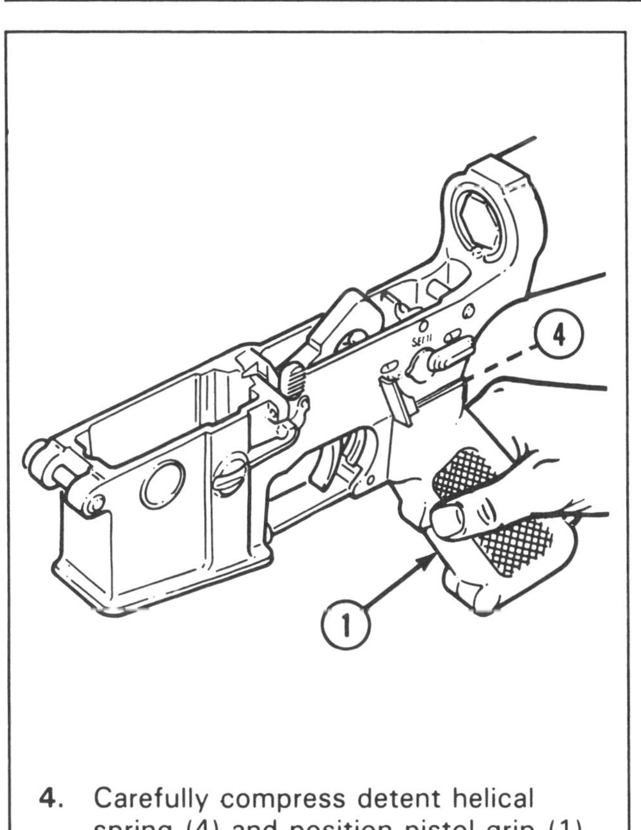

4. Carefully compress detent helical spring (4) and position pistol grip (1).

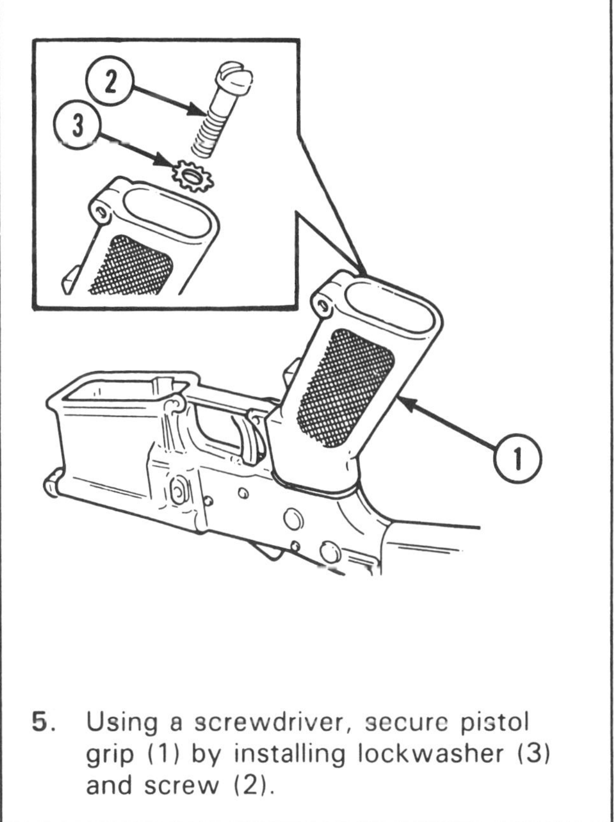

5. Using a screwdriver, secure pistol grip (1) by installing lockwasher (3) and screw (2).

4-2. LOCK PLATE (CONT).

b. REMOVAL

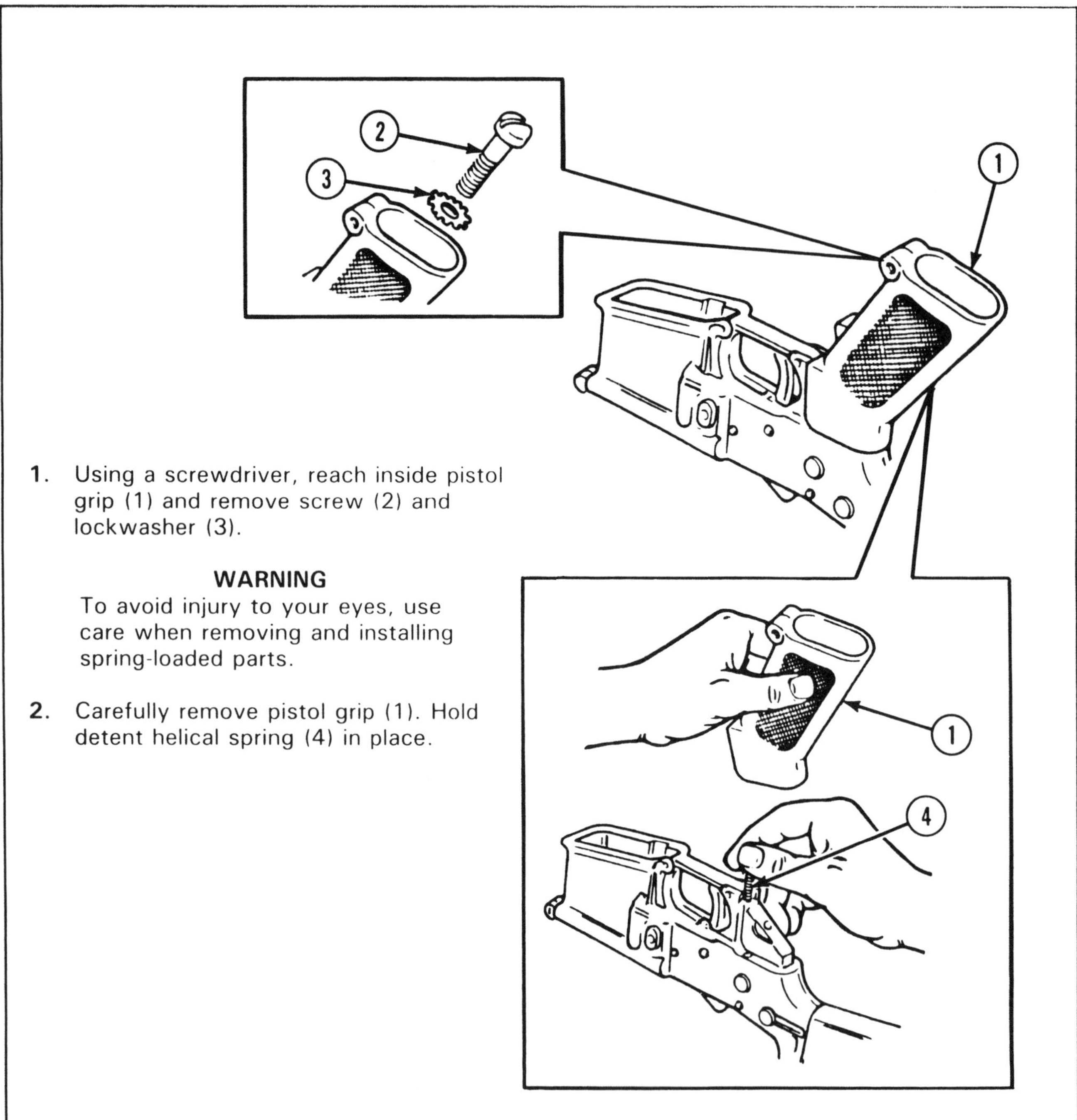

1. Using a screwdriver, reach inside pistol grip (1) and remove screw (2) and lockwasher (3).

WARNING

To avoid injury to your eyes, use care when removing and installing spring-loaded parts.

2. Carefully remove pistol grip (1). Hold detent helical spring (4) in place.

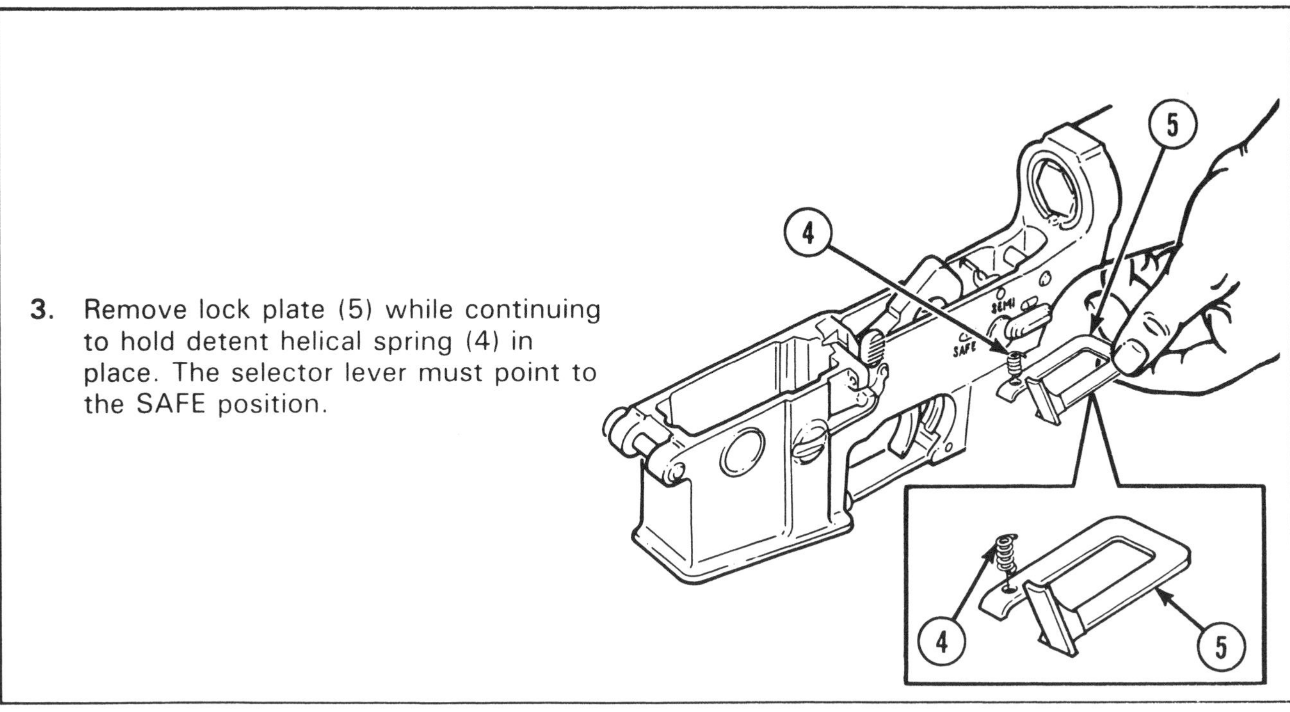

3. Remove lock plate (5) while continuing to hold detent helical spring (4) in place. The selector lever must point to the SAFE position.

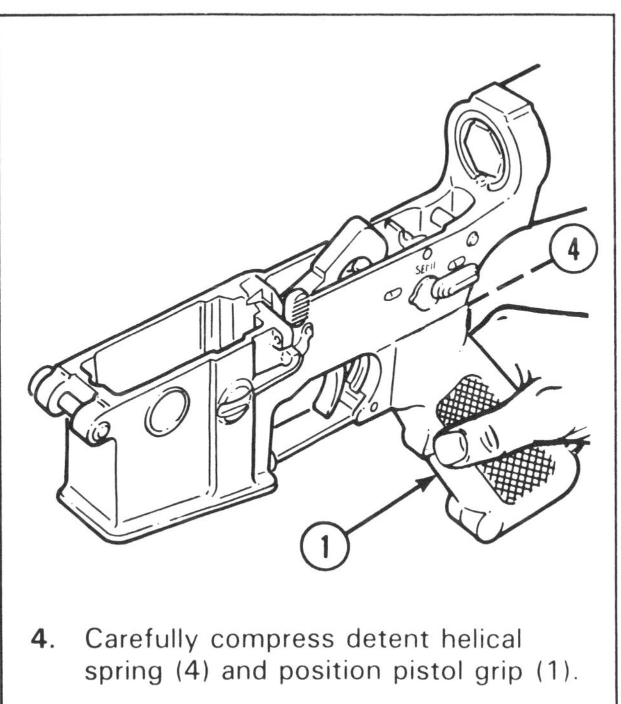

4. Carefully compress detent helical spring (4) and position pistol grip (1).

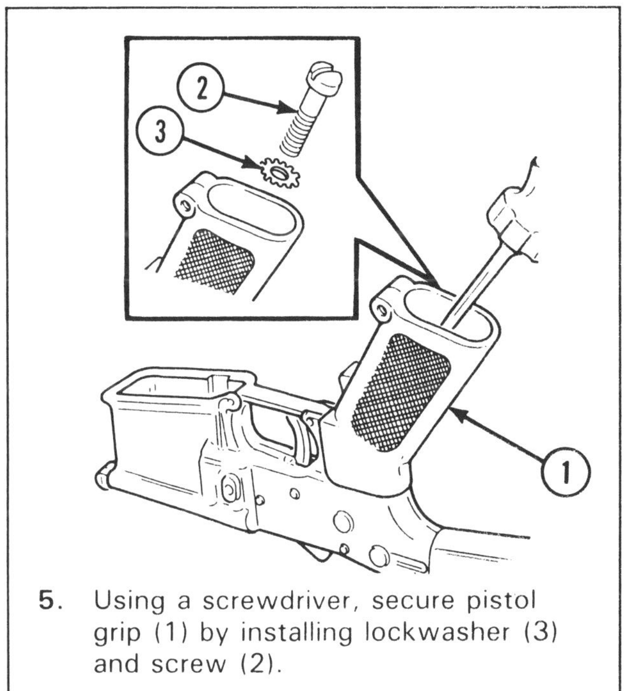

5. Using a screwdriver, secure pistol grip (1) by installing lockwasher (3) and screw (2).

c. **INSPECTION**

Inspect lock plate for serviceability and broken arm. Replace if unserviceable or if arm is broken off.

4-3. TOP SLING ADAPTER

This task covers:

    a. Installation   **c.** Inspection
    b. Removal

INITIAL SETUP

Materials/Parts
    Top sling adapter kit PN 8448471

References
    TM 9-1005-319-10

a. INSTALLATION

1. Refer to TM 9-1005-319-10 to remove sling.

RIFLE ONLY

2. Install top sling adapter strap (1) through sling swivel (2) and tie.

CARBINE ONLY

2.1. Install top sling adapter strap (1) through sling opening (2.1) and tie.

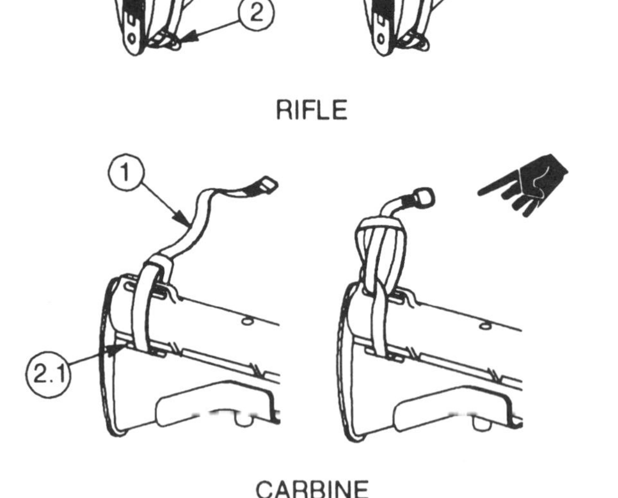

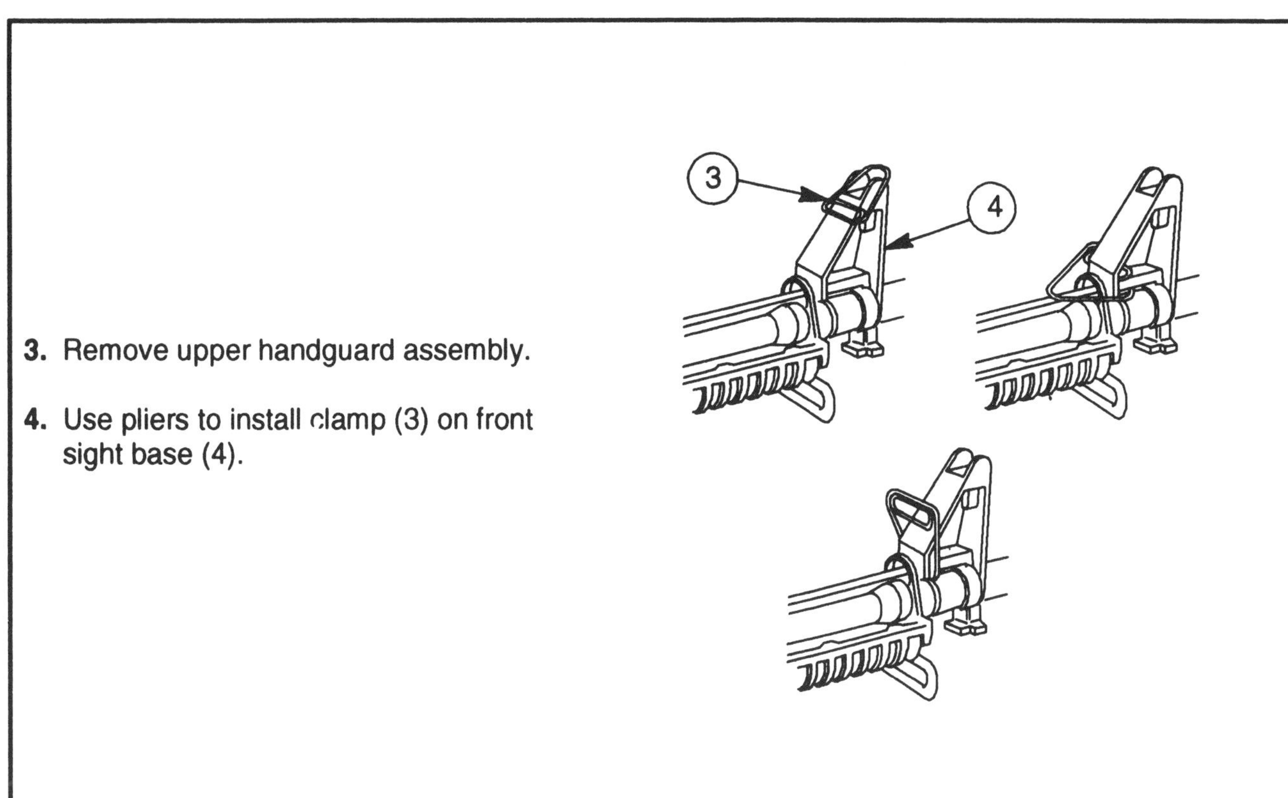

3. Remove upper handguard assembly.

4. Use pliers to install clamp (3) on front sight base (4).

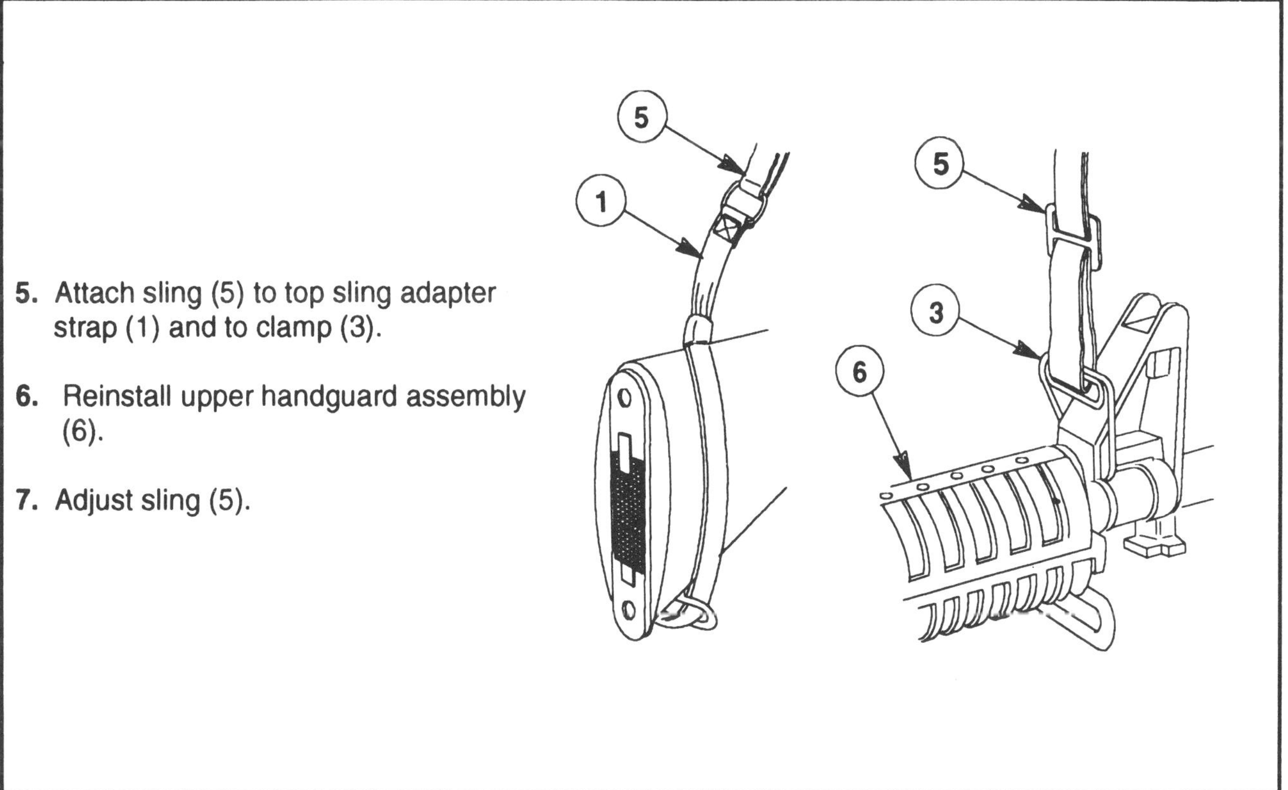

5. Attach sling (5) to top sling adapter strap (1) and to clamp (3).

6. Reinstall upper handguard assembly (6).

7. Adjust sling (5).

4-3. TOP SLING ADAPTER (CONT).

b. REMOVAL

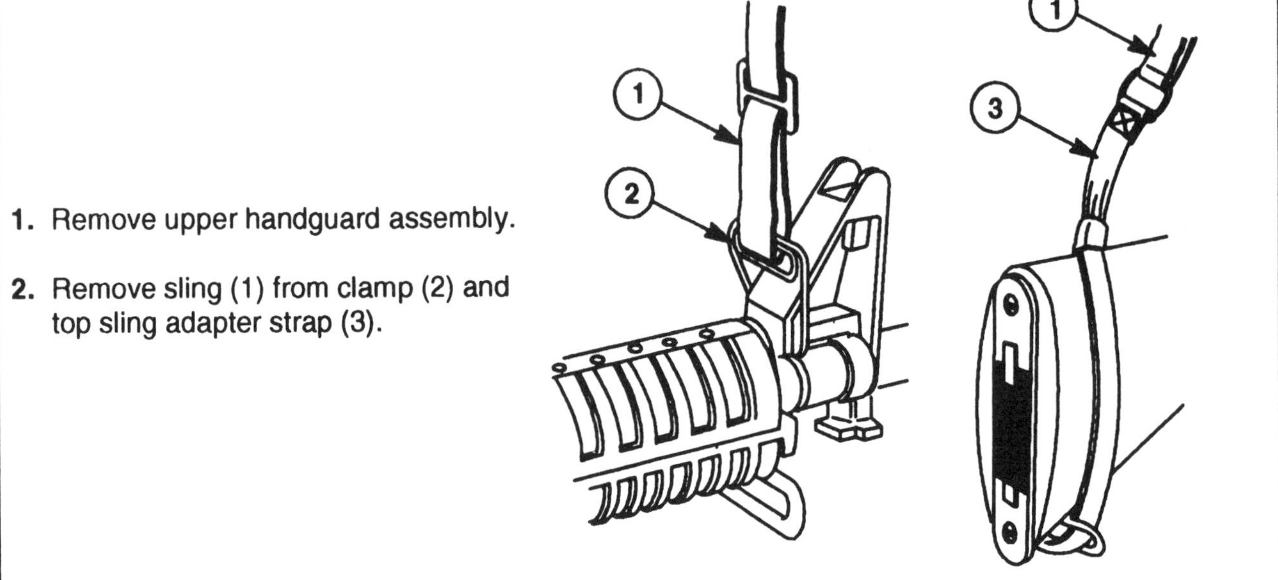

1. Remove upper handguard assembly.

2. Remove sling (1) from clamp (2) and top sling adapter strap (3).

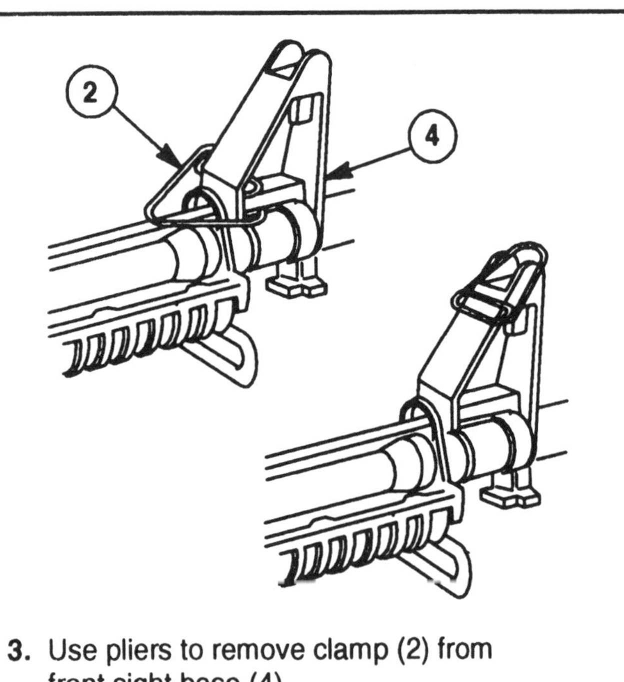

3. Use pliers to remove clamp (2) from front sight base (4).

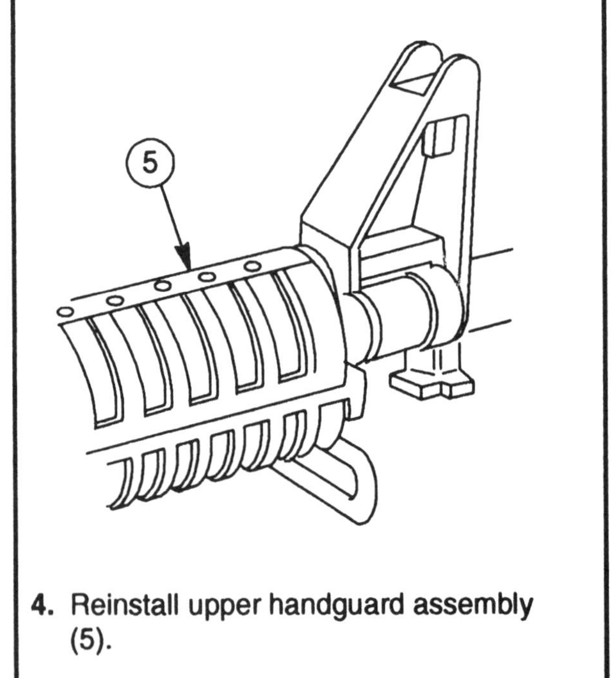

4. Reinstall upper handguard assembly (5).

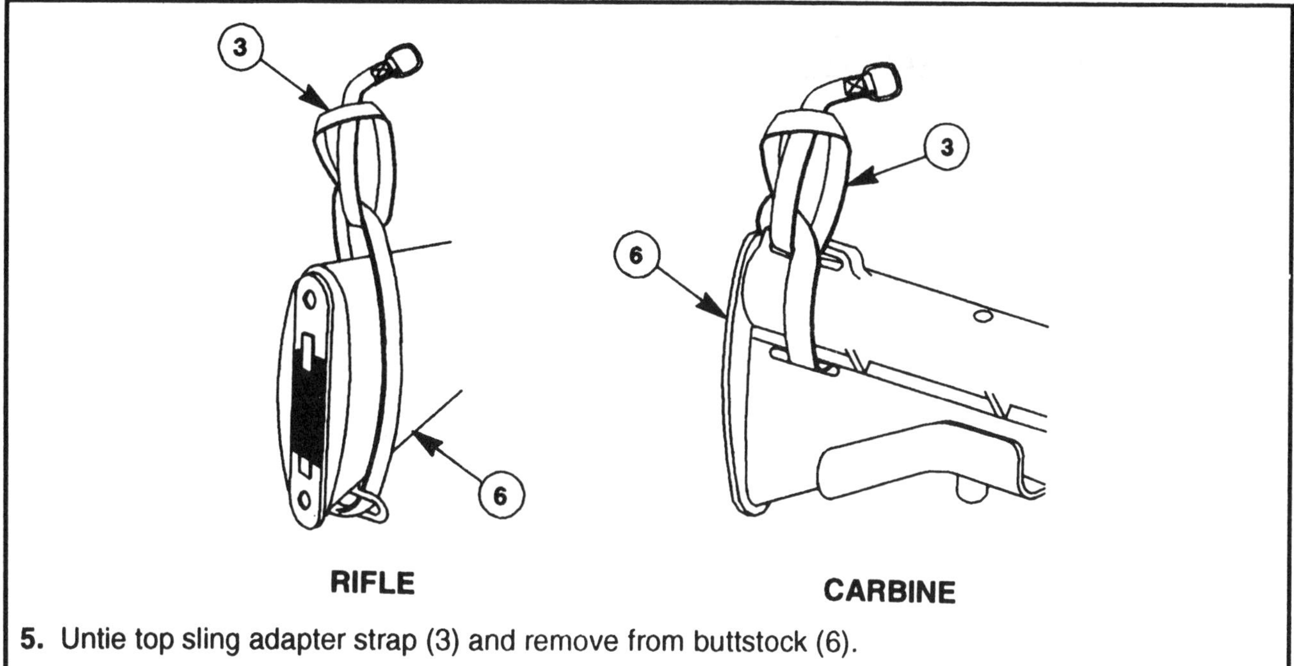

5. Untie top sling adapter strap (3) and remove from buttstock (6).

6. Refer to TM 9-1005-319-10 and install sling.

4-3. TOP SLING ADAPTER (CONT).

c. INSPECTION

Visually inspect top sling adapter strap and replace if it is badly worn or damaged.

4-4. BLANK FIRING ATTACHMENTS M15A2 AND M23.

This task covers:

a. Installation
b. Removal
c. Cleaning

d. Inspection
e. Repainting
f. Replacement

INITIAL SETUP

Materials/Parts
Cleaner, lubricant, and preservative (CLP) (item 9, app D)
Coating compound, fluorescent (item 14, app D)

General Safety Instructions
Do not keep live ammunition near the work area.

Only blank cartridge M200 is to be used when the blank firing attachment is attached to the rifle.
Do not fire blank ammunition at a representative enemy at distances of less than 20 feet (6.10 m). The unburned propellant grains can cause injury within this distance.

a. INSTALLATION

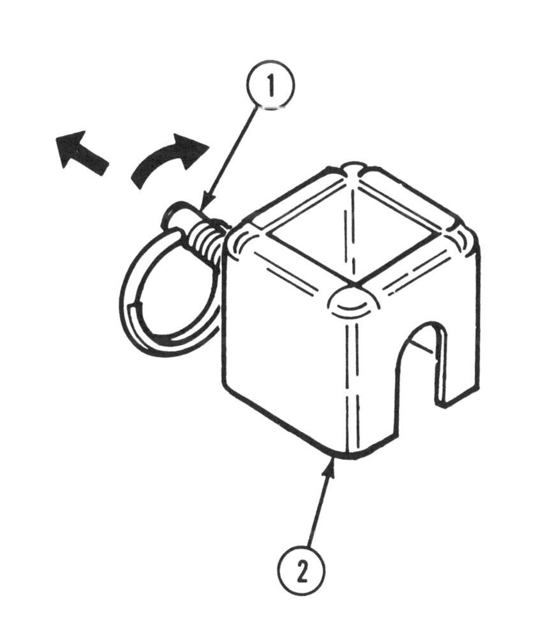

1. Unscrew and pull slide (1) all the way out on blank firing attachment (2).

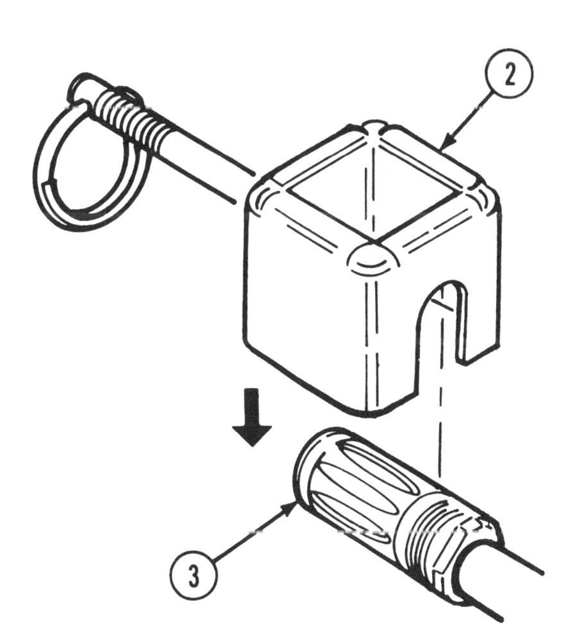

2. Hook blank firing attachment (2) behind the first groove of the compensator (3).

4-4. BLANK FIRING ATTACHMENTS M15A2 AND M23 (CONT).

a. INSTALLATION (CONT)

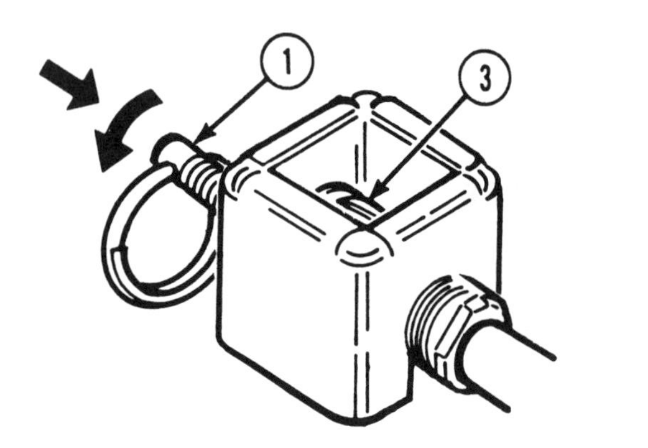

CAUTION

Do not use tools to tighten the blank firing attachment. **USE HANDS ONLY.**

3. Push slide (1) into compensator (3) and hand tighten.

NOTE

Check for tightness after firing approximately 50 blank rounds.

b. REMOVAL

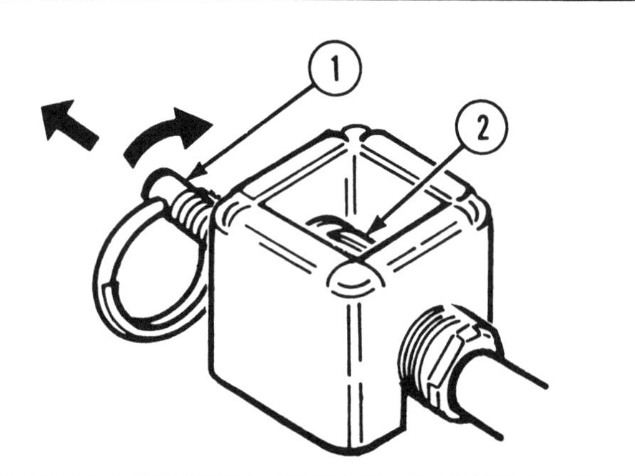

CAUTION

Do not use tools to loosen the blank firing attachment. **USE HANDS ONLY.**

1. Unscrew slide (1) to remove from compensator (2).

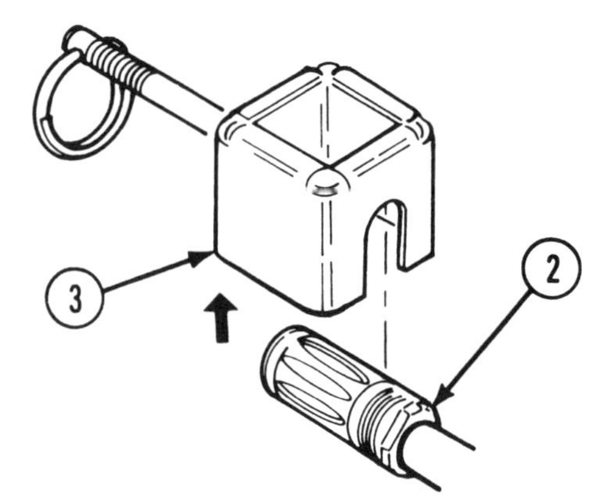

2. Unhook blank firing attachment (3) from behind the first groove of the compensator (2).

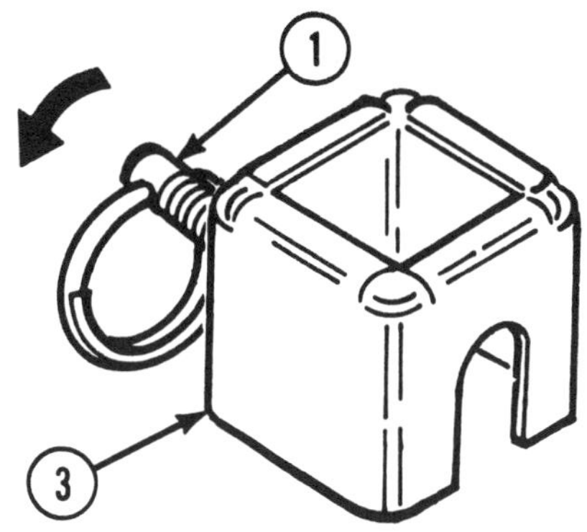

3. Screw slide (1) all the way in on blank firing attachment (3).

---

c. CLEANING

Clean blank firing attachment with CLP, wipe dry, and coat with CLP.

d. INSPECTION

Inspect blank firing attachment for cracks or distortion. Be sure the parts in the slide are clear and clean. If blank firing attachment is cracked or distorted, it is unserviceable.

e. REPAINTING

Repaint blank firing attachment using fluorescent coating compound. Painting is the only repair authorized.

f. REPLACEMENT

Replace blank firing attachment if unserviceable.

APPENDIX A
REFERENCES

A-1. TECHNICAL INSTRUCTIONS/TECHNICAL MANUALS/TECHNICAL ORDERS.

TI 05538A-15/2A . . . . Early Warning Indications of Malfunctions

TI 05538A-35/12A . . . . Checking Muzzle and Breech Erosion in Barrel

TI 8005-24/20 . . . . Prefire Inspection, Small Arms Weapons, Ordnance Material

TM 9-1005-237-23&P . . . . Organizational and Direct Support Maintenance Manual (Including Repair Parts and Special Tools List) for Bayonet-Knife M6 and M7, with Bayonet Knife Scabbard M10

TM 9-1005-319-10 . . . . . . Operator's Manual for 5.56mm, M16A2 Rifle

TM 9-1010-221-24&P . . . . Organizational, Direct Support and General Support Maintenance Manual Including Repair Parts and Special Tools List for Launcher, Grenade: 40mm M203 W/E (NSN 1010-00-179-6447)

TM 9-6920-363-12&P . . . . M261 Conversion Kit

TM 750-244-7 . . . . Procedures for Destruction of Equipment in Federal Supply Classifications 1000, 1005, 1015, 1020, 1025, 1030, 1055, 1090, and 1095 to Prevent Enemy Use

TM 4700-15/1 . . . . Equipment Record Procedures

TO 00-35D-54 . . . . Materiel Deficiency Reporting and Investigating System

TO 11A-13-10-7 . . . . Storage and Maintenance Procedures for Small Arms Ammunition

TO 11W-1-10 . . . . Historical Data Recording of Inspection, Maintenance, and Firing Data for Ground Weapons

TO 33K-1-100 . . . . (TMDE) Interval Calibration and Repair Reference Guide and Work Unit Code Manual

A-2. FIELD MANUALS.

FM 21-11 . . . . First Aid for Soldiers

FM 23-9 . . . . M16A1 and M16A2 Rifle and Rifle Marksmanship

A-3. PAMPHLETS.

DA PAM 25-30 . . . . Consolidated Index of Army Publications and Blank Forms

DA PAM 738-750 . . . . The Army Maintenance Management System (TAMMS)

A-4. RELATED PUBLICATIONS.

AFR 50-36 Vol 1 . . . . Combat Arms Training and Maintenance Program Management

AMC-P 310-9 . . . . Equipment Publications Listing

AR 725-50 . . . . Requisitioning, Receipt, and Issue System

AR 750-1 . . . . Army Materiel Policies

CTA 8-100 . . . . Army Medical Department Expendable/Durable Items

CTA 50-970 . . . . Expendable/Durable Items (Except: Medical, Class V, Repair
Parts and Heraldic Items)

FMFM 1-3A . . . . Field Firing Techniques

SB 746-1 . . . . Publications for Packing Army General Supplies

SPI 00-856-6885 . . . . Special Packaging Instructions for M16 Rifle

A-5. FORMS.

AFTO Form 22 . . . . Technical Order System Publications Improvement Report

AFTO Form 105 . . . . Inspection, Maintenance, and Firing Data for Ground Weapons

DA Form 2028 . . . . Recommended Changes to Publications and Blank Forms

DA Form 2028-2 . . . . Recommended Changes to Equipment Technical Manuals

DA Form 2404 . . . . Equipment Inspection and Maintenance Worksheet

DA Form 2408-9 . . . . Equipment Control Record

DD Form 314 . . . . Preventive Maintenance Schedule and Record

NAVMC 10772 . . . . Recommended Changes to Technical Publications

NAVMC 11003 . . . . Ordnance Serialized Items Subsidiary Records

SF 364 . . . . Report of Discrepancy (ROD)

SF 368 . . . . Product Quality Deficiency Report

A-6. SUPPLY CATALOGS/SUPPLY LISTS.

SC 4933-95-CL-A11 . . . . . . Shop Set, Small Arms Field Maintenance

SC 5180-95-CL-A07 . . . . . . Tool Kit, Small Arms Repair

APPENDIX B

MAINTENANCE ALLOCATION CHART

Section I. INTRODUCTION

B-1. GENERAL.

a.   This section provides a general explanation of all maintenance and repair functions authorized at various maintenance levels.

b.   The Maintenance Allocation Chart (MAC) in section II designates overall authority and responsibility for the performance of maintenance functions on the identified end item or component. The application of the maintenance functions to the end item or component will be consistent with the capacities and capabilities of the designated maintenance levels.

c.   Section III lists the tools and test equipment (both special tools and common tool sets) required for each maintenance function as referenced from section II.

d.   Section IV contains supplemental instructions and explanatory notes for a particular maintenance function.

B-2. MAINTENANCE FUNCTIONS.

Maintenance functions will be limited to and defined as follows: (except for ammunition MAC[1])

a.   **Inspect**. To determine the serviceability of an item by comparing its physical, mechanical, and/or electrical characteristics with established standards through examination (e.g., by sight, sound, or feel).

b.   **Test**. To verify serviceability by measuring the mechanical, pneumatic, hydraulic, or electrical characteristics of an item and comparing those characteristics with prescribed standards.

c.   **Service**. Operations required periodically to keep an item in proper operating condition, i.e., to clean (includes decontaminate, when required), to preserve, to drain, to paint, or to replenish fuel, lubricants, chemical fluids, or gases.

d.   **Adjust**. To maintain or regulate, within prescribed limits, by bringing into proper or exact position, or by setting the operating characteristics to specified parameters.

---

[1]Exception is authorized for ammunition MAC to permit the redesignation/redefinition of maintenance function headings to more adequately identify ammunition maintenance functions. The heading designations and definitions will be included in the appropriate technical manual for each category of ammunition.

B-2. MAINTENANCE FUNCTIONS (CONT).

e. Align. To adjust specified variable elements of an item to bring about optimum or desired performance.

NOTE

Air Force Precision Measurement Equipment Laboratory (PMEL) person will calibrate small arms inspection gages in accordance with TO 33K-1-100.

f. Calibrate. To determine and cause corrections to be made or to be adjusted on instruments or test, measuring, and diagnostic equipments used in precision measurement. Consists of comparisons of two instruments, one of which is a certified standard of known accuracy, to detect and adjust any discrepancy in the accuracy of the instrument being compared.

g. Remove/Install. To remove and install the same item when required to perform service or other maintenance functions. Install may be the act of emplacing, seating, or fixing into position a spare, repair part, or module (component or assembly) in a manner to allow the proper functioning of an equipment or system.

h. Replace. To remove an unserviceable item and install a serviceable counterpart in its place. ''Replace'' is authorized by the MAC and is shown as the 3d position code of the SMR code.

i. Repair. The application of maintenance services[2], including fault location/troubleshooting[3], removal/installation, and disassembly/assembly[4] procedures, and maintenance actions[5] to identify troubles and restore serviceability to an item by correcting specific damage, fault, malfunction, or failure in a part, subassembly, module (component or assembly), end item, or system.

j. Overhaul. That maintenance effort (service/action) prescribed to restore an item to a completely serviceable/operational condition as required by maintenance standards in appropriate technical publications (i.e., DMWR). Overhaul is normally the highest degree of maintenance performed by the Army. Overhaul does not normally return an item to like new condition.

k. Rebuild. Consists of those services/actions necessary for the restoration of unserviceable equipment to a like new condition in accordance with original manufacturing standards. Rebuild is the highest degree of materiel maintenance applied to Army equipment. The rebuild operation includes the act of returning to zero those age measurements (hours/miles, etc.) considered in classifying Army equipment/components.

---

[2]Services—inspect, test, service, adjust, align, calibrate, and/or replace.
[3]Fault locate/troubleshoot—The process of investigating and detecting the cause of equipment malfunctioning; the act of isolating a fault within a system or unit under test (UUT).
[4]Disassemble/assemble—encompasses the step-by-step taking apart (or breakdown) of a spare/functional group coded item to the level of its least componency identified as maintenance significant (i.e., assigned an SMR code) for the level of maintenance under consideration.
[5]Actions—welding, grinding, riveting, straightening, facing, remachinery, and/or resurfacing.

B-3. EXPLANATION OF COLUMNS IN THE MAC, SECTION II.

a. Column 1, Group Number. Column 1 lists functional group code numbers, the purpose of which is to identify maintenance significant components, assemblies, subassemblies, and modules with the next higher assembly. End item group number shall be "00".

b. Column 2, Component/Assembly. Column 2 contains the names of components, assemblies, subassemblies, and modules for which maintenance is authorized.

c. Column 3, Maintenance Function. Column 3 lists the functions to be performed on the item listed in Column 2. (For detailed explanation of these functions, see paragraph B-2.)

d. Column 4, Maintenance Level. Column 4 specifies, by the listing of a work time figure in the appropriate subcolumn(s), the level of maintenance authorized to perform the function listed in Column 3. This figure represents the active time required to perform that maintenance function at the indicated level of maintenance. If the number or complexity of the tasks within the listed maintenance function vary at different maintenance levels, appropriate work time figures will be shown for each level. The work time figure represents the average time required to restore an item (assembly, subassembly, component, module, end item, or system) to a serviceable condition under typical field operating conditions. This time includes preparation time (including any necessary disassembly/assembly time), troubleshooting/fault location time, and quality assurance/quality control time in addition to the time required to perform the specific tasks identified for the maintenance functions authorized in the maintenance allocation chart. The symbol designations for the various maintenance levels are as follows:

    C . . . .   Operator or Crew
    O . . . .   Unit Maintenance
    F . . . .   Direct Support Maintenance
    H . . . .   General Support Maintenance
    L . . . .   Specialized Repair Activity (SRA)[6]
    D . . . .   Depot Maintenance

e. Column 5, Tools and Equipment. Column 5 specifies, by code, those common tool sets (not individual tools) and special tools, TMDE, and support equipment required to perform the designated function.

f. Column 6, Remarks. This column shall, when applicable, contain a letter code, in alphabetic order, which shall be keyed to the remarks contained in Section IV.

---

[6]This maintenance level is not included in Section II, column (4) of the Maintenance Allocation Chart. To identify functions to this level of maintenance, enter a work time figure in the "H" column of Section II, column (4), and use an associated reference code in the Remarks column (6). Key the code to Section IV, Remarks, and explain the SRA complete repair application there. The explanatory remark(s) shall reference the specific Repair Parts and Special Tools List (RPSTL) TM which contains additional SRA criteria and the authorized spare/repair parts.

B-4. EXPLANATION OF COLUMNS IN TOOL AND TEST EQUIPMENT REQUIREMENTS, SECTION III.

a. **Column 1, Reference Code.** The tool and test equipment reference code correlates with a code used in the MAC, Section II, Column 5.

b. **Column 2, Maintenance Level.** The lowest level of maintenance authorized to use the tool or test equipment.

c. **Column 3, Nomenclature.** Name or identification of the tool or test equipment.

d. **Column 4, National Stock Number.** The National stock number of the tool or test equipment.

e. **Column 5, Tool Number.** The manufacturer's part number.

B-5. EXPLANATION OF COLUMNS IN REMARKS, SECTION IV.

a. **Column 1, Reference Code.** The code recorded in column 6, Section II.

b. **Column 2, Remarks.** This column lists information pertinent to the maintenance function being performed as indicated in the MAC, Section II.

Section II. MAINTENANCE ALLOCATION CHART
FOR
M16A2 RIFLE AND M4 CARBINE

(1) Group Number	(2) Component/ Assembly	(3) Maintenance Function	(4) Maintenance Level					(5) Tools and EQPT	(6) Remarks
			Unit		DS	GS	Depot		
			C	O	F	H	D		
00	M16A2 5.56MM RIFLE AND M4 5.56MM CARBINE	Inspect	0.1	0.3	0.7			2	
		Test			0.2			2,3	
		Service	0.2	0.3	0.3				
		Replace		0.1					
		Overhaul					**		
01	BOLT CARRIER ASSEMBLY	Inspect	0.1	0.1	0.1				
		Service	0.1	0.1					
		Remove/ Install	0.1						
		Replace			0.1			1,2	
		Repair		0.1	0.1			3	
0101	Bolt Assembly	Inspect	0.1	0.1	0.1				
		Test			0.1			2	
		Service	0.1						
		Remove/ Install	0.1						
		Replace			0.1			2	
		Repair		0.1	0.1			3	
0102	Key and Bolt Carrier Assembly	Inspect	0.1		0.1				
		Service	0.1						
		Remove/ Install	0.1						
		Replace			0.1			1	
		Repair		0.1	0.1			1,3	
02	HANDLE ASSEMBLY	Inspect	0.1						
		Service	0.1						
		Remove/ Install	0.1						
		Replace		0.1					
		Repair		0.1				3	

**Worktimes are included in DMWR 9-1005-249

MAINTENANCE ALLOCATION CHART (CONT)

(1) Group Number	(2) Component/ Assembly	(3) Maintenance Function	(4) Maintenance Level					(5) Tools and EQPT	(6) Remarks
			Unit		DS	GS	Depot		
			C	O	F	H	D		
03	UPPER RECEIVER AND BARREL ASSEMBLY	Inspect	0.1	0.2	0.1				
		Test			0.2			2	
		Service	0.2		0.1			1	
		Remove/ Install	0.1						
		Replace			0.5			1,2	
		Repair		0.2	0.5			1,2,3	A,B
0301	M16A2 Barrel Assembly and M4 Replacement Barrel	Inspect		0.1	0.1				
		Replace			0.3			1,2	
		Repair		0.1				3	
0302	Upper Receiver Assembly	Inspect			0.1				
		Replace			0.5			1,2	
		Repair			0.3			1,3	
030201	Forward Assist Assembly	Inspect			0.1				
		Replace			0.1				
		Repair			0.1			1,3	
030202	Rear Sight Assembly	Inspect			0.1				
		Replace			0.3			1,3	
		Repair			0.2			1,3	
04	LOWER RECEIVER AND BUTTSTOCK ASSEMBLY	Inspect	0.1	0.1	0.1				
		Test			0.1			2	
		Service	0.2	0.2					
		Repair		0.2	0.3			1,2,3	C,D,E
0401 and 0401A	Buttstock Assembly	Inspect		0.1					
		Remove/ Install		0.1				3	
		Replace		0.1				3	
		Repair		0.1				3	

MAINTENANCE ALLOCATION CHART (CONT)

(1) Group Number	(2) Component/ Assembly	(3) Maintenance Function	(4) Maintenance Level					(5) Tools and EQPT	(6) Remarks
			Unit		DS	GS	Depot		
			C	O	F	H	D		
0402	Hammer Assembly	Inspect			0.1				
		Remove/ Install			0.1				
		Replace			0.1				
		Repair			0.1				
0403	Trigger Assembly	Inspect			0.1				
		Remove/ Install			0.1			1,3	
		Replace			0.1			1,3	
		Repair			0.1			1,3	
040301	Trigger Sub-assembly	Inspect			0.1				
		Replace			0.1			1,3	
		Repair			0.1			1,3	
0404 and 0404A	Lower Receiver and Receiver Extension Assembly	Inspect			0.1				
		Test			0.1			2	
		Repair			0.2			1,3	

Section III. TOOL AND TEST EQUIPMENT REQUIREMENTS
FOR
M16A2 RIFLE AND M4 CARBINE

(1) Tool/Test Equipment Ref. Code	(2) Maintenance Level	(3) Nomenclature	(4) National/NATO Stock Number	(5) Tool Number
1 (ARMY ONLY)	F	Shop Set, Small Arms; Field Maintenance, Basic Less Power	4933-00-754-0664	SC 4933-95-CL-A11
2 (ARMY ONLY)	F	Tool and Gage Set, DS/GS Maintenance for 5.56mm Rifle, M16 Series	4933-00-056-7106	8426685
3 (ARMY ONLY)	O	Tool Kit, Small Arms Repairman	5180-00-357-7770	SC 5180-95-CL-A07
4 (A.F. ONLY)	F	Torque Wrench, ft-lb	5120-00-640-6365	A-A-411
5 (A.F. ONLY)	F	Torque Wrench, in.-lb	5120-00-230-6380	T-E-12A
6 (A.F. ONLY)	F	Trigger Weights	4933-00-647-3696	7274758

Section IV. REMARKS

Reference Code	Remarks
A	Tool, Front Sight Post Removal and Installation
B	Depressor, Front Sight Detent
C	Tool, Pivot Pin Removal
D	Tool, Pivot Pin Installation
E	Pin, Slave

APPENDIX C
REPAIR PARTS AND SPECIAL TOOLS LIST

Section I. INTRODUCTION

C-1. SCOPE.

This RPSTL lists and authorizes spares and repair parts; special tools; special test, measurement, and diagnostic equipment (TMDE); and other special support equipment required for performance of unit and direct support maintenance of the M16A2 rifle. It authorizes the requisitioning, issue, and disposition of spares, repair parts, and special tools as indicated by the source, maintenance, and recoverability (SMR) codes.

C-2. GENERAL.

In addition to Section I, Introduction, this Repair Parts and Special Tools List is divided into the following sections:

   a. **Section II—Repair Parts List**. A list of spares and repair parts authorized by this RPSTL for use in the performance of maintenance. The list also includes parts which must be removed for replacement of the authorized parts. Parts lists are composed of functional groups in ascending alphanumeric sequence, with the parts in each group listed in ascending figure and item number sequence. Bulk materials are listed by item name in FIG BULK at the end of the section. Repair parts kits or sets are listed separately in their own functional group within Section II. Repair parts for repairable special tools are also listed in the section.

   b. **Section III—Special Tools List**. A list of special tools, special TMDE, and other special support equipment authorized by this RPSTL (as indicated by Basis of Issue (BOI) information in DESCRIPTION AND USABLE ON CODE (UOC) column) for the performance of maintenance.

   c. **Section IV—Cross-Reference Indexes**. A list, in National item identification number (NIIN) sequence, of all National stock numbered items appearing in the listings, followed by a list in alphanumeric sequence of all part numbers appearing in the listing. National stock numbers and part numbers are cross-referenced to each illustration figure and item number in alphanumeric sequence and cross-references NSN, CAGEC, and part number.

C-3. EXPLANATION OF COLUMNS (SECTIONS II AND III).

   a. **ITEM NO. (Column (1))**. Indicates the number used to identify items called out in the illustration.

   b. **SMR CODE (Column (2))**. The Source, Maintenance, and Recoverability (SMR) code is a 5-position code containing supply/requisitioning information, maintenance level authorization criteria, and disposition instruction, as shown in the following breakout:

C-3. EXPLANATION OF COLUMNS (SECTION II AND III) (CONT).

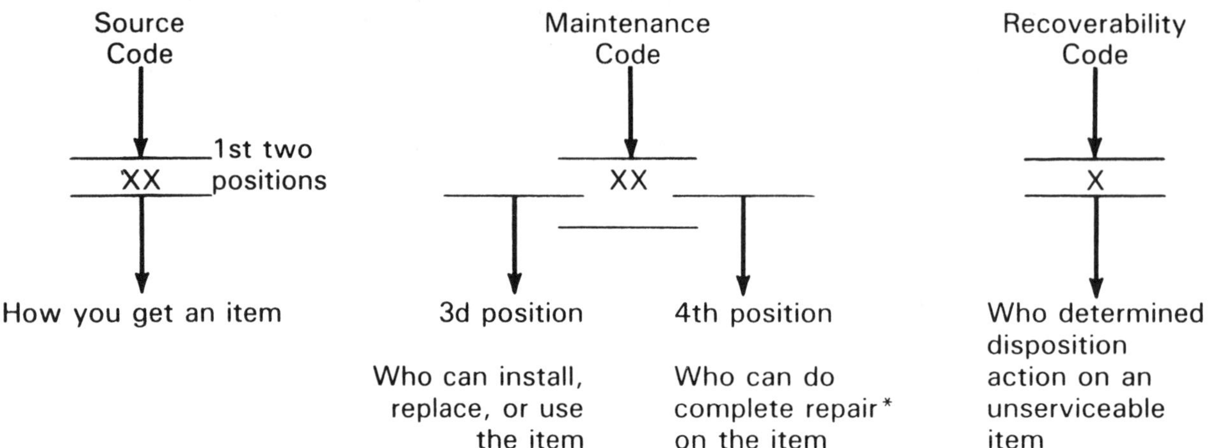

*Complete Repair: Maintenance capacity, capability, and authority to perform all corrective maintenance tasks of the ''Repair'' function in a use/user environment in order to restore serviceability to a failed item.

(1) **Source Code.** The source code tells you how to get an item needed for maintenance, repair, or overhaul of an end item/equipment. Explanations of source codes follow:

Code   **Explanation**

PA
PB
PC**
PD
PE
PF
PG

Stocked items; use the applicable NSN to request/requisition items with these source codes. They are authorized to the level indicated by the code entered in the 3d position of the SMR code.

**NOTE: Items coded PC are subject to deterioration.

KD
KF
KB

Items with these codes are not to be requested/requisitioned individually. They are part of a kit which is authorized to the maintenance level indicated in the 3d position of the SMR code. The complete kit must be requisitioned and applied.

MO—(Made at Unit/ AVUM Level)
MF—(Made at DS/ AVIM Level)
MH—(Made at GS Level)
ML—(Made at Specialized Repair Act (SRA))
MD—(Made at Depot)

Items with these codes are not to be requested/requisitioned individually. They must be made from bulk material which is identified by the part number in the DESCRIPTION AND USABLE ON CODE (UOC) column and listed in the Bulk Material group of the repair parts list in this RPSTL. If the item is authorized to you by the 3d position code of the SMR code, but the source code indicates it is made at a higher level, order the item from the higher level of maintenance.

Code   **Explanation**

AO — (Assembled by
     unit/AVUM Level)
AF — (Assembled by
     DS/AVIM Level)
AH — (Assembled by
     GS Level)
AL — (Assembled by
     SRA)
AD — (Assembled by
     Depot)

Items with these codes are not to be requested/requisitioned individually. The parts that make up the assembled item must be requisitioned or fabricated and assembled at the level of maintenance indicated by the source code. If the 3d position code of the SMR code authorizes you to replace the item, but the source code indicates the item is assembled at a higher level, order the item from the higher level of maintenance.

XA — Do not requisition an ''XA''-coded item. Order its next higher assembly. (Also, refer to the NOTE below.)

XB — If an ''XB'' item is not available from salvage, order it using the CAGEC and part number given.

XC — Installation drawing, diagram, instruction sheet, field service drawing, that is identified by manufacturer's part number.

XD — Item is not stocked. Order an ''XD''-coded item through normal supply channels using the CAGEC and part number given, if no NSN is available.

NOTE

Cannibalization or controlled exchange, when authorized, may be used as a source of supply for items with the above source codes, except for those source coded ''XA'' or those aircraft support items restricted by requirements of AR 750-1.

(2) **Maintenance Code.** Maintenance codes tell you the level(s) of maintenance authorized to USE and REPAIR support items. The maintenance codes are entered in the third and fourth positions of the SMR code as follows:

(a) The maintenance code entered in the third position tells you the lowest maintenance level authorized to remove, replace, and use an item. The maintenance code entered in the third position will indicate authorization to one of the following levels of maintenance.

Code   **Application/Explanation**

C — Crew or operator maintenance done within unit or aviation unit maintenance.

O — Unit or aviation unit level can remove, replace, and use the item.

F — Direct support or aviation intermediate level can remove, replace, and use the item.

H — General support level can remove, replace, and use the item.

L — Specialized repair activity can remove, replace, and use the item.

D — Depot level can remove, replace, and use the item.

C-3. EXPLANATION OF COLUMNS (SECTIONS II AND III) (CONT).

(b)   The maintenance code entered in the fourth position tells whether or not the item is to be repaired and identifies the lowest maintenance level with the capability to do complete repair (i.e., perform all authorized repair functions). (NOTE: Some limited repair may be done on the item at a lower level of maintenance, if authorized by the Maintenance Allocation Chart (MAC) and SMR codes.) This position will contain one of the following maintenance codes.

Code	Application/Explanation
O	-Unit or aviation unit is the lowest level that can do complete repair of the item.
F	-Direct support or aviation intermediate is the lowest level that can do complete repair of the item.
H	-General support is the lowest level that can do complete repair of the item.
L	-Specialized repair activity is the lowest level that can do complete repair of the item.
D	-Depot is the lowest level that can do complete repair of the item.
Z	-Nonreparable. No repair is authorized.
B	-No repair is authorized. (No parts or special tools are authorized for the maintenance of a ''B'' coded item.) However, the item may be reconditioned by adjusting, lubricating, etc., at the user level.

(3)   Recoverability Code. Recoverability codes are assigned to items to indicate the disposition action on unserviceable items. The recoverability code is entered in the fifth position of the SMR code as follows:

Recoverability Codes	Application/Explanation
Z	-Nonreparable item. When unserviceable, condemn and dispose of the item at the level of maintenance shown in 3d position of SMR code.
O	-Reparable item. When uneconomically reparable, condemn and dispose of the item at unit or aviation unit level.
F	-Reparable item. When uneconomically reparable, condemn and dispose of the item at the direct support or aviation intermediate level.
H	-Reparable item. When uneconomically reparable, condemn and dispose of the item at the general support level.

Recoverability Codes	Application/Explanation
D	-Reparable item. When beyond lower level repair capability, return to depot. Condemnation and disposal of item not authorized below depot level.
L	-Reparable item. Condemnation and disposal not authorized below Specialized Repair Activity (SRA).
A	-Item requires special handling or condemnation procedures because of specific reasons (e.g., precious metal content, high dollar value, critical material, or hazardous material). Refer to appropriate manuals/directives for specific instructions.

c. CAGEC (Column (3)). The Contractor and Government Entity Code (CAGEC) is a 5-digit numeric code which is used to identify the manufacturer, distributor, or Government agency, etc., that supplies the item.

d. PART NUMBER (Column (4)). Indicates the primary number used by the manufacturer (individual, company, firm, corporation, or Government activity), which controls the design and characteristics of the item by means of its engineering drawings, specifications standards, and inspection requirements to identify an item or range of items.

NOTE

When you use an NSN to requisition an item, the item you receive may have a different part number from the part ordered.

e. DESCRIPTION AND USABLE ON CODE (UOC) (Column (5)). This column includes the following information:

(1) The Federal item name and, when required, a minimum description to identify the item.

(2) The physical security classification of the item is indicated by the parenthetical entry which is a physical security classification abbreviation, e.g., Phy Sec C1 (C)—Confidential, Phy Sec C1 (S)—Secret, Phy Sec C1 (T)—Top Secret.

(3) Items that are included in kits and sets are listed below the name of the kit or set.

(4) Spare/repair parts that make up an assembled item are listed immediately following the assembled item line entry.

(5) Part numbers for bulk materials are referenced in this column in the line item entry for the item to be manufactured/fabricated.

(6) When the item is not used with all serial numbers of the same model, the effective serial numbers are shown on the last line(s) of the description (before UOC).

(7) The usable on code, when applicable (see paragraph 5, Special Information).

(8) In the Special Tools List section, the basis of issue (BOI) appears as the last line(s) in the entry for each special tool, special TMDE, and other special support equipment. When density of equipment supported exceeds density spread indicated in the basis of issue, the total authorization is increased proportionately.

C-3. EXPLANATION OF COLUMNS (SECTIONS II AND III) (CONT).

(9) The statement "END OF FIGURE" appears just below the last item description in Column 5 for a given figure in both Section II and Section III.

f. QTY (Column (6)). The QTY (quantity per figure column) indicates the quantity of the item used in the breakout shown on the illustration figure, which is prepared for a functional group, subfunctional group, or an assembly. A "V" appearing in this column in lieu of a quantity indicates that the quantity is variable and the quantity may vary from application to application.

C-4. EXPLANATION OF COLUMNS (SECTION IV).

a. NATIONAL STOCK NUMBER (NSN) INDEX.

(1) STOCK NUMBER column. This column lists the NSN by National Item Identification Number (NIIN) sequence. The NIIN consists of the last nine digits of the NSN (i.e., 5305-$\underline{01\text{-}674\text{-}1467}$). When using this column to locate an item, ignore the first 4 digits of the NSN. However, the complete NSN should be used when ordering items by stock number.

(2) FIG. column. This column lists the number of the figure where the item is identified/located. The figures are in numerical order in Section II and Section III.

(3) ITEM column. The item number identifies the item associated with the figure listed in the adjacent FIG. column. This item is also identified by the NSN listed on the same line.

b. PART NUMBER INDEX. Part numbers in this index are listed by part number in ascending alphanumeric sequence (i.e., vertical arrangement of letter and number combination which places the first letter or digit of each group in order A through Z, followed by the numbers 0 through 9 and each following letter or digit in like order).

(1) CAGEC column. The Contractor and Government Entity Code (CAGEC) is a 5-digit numeric code used to identify the manufacturer, distributor, or Government agency, etc., that supplies the item.

(2) PART NUMBER column. Indicates the primary number used by the manufacturer (individual, firm, corporation, or Government activity), which controls the design and characteristics of the item by means of its engineering drawings, specifications standards, and inspection requirements to identify an item or range of items.

(3) STOCK NUMBER column. This column lists the NSN for the associated part number and manufacturer identified in the PART NUMBER and CAGEC columns to the left.

(4) FIG. column. This column lists the number of the figure where the item is identified/located in Sections II and III.

(5) ITEM column. The item number is that number assigned to the item as it appears in the figure referenced in the adjacent figure number column.

c. FIGURE AND ITEM NUMBER INDEX.

(1) FIG. column. This column lists the number of the figure where the item is identified/located in Sections II and III.

(2) ITEM column. The item number is that number assigned to the item as it appears in the figure referenced in the adjacent figure number column.

(3) STOCK NUMBER column. This column lists the NSN for the item.

(4) CAGEC column. The Contractor and Government Entity Code (CAGEC) is a 5-digit numeric code used to identify the manufacturer, distributor, or Government agency, etc., that supplies the item.

(5) PART NUMBER column. Indicates the primary number used by the manufacturer (individual, firm, corporation, or Government activity), which controls the design and characteristics of the item by means of its engineering drawings, specifications standards, and inspection requirements to identify an item or range of items.

C-5. SPECIAL INFORMATION.

a. ASSEMBLY INSTRUCTION. Detailed assembly instructions for items source coded to be assembled from component spare/repair parts are found in Chapter 3. Items that make up the assembly are listed immediateluy following the assembly item entry or reference is made to an applicable figure.

b. ASSOCIATED PUBLICATIONS. The publications listed below pertain to the M16A2 rifle and its components:

Publication	Short Title
TM 9-1005-319-10	M16A2 Rifle & M4 Carbine

c. USABLE ON CODE. The usable on code appears in the lower left corner of the DESCRIPTION column heading, Usable on codes are shown as "UOC..." in DESCRIPTION column (justified left) on the first line applicable item description/nomenclature. Uncoded items are applicable to all models. Identification of the usable on codes used in the RPSTL are:

Code	Used On
AR8	M16A2 Rifle
AS1	M4 Carbine

C-6. HOW TO LOCATE REPAIR PARTS.

a. When National Stock Number or Part Number is Not Known.

(1) First. Using the table of contents, determine the assembly group or subassembly group to which the item belongs. This is necessary since figures are prepared for assembly groups and subassembly groups, and listings are divided into the same groups.

C-6. HOW TO LOCATE REPAIR PARTS (CONT).

(2) Second. Find the figure covering the assembly group or subassembly group to which the item belongs.

(3) Third. Identify the item on the figure and use the Figure and Item Number Index to find the NSN.

b. When National Stock Number or Part Number is Known.

(1) First. Using the National Stock Number or the Part Number Index, find the pertinent National Stock Number or Part Number. The NSN index is in National Item Identifcation Number (NIIN) sequence. (See C-4.a.(1).) The part numbers in the Part Number Index are listed in ascending alphanumeric sequence. (See C-4.b.) Both indexes cross-reference you to the illustration/figure and item number of the item you are looking for.

(2) Second. Turn to the figure and item number, verify that the item is the one you're looking for, then locate the item number in the repair parts list for the figure.

C-7. ABBREVIATIONS.

Not applicable.

Section II. REPAIR PARTS LIST

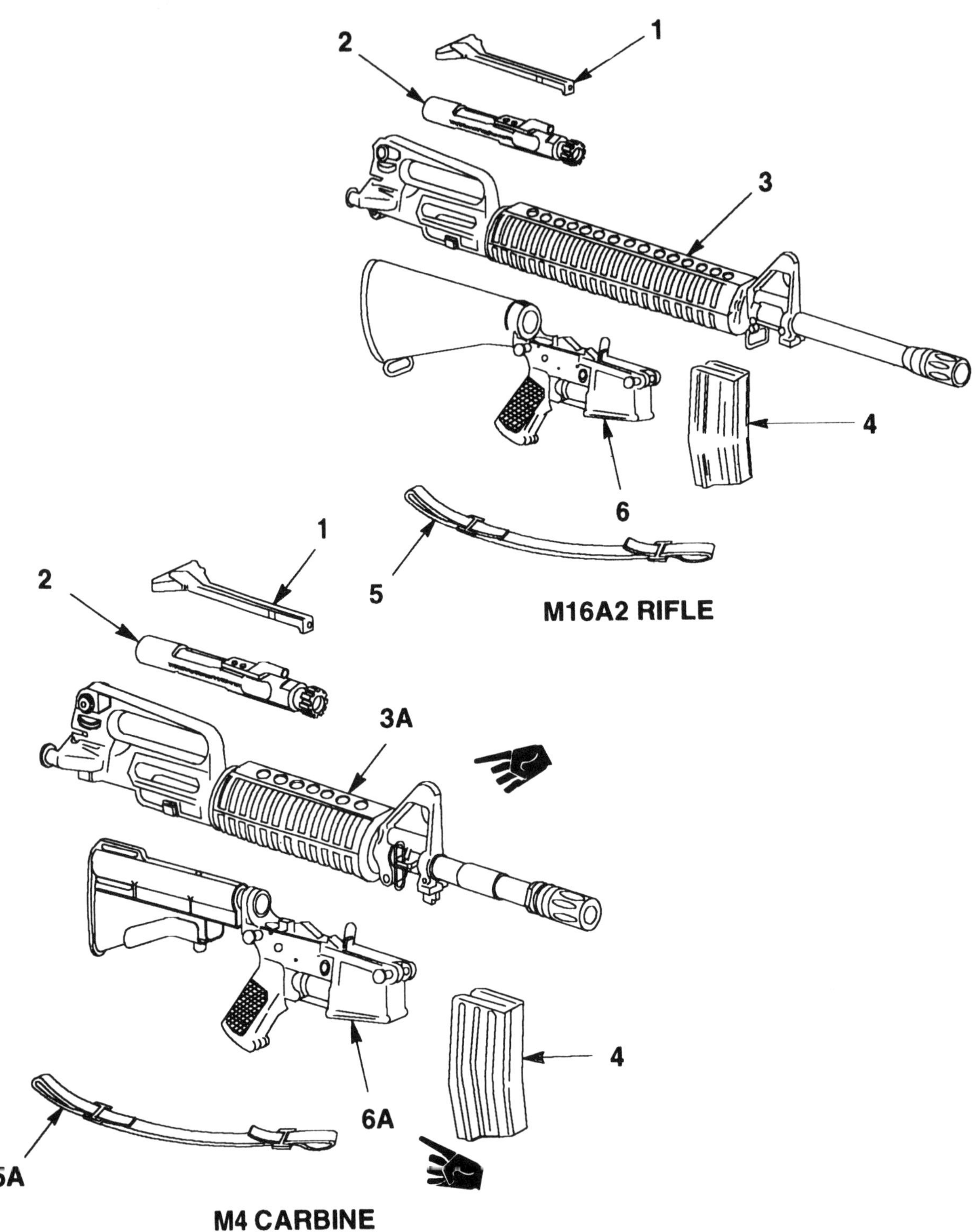

Figure C-1. M16A2 Rifle 5.56mm 9349000 and
M4 Carbine 5.56mm 9390000.

(1)	(2)	(3)	(4)	(5)	(6)
ITEM NO	SMR CODE	CAGEC	PART NUMBER	DESCRIPTION AND USABLE ON CODES(UOC)	QTY
				GROUP 00 M16A2 RIFLE 5.56MM 9349000 AND M4 CARBINE 5.56MM 9390000 FIG. C-1. M16A2 RIFLE AND M4 CARBINE	
1	PAOOO	19204	8448517	HANDLE ASSEMBLY, CHARGING..........	1
2	AFFFF	19204	8448501	BOLT CARRIER ASSEMBLY..............	1
3	AFFFF	19200	9349050	UPPER RECEIVER AND BARREL ASSEMBLY (M16A2).............................. UOC:AR8	1
3A	AFFFF	19200	9390001	UPPER RECEIVER AND BARREL ASSEMBLY (M4)................................. UOC:AS1	1
4	PACZZ	19200	8448670	MAGAZINE,CARTRIDGE (30 ROUND)......	1
5	PACZZ	19204	12624561	SLING,SMALL ARMS (M16A2).......... UOC:AR8	1
5A	PACZZ	19200	12011996	SLING,SMALL ARMS (M4)............. UOC:AS1	1
6	XAFFA	19200	9349100	LOWER RECEIVER AND BUTTSTOCK ASSEMBLY (M16A2).................... UOC:AR8	1
6A	XAFFA	19200	9390011	LOWER RECEIVER AND BUTTSTOCK ASSEMBLY (M4)....................... UOC:AS1	1

END OF FIGURE

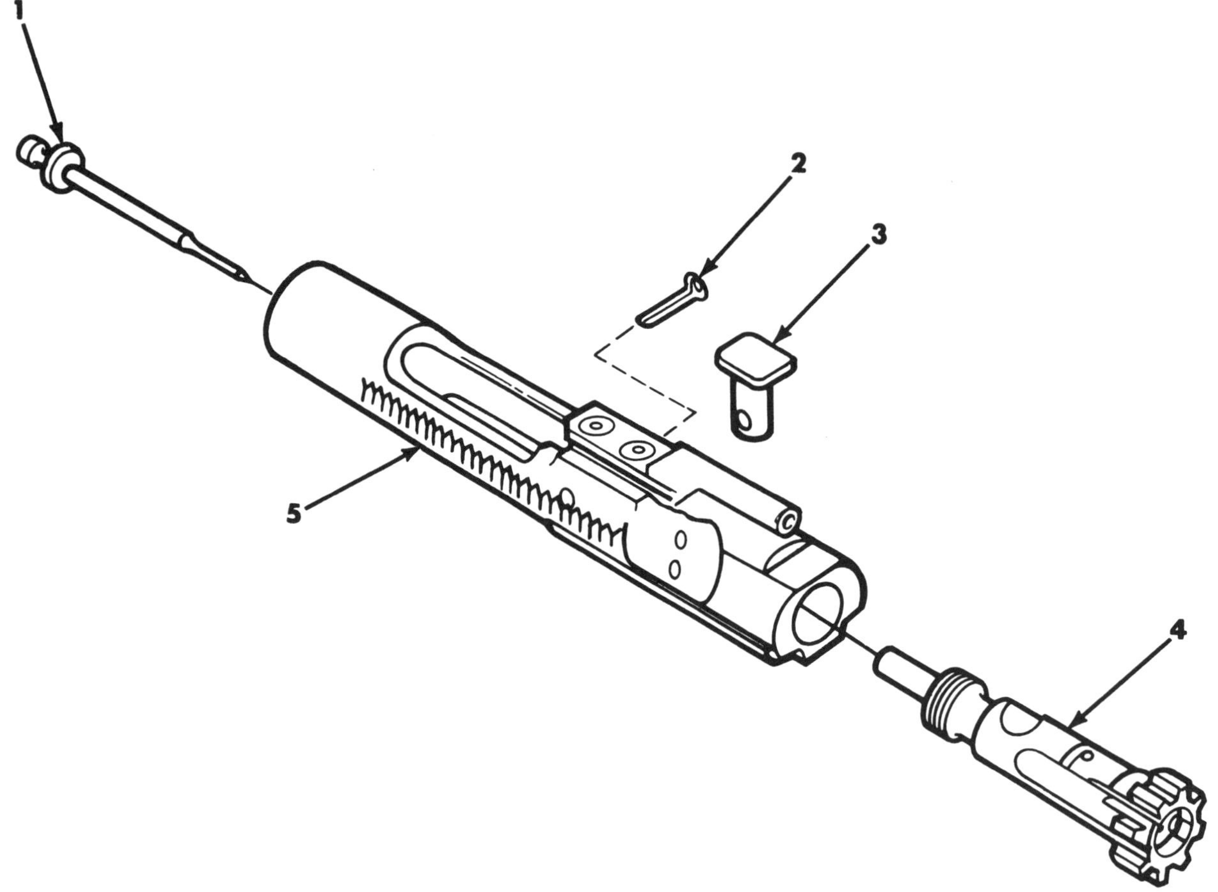

Figure C-2. Bolt Carrier Assembly 8448501.

(1) ITEM NO	(2) SMR CODE	(3) CAGEC	(4) PART NUMBER	(5) DESCRIPTION AND USABLE ON CODES (UOC)	(6) QTY
				GROUP 01 BOLT CARRIER ASSEMBLY 8448501 FIG. C-2. BOLT CARRIER ASSEMBLY	
1	PAFZZ	19204	8448503	PIN,FIRING	1
2	PAOZZ	19204	8448504	PIN,FIRING PIN RETAINING	1
3	PAOZZ	19204	8448502	PIN,GROOVED,HEADED,BOLT CAM	1
4	PAFFF	19200	8448509	**BOLT ASSEMBLY**	1 ▮
5	AFFFF	19200	8448505	KEY AND BOLT CARRIER ASSEMBLY	1

END OF FIGURE

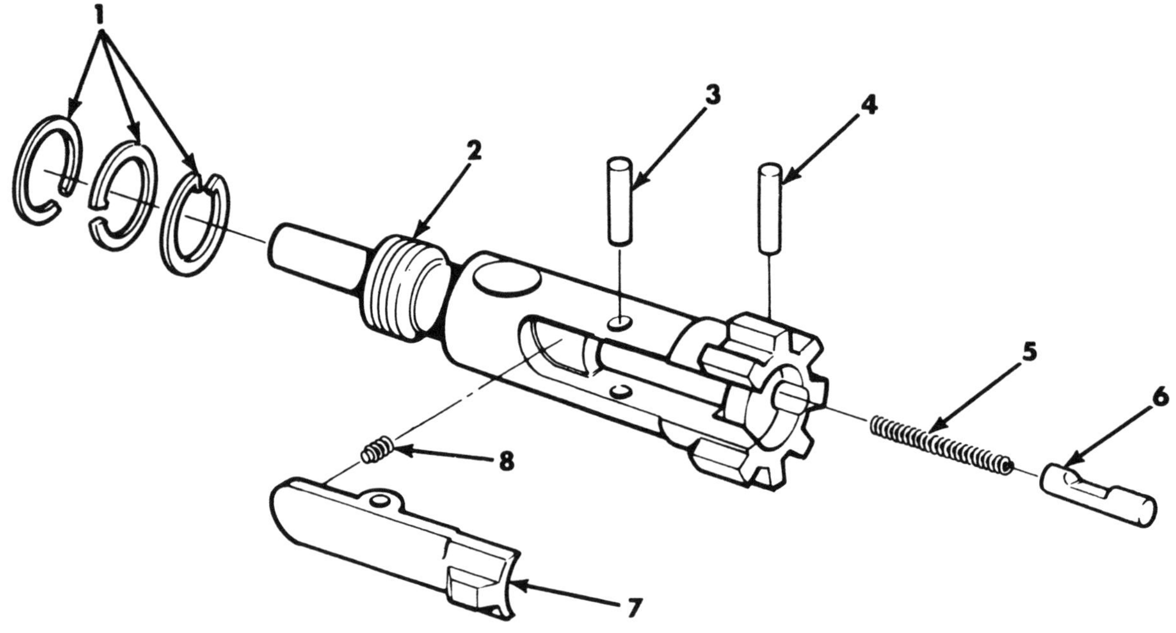

Figure C-3. Bolt Assembly 8448509.

(1) ITEM NO	(2) SMR CODE	(3) CAGEC	(4) PART NUMBER	(5) DESCRIPTION AND USABLE ON CODES (UOC)	(6) QTY
				GROUP 0101 BOLT ASSEMBLY 8448509 FIG. C-3. BOLT ASSEMBLY	
1	PAFZZ	19204	8448511	RING,BOLT	3
2	XAFZZ	19200	8448510	BOLT	1
3	PAOZZ	19204	8448513	PIN,EXTRACTOR	1
4	PAOZZ	96906	MS16562-98	PIN,SPRING ,EJECTOR	1
5	PAOZZ	19204	8448516	SPRING,HELICAL ,EJECTOR	1
6	PAOZZ	19204	8448515	EJECTOR,CARTRIDGE	1
7	PAOZZ	19204	8448512	EXTRACTOR,CARTRIDGE	1
8	PAOZZ	19200	8448755	SPRING ASSEMBLY ,EXTRACTOR	1

END OF FIGURE

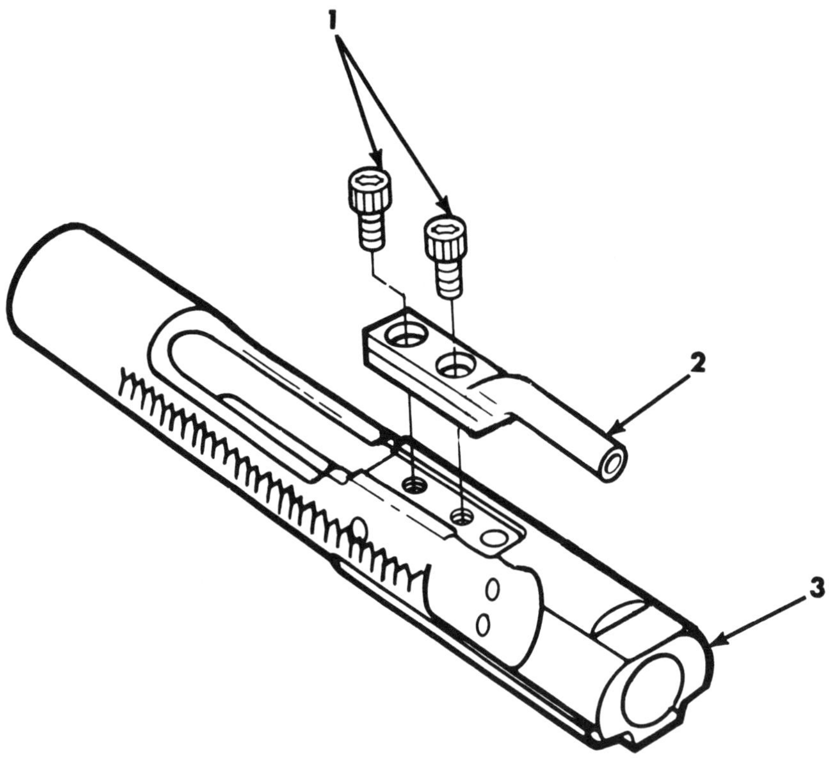

Figure C-4. Key and Bolt Carrier Assembly 8448505.

(1) ITEM NO	(2) SMR CODE	(3) CAGEC	(4) PART NUMBER	(5) DESCRIPTION AND USABLE ON CODES (UOC)	(6) QTY
				GROUP 0102 KEY AND BOLT CARRIER ASSEMBLY 8448505	
				FIG. C-4. KEY AND BOLT CARRIER ASSEMBLY	
1	PAFZZ	19204	8448508	SCREW,CARRIER AND KEY	2
2	PAFZZ	19200	8448506	KEY,BOLT CARRIER	1
3	PAFZZ	19200	8448507	CARRIER,BOLT ,BOLT CARRIER	1

END OF FIGURE

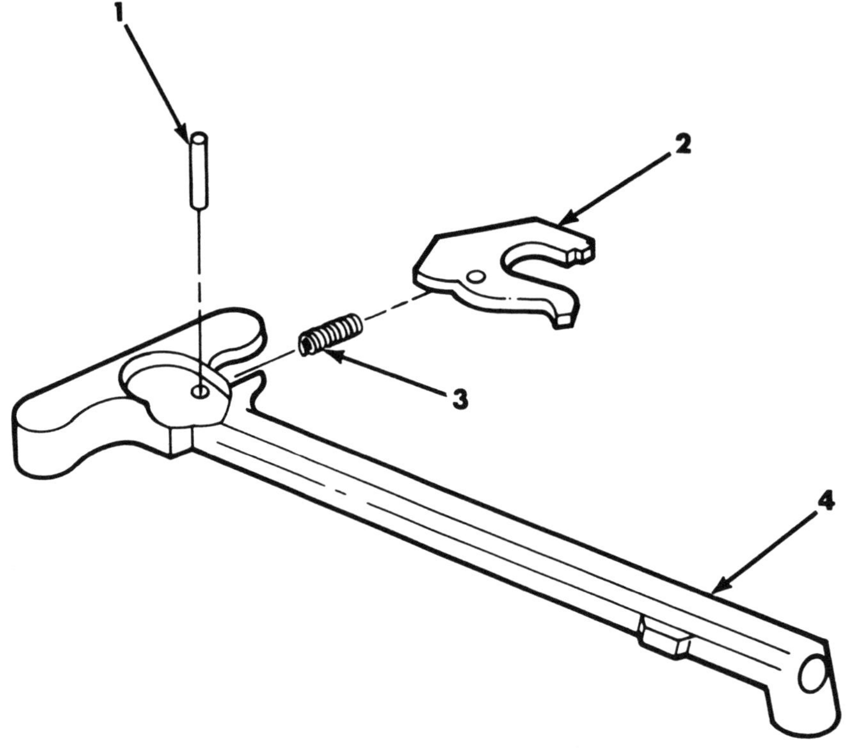

Figure C-5. Handle Assembly 8448517.

(1) ITEM NO	(2) SMR CODE	(3) CAGEC	(4) PART NUMBER	(5) DESCRIPTION AND USABLE ON CODES(UOC)	(6) QTY
				GROUP 02 HANDLE ASSEMBLY	
				8448517	
				FIG. C-5. HANDLE ASSEMBLY	
1	PAOZZ	19204	8448521-2	PIN,SPRING, CHARGING HANDLE........	1
2	PAOZZ	19200	8448519	LATCH,CHARGING HANDLE..............	1
3	PAOZZ	19204	8448520	SPRING,HELICAL, COMPRESSION, CHARGING HANDLE.....................	1
4	XAOZZ	19204	8448518	HANDLE,CHARGING.....................	1

END OF FIGURE

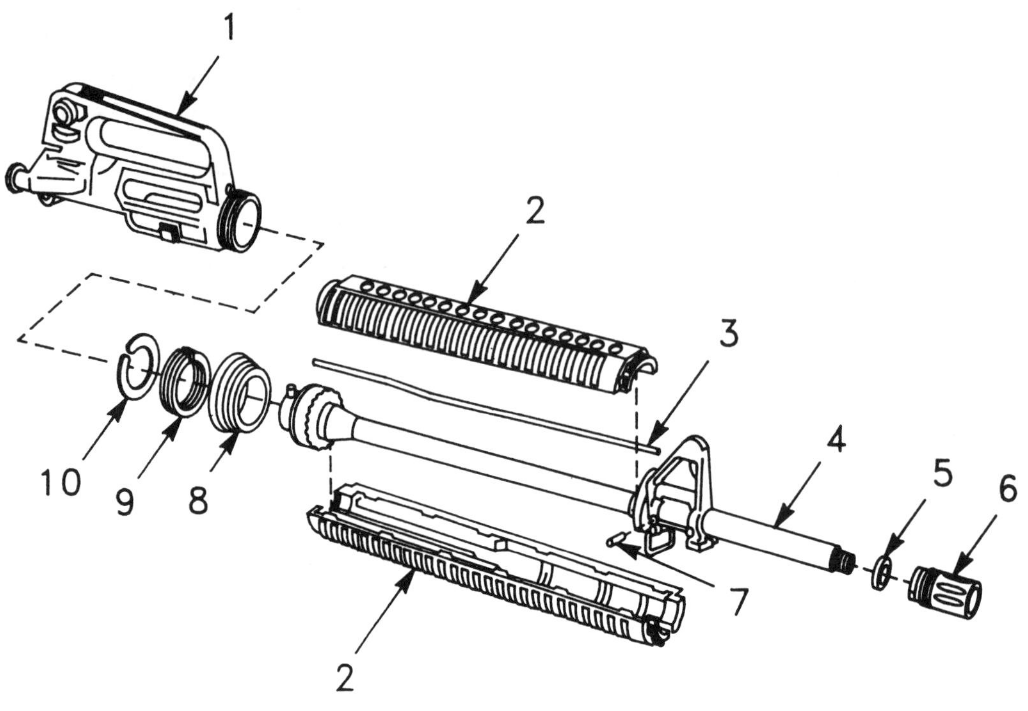

M16A2 Rifle

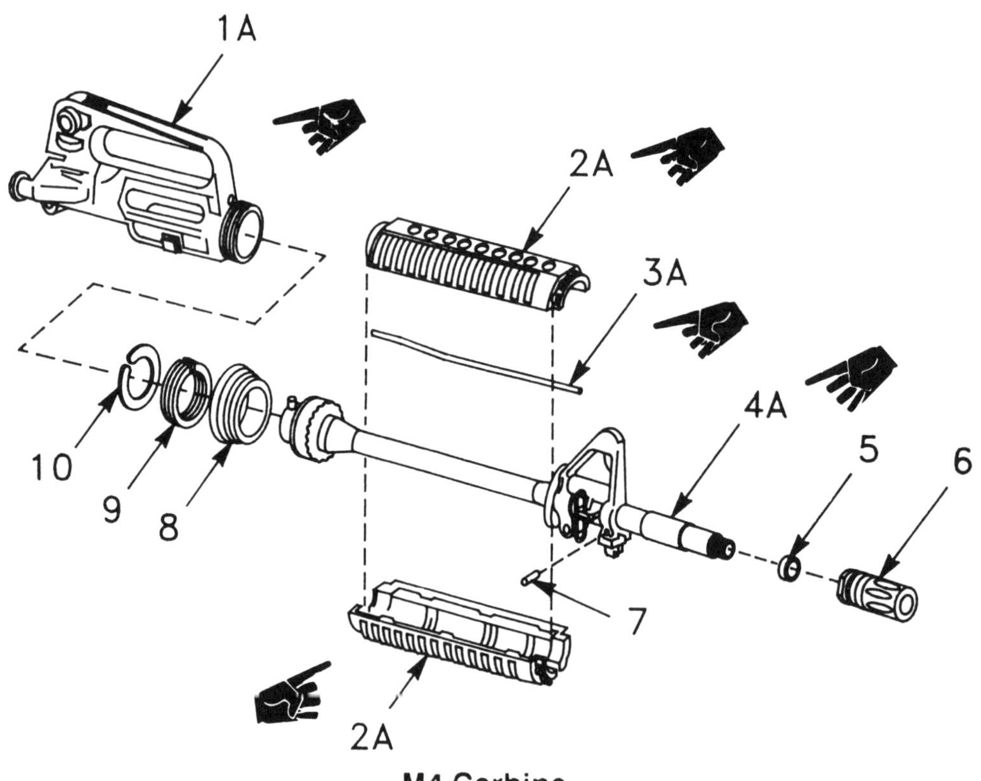

M4 Carbine

Figure C-6. Upper Receiver and Barrel Assembly
M16A2 9349050 and M4 9390001.

(1) ITEM NO	(2) SMR CODE	(3) CAGEC	(4) PART NUMBER	(5) DESCRIPTION AND USABLE ON CODES(UOC)	(6) QTY
				GROUP 03 UPPER RECEIVER AND BARREL ASSEMBLY M16A2 9349050 AND M4 9390001 FIG. C-6. UPPER RECEIVER AND BARREL ASSEMBLY	
1	AFFFF	19200	9349062	UPPER RECEIVER ASSEMBLY (M16A2).... UOC:AR8	1
1A	AFFFF	19200	9390029	UPPER RECEIVER ASSEMBLY (M4)....... UOC:AS1	1
2	PAOZZ	19200	9349059	HANDGUARD ASSEMBLY (M16A2)......... UOC:AR8	2
2A	PAOZZ	19200	9390003	GUARD,HAND,GUN (M4)................ UOC:AS1	2
3	PAFZZ	19200	8448567	TUBE,BENT,METALLIC (M16A2)........ UOC:AR8	1
3A	PAFZZ	19200	9390016	TUBE,BENT,METALLIC (M4)........... UOC:AS1	1
4	PAFFF	19200	9349124	BARREL ASSEMBLY (M16A2)........... UOC:AR8	1
4A	PAFFF	19200	9390007	BARREL AND FRONT SIGHT ASSEMBLY, REPLACEMENT (M4).................... UOC:AS1	1
5	PAFZZ	19200	9349052	SHIM COMPENSATOR...................	1
6	PAFZZ	19200	9349051	COMPENSATOR.........................	1
7	PAFZZ	96906	MS16562-106	PIN,SPRING,GAS TUBE.................	1
8	PAFZZ	19204	8448712	RING,SLIP,HAND GUARD...............	1
9	PAFZZ	19204	8448555	SPRING,SLIP RING, HANDGUARD UPPER RECEIVER.............................	1
10	PAFZZ	96906	MS16626-3137	RING,RETAINING......................	1

END OF FIGURE

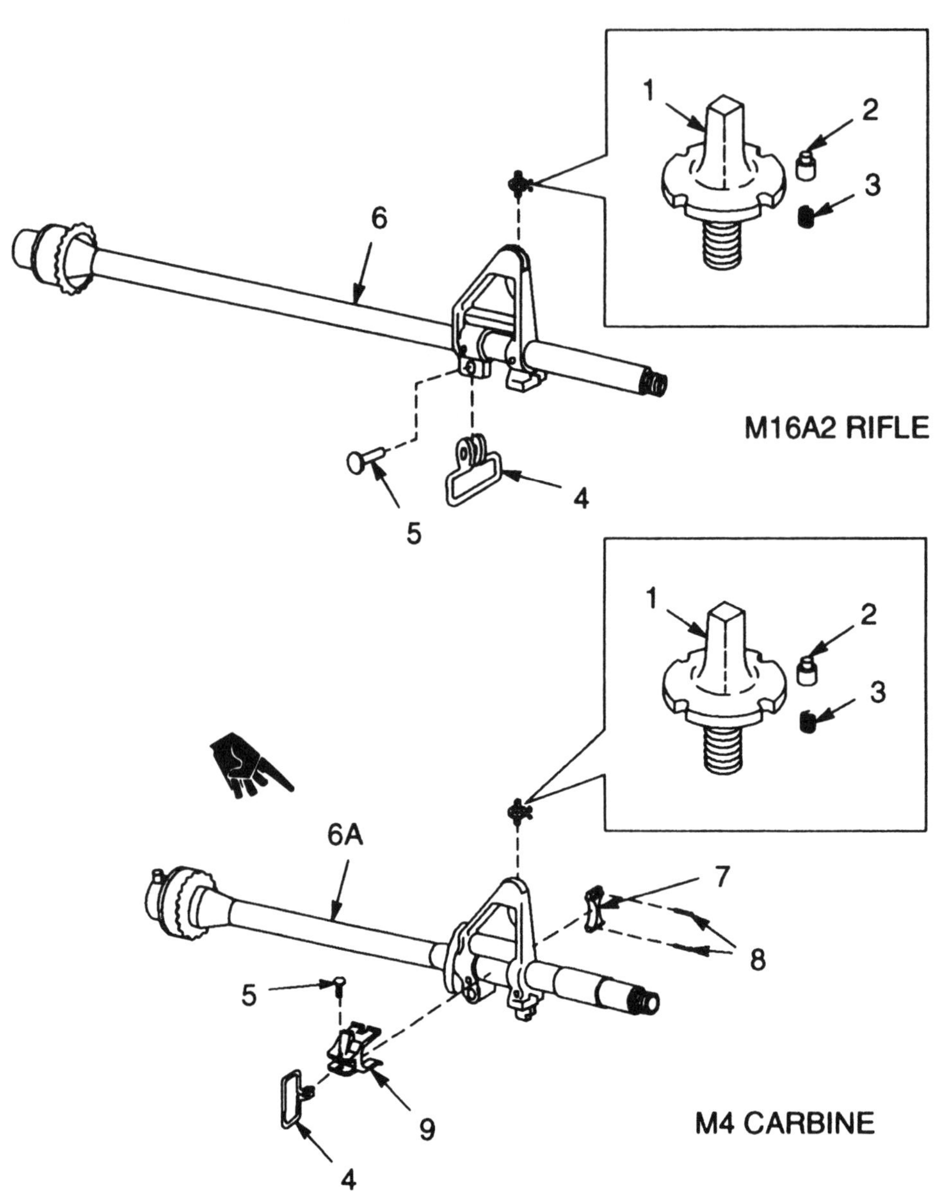

Figure C-7. M16A2 Barrel Assembly 9349124 and
M4 Replacement Barrel and Front Sight Assembly 9390007.

(1) ITEM NO	(2) SMR CODE	(3) CAGEC	(4) PART NUMBER	(5) DESCRIPTION AND USABLE ON CODES(UOC)	(6) QTY
				GROUP 0301 M16A2 BARREL ASSEMBLY 9349124 M4 REPLACEMENT BARREL AND FRONT SIGHT ASSEMBLY 9390007 FIG. C-7. M16A2 BARREL ASSEMBLY AND M4 REPLACEMENT BARREL AND FRONT SIGHT ASSEMBLY	
1	PAOZZ	19200	9349056	POST,FRONT SIGHT....................	1
2	PAOZZ	19204	8448573	PIN,SHOULDER, HEADLESS (DETENT, FRONT SIGHT)........................	1
3	PAOZZ	19204	8448574	SPRING,HELICAL, COMPRESSION, FRONT SIGHT................................	1
4	PAOZZ	19204	8448571	SWIVEL,SLING,SMALL..................	1
5	PAOZZ	19204	8448697	RIVET,TUBULAR......................	1
6	XAFZZ	19200	9349054	BARREL AND BARREL EXTENSION ASSEMBLY (M16A2).................... UOC:AR8	1
6A	XAFZZ	19200	9390009	BARREL AND BARREL EXTENSION ASSEMBLY (M4)....................... UOC:AS1	1
7	PAOZZ	19200	12598618	CLAMP,RIM CLENCHING (M4).......... UOC:AS1	1
8	PAOZZ	96906	MS39086-41	PIN,SPRING (M4)................... UOC:AS1	2
9	PAOZZ	19200	12598617	MOUNT,SWIVEL (M4)................. UOC:AS1	1

END OF FIGURE

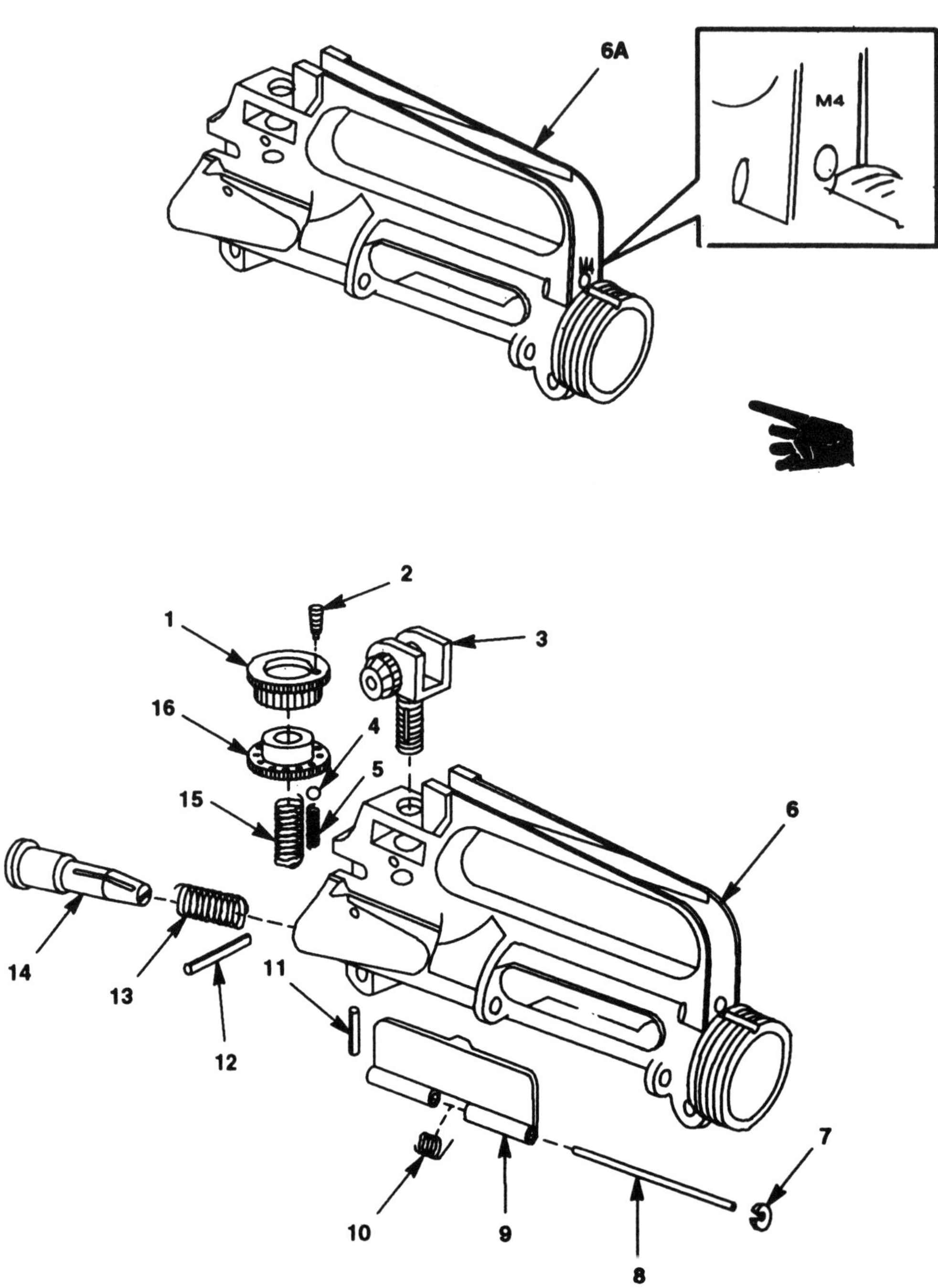

Figure C-8. Upper Receiver Asesembly M16A2 9349062 and M4 9390029

(1) ITEM NO	(2) SMR CODE	(3) CAGEC	(4) PART NUMBER	(5) DESCRIPTION AND USABLE ON CODES(UOC)	(6) QTY
				GROUP 0302 UPPER RECEIVER ASSEMBLY 9349062 (M16A2) - 9390029 (M4) FIG. C-8. UPPER RECEIVER ASSEMBLY	
1	PAFZZ	19200	9349066	INDEX,ELEVATION......................	1
2	PAFZZ	19200	9349065	SCREW,INDEX...........................	1
3	AFFFF	19200	9349072	REAR SIGHT ASSEMBLY..................	1
4	PAFZZ	96906	MS19060-4808	BALL,BEARING.........................	1
5	PAFZZ	19200	9349069	SPRING,HELICAL, COMPRESSION, INDEX.	1
6	PAFZZ	19200	9349063	RECEIVER,UPPER (M16A2)............. UOC:AR8	1
6A	PAFZZ	19200	9390028	RECEIVER,CARTRIDGE UPPER, (M4)..... UOC:AS1	1
7	PAOZZ	96906	MS16632-3012	RING,RETAINING, COVER..............	1
8	PAOZZ	19204	8448533	PIN,GROOVED, HEADLESS COVER........	1
9	PAOZZ	19204	8448525	COVER,EJECTION PORT..................	1
10	PAOZZ	19204	8448532	SPRING,HELICAL, TORSION, COVER.....	1
11	PAFZZ	96906	MS16562-121	PIN,SPRING FORWARD ASSIST..........	1
12	PAFZZ	96906	MS16562-121	PIN,SPRING...........................	1
13	PAFZZ	19200	8448540	SPRING,HELICAL, COMPRESSION FORWARD ASSIST.......................	1
14	AFFFF	19204	9349086	FORWARD ASSIST ASSEMBLY............	1
15	PAFZZ	19200	9349070	SPRING,HELICAL, COMPRESSION, ELEVATION............................	1
16	PAFZZ	19200	9349067	KNOB,ELEVATION.......................	1

END OF FIGURE

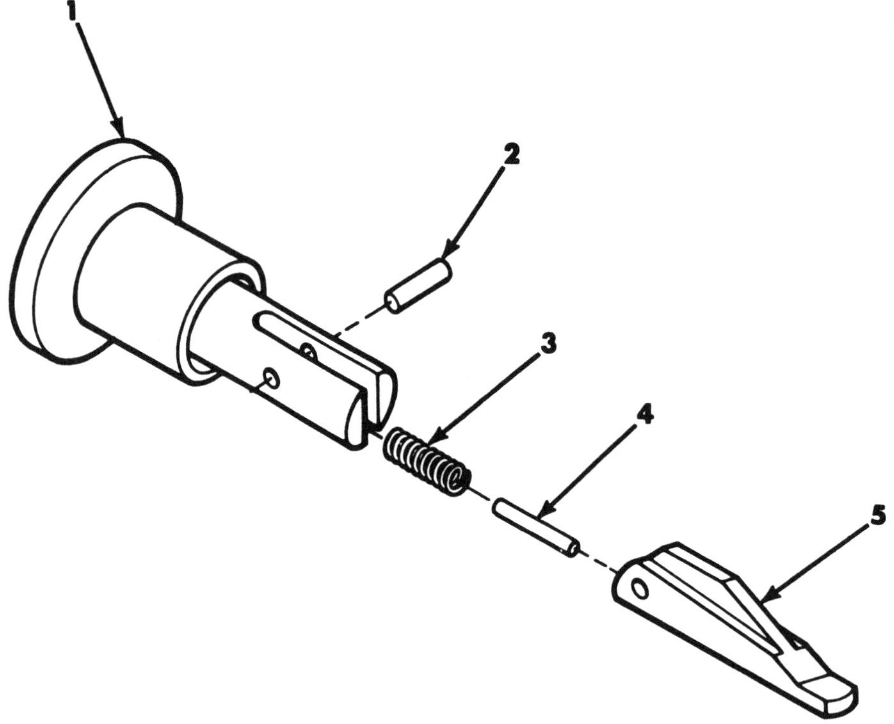

Figure C-9. Forward Assist Assembly 9349086.

(1) ITEM NO	(2) SMR CODE	(3) CAGEC	(4) PART NUMBER	(5) DESCRIPTION AND USABLE ON CODES (UOC)	(6) QTY
				GROUP 030201 FORWARD ASSIST ASSEMBLY 9349086 FIG. C-9. FORWARD ASSIST ASSEMBLY	
1	PAFZZ	19200	9349085	PLUNGER ASSEMBLY	1
2	PAFZZ	13629	8448521-2	PIN,SPRING ,PAWL	1
3	PAFZZ	19200	8448542	SPRING,HELICAL, PAWL	1
4	PAFZZ	19204	8448544	DETENT,PAWL	1
5	PAFZZ	19204	8448543	PAWL,FORWARD ASSIST	1

END OF FIGURE

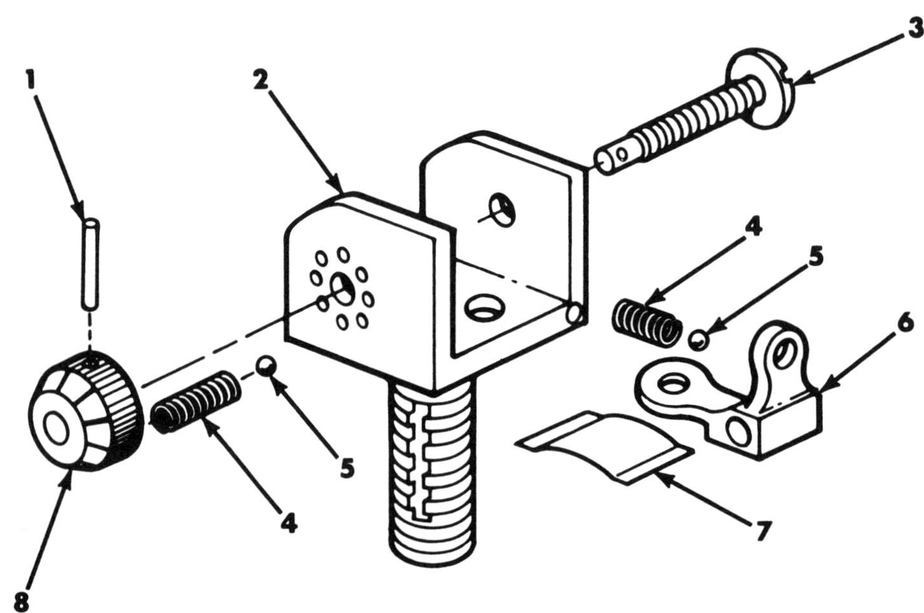

Figure C-10. Rear Sight Assembly 9349072.

(1) ITEM NO	(2) SMR CODE	(3) CAGEC	(4) PART NUMBER	(5) DESCRIPTION AND USABLE ON CODES(UOC)	(6) QTY
				GROUP 030202 REAR SIGHT ASSEMBLY 9349072 FIG. C-10. REAR SIGHT ASSEMBLY	
1	PAFZZ	96906	MS16562-103	PIN,SPRING, WINDAGE................	1
2	PAFZZ	19200	9349074	BASE,REAR SIGHT.....................	1
3	PAFZZ	19200	9349076	SCREW,EXTERNALLY....................	1
4	PAFZZ	19200	9349069	SPRING,HELICAL, COMPRESSION, REAR SIGHT................................	2
5	PAFZZ	96906	MS19060-4808	BALL,BEARING........................	2
6	PAFZZ	19200	9349075	APERTURE,SIGHT......................	1
7	PAFZZ	19200	12011987	SPRING,FLAT REAR SIGHT.............	1
8	PAFZZ	19200	9349077	KNOB,WINDAGE........................	1

END OF FIGURE

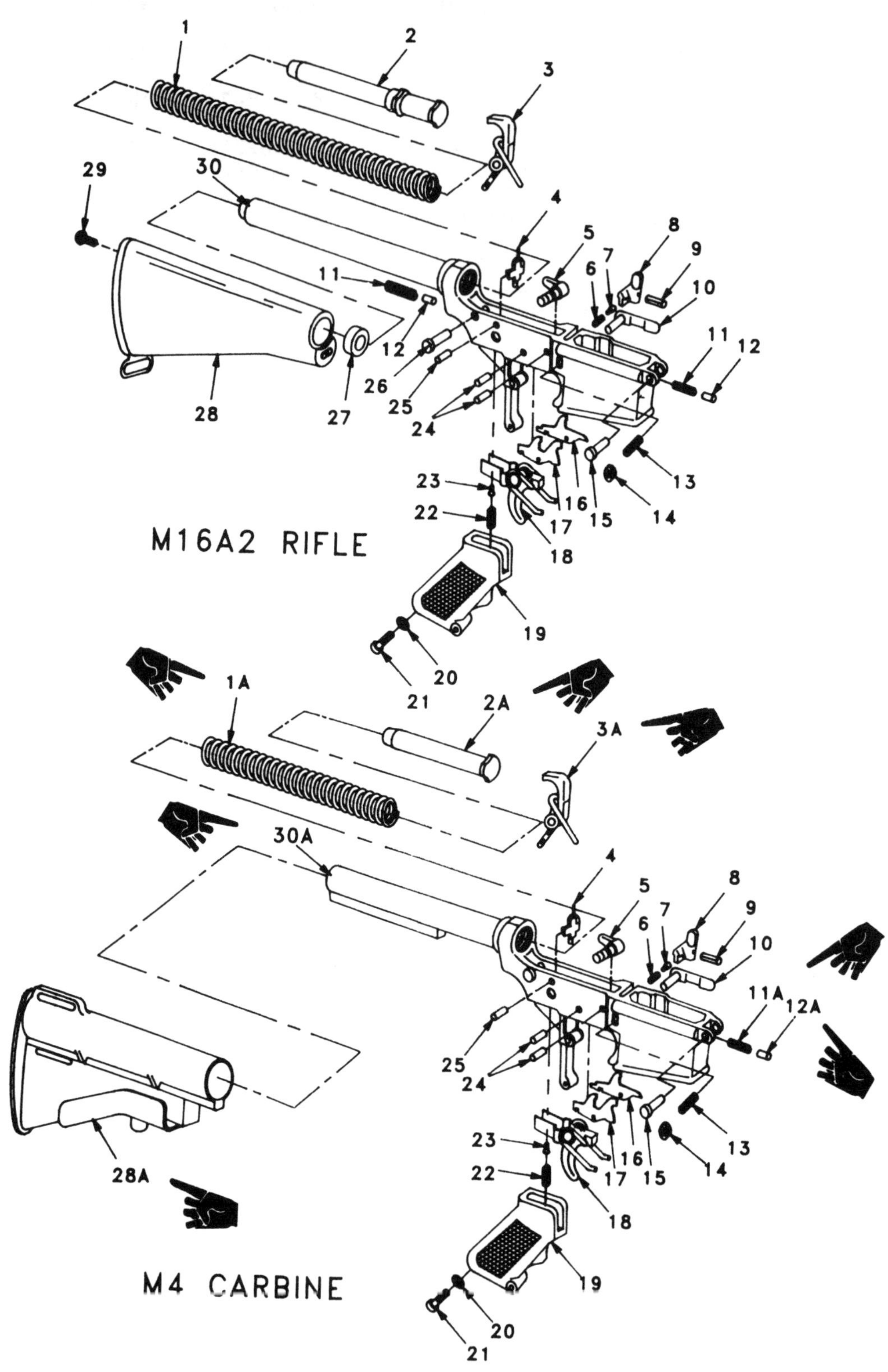

Figure C-11. Lower Receiver and Buttstock Assembly
M16A2 9349100 and M4 9390011.

(1) ITEM NO	(2) SMR CODE	(3) CAGEC	(4) PART NUMBER	(5) DESCRIPTION AND USABLE ON CODES(UOC)	(6) QTY
				GROUP 04 LOWER RECEIVER AND BUTTSTOCK ASSEMBLY M16A2 9349100 AND M4 9390011 FIG. C-11. LOWER RECEIVER AND BUTTSTOCK ASSEMBLY	
1	PAOZZ	19204	8448629	SPRING,HELICAL, COMPRESSION ACTION (M16A2)................................	1
1A	PAOZZ	19200	9390022	SPRING,HELICAL, COMPRESSION (M4)... UOC:AS1	1
2	PAOZZ	19200	8448615	BUFFER ASSEMBLY (M16A2)............ UOC:AR8	1
2A	PAOZZ	19200	9390023	BUFFER ASSEMBLY (M4)............... UOC:AS1	1
3	AFFFF	19200	9349106	HAMMER ASSEMBLY (M16A2)............ UOC:AR8	1
3A	AFFFF	19200	9390032	HAMMER ASSEMBLY (M4)............... UOC:AS1	1
4	PAFZZ	19200	8448595	SEAR................................	1
5	PAFZZ	19200	9381367	SELECTOR,FIRE CONTROL..............	1
6	PAFZZ	19204	8448633	SPRING,HELICAL, COMPRESSION, BOLT CATCH...............................	1
7	PAFZZ	19204	8448634	PLUNGER,BOLT CATCH..................	1
8	PAFZZ	19200	8448628	CATCH,BOLT..........................	1
9	PAFZZ	96906	MS16562-119	PIN,SPRING,BOLT CATCH.............	1
10	PAFZZ	19204	8448638	CATCH,MAGAZINE......................	1
11	PAOZZ	19204	8448586	SPRING,HELICAL, COMPRESSION, TAKE DOWN/PIVOT PIN (M16A2)..............	2
11A	PAOZZ	19204	8448586	SPRING,HELICAL, COMPRESSION, TAKE DOWN/PIVOT PIN (M4)................. UOC:AS1	1
12	PAOZZ	19204	8448585	PIN,STRAIGHT, HEADLESS DETENT, TAKEDOWN PIN (M16A2)................	2
12A	PAOZZ	19204	8448585	PIN,STRAIGHT HEADLESS DETENT, TAKEDOWN PIN (M4)................... UOC:AS1	1
13	PAFZZ	19204	8448637	SPRING,HELICAL, COMPRESSION, MAGAZINE CATCH......................	1
14	PAFZZ	19204	8448636	BUTTON,MAGAZINE CATCH.............	1
15	PAOZZ	19204	8448621	PIN,GROOVED,HEADED (PIVOT PIN).....	1
16	PAFZZ	19200	9349114	LEVER,LOCK-RELEASE, SEMI...........	1
17	PAFZZ	19200	9349113	LEVER,LOCK-RELEASE, BURST..........	1
18	AFFFF	19200	9349115	TRIGGER ASSEMBLY.................... UOC:AR8	1
19	PAOZZ	19200	9349127	GRIP,RIFLE PLASTIC,BLACK...........	1
20	PAOZZ	96906	MS35335-61	WASHER,LOCK,RIFLE GRIP.............	1
21	PAOZZ	88044	AN501D416-18	SCREW,MACHINE,RIFLE GRIP...........	1
22	PAOZZ	19204	8448516	SPRING,HELICAL, COMPRESSION SAFETY.	1
23	PAOZZ	19204	8448631	DETENT,SAFETY.......................	1
24	PAFZZ	19204	8448609	PIN,GROOVED, HEADLESS TRIGGER AND HAMMER..............................	2
25	PAFZZ	19204	8448599	PIN,GROOVED, HEADLESS,AUTOMATIC	1

(1) ITEM NO	(2) SMR CODE	(3) CAGEC	(4) PART NUMBER	(5) DESCRIPTION AND USABLE ON CODES(UOC)	(6) QTY
27	PAOZZ	19200	12597640	SPACER,STEPPED (M16A2)............. UOC:AR8	1
28	PAOOO	19200	9349119	BUTTSTOCK ASSEMBLY, (M16A2)........ UOC:AR8	1
28A	AOOOO	19200	9390012	BUTTSTOCK ASSEMBLY (M4)........... UOC:AS1	1
29	PAOZZ	19200	9349128	SCREW,MACHINE, BUTTCAP (M16A2).... UOC:AR8	1
30	XAFFA	19200	9349101	LOWER RECEIVER AND RECEIVER EXTENSION ASSEMBLY (M16A2).......... UOC:AR8	1
30A	XAFFA	99999	NPN	LOWER RECEIVER ASSEMBLY (M4)....... UOC:AS1	1

END OF FIGURE

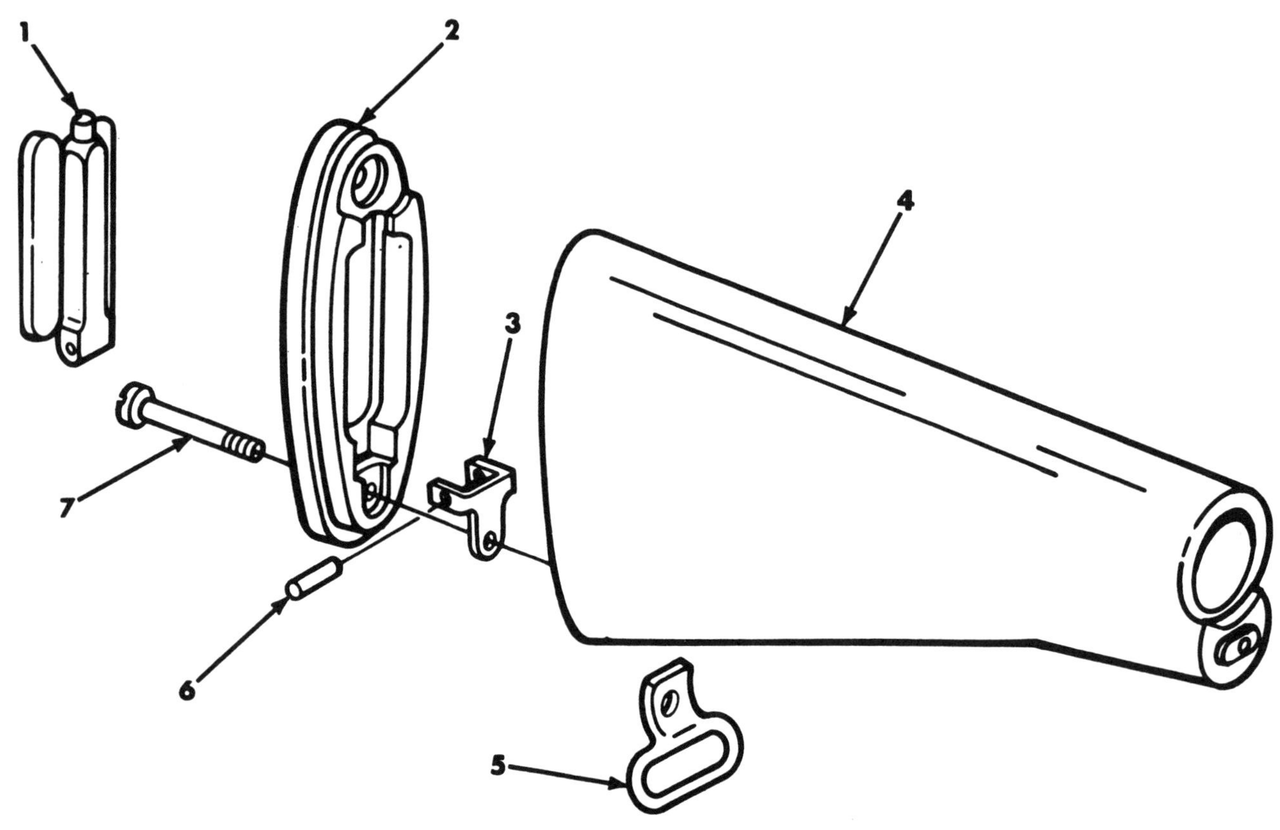

Figure C-12. M16A2 Buttstock Assembly 9349119.

(1) ITEM NO	(2) SMR CODE	(3) CAGEC	(4) PART NUMBER	(5) DESCRIPTION AND USABLE ON CODES(UOC)	(6) QTY
				GROUP 0401 M16A2 BUTTSTOCK ASSEMBLY 9349119 FIG. C-12. M16A2 BUTTSTOCK ASSEMBLY	
1	PAOZZ	19200	9381380	DOOR ASSEMBLY,THUMB................. UOC:AR8	1
2	PAOZZ	19200	9349130	PLATE,BUTT,SHOULDER GUN STOCK...... UOC:AR8	1
3	PAOZZ	19200	8448653	HINGE,ACCESS DOOR BUTT PLATE....... UOC:AR8	1
4	XAOZZ	19200	9349121	BUTTSTOCK........................... UOC:AR8	1
5	PAOZZ	19204	8448652	SWIVEL,SLING,SMALL.................. UOC:AR8	1
6	PAOZZ	19204	8448655	PIN,STRAIGHT, HEADLESS, ACCESS DOOR UOC:AR8	1
7	PAOZZ	19200	9349120	SCREW,MACHINE, BUTT PLATE.......... UOC:AR8	1

END OF FIGURE

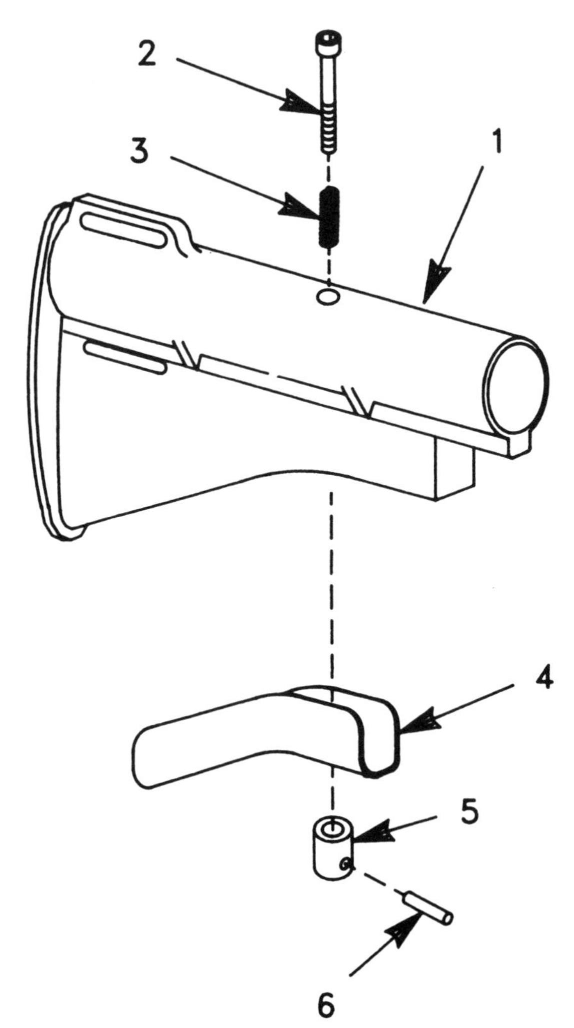

Figure C12A. M4 Buttstock Assembly 9390012.

(1) ITEM NO	(2) SMR CODE	(3) CAGEC	(4) PART NUMBER	(5) DESCRIPTION AND USABLE ON CODES(UOC)	(6) QTY
				GROUP 0401A M4 BUTTSTOCK ASSEMBLY 9390012 FIG. C12A. M4 BUTTSTOCK ASSEMBLY	
1	PAOZZ	19200	9390013	STOCK,GUN,SHOULDER (M4)............ UOC:AS1	1
2	PAOZZ	19200	9390025	PIN,SHOULDER, HEADLESS (M4)........ UOC:AS1	1
3	PAOZZ	19200	9390027	SPRING,HELICAL, COMPRESSION (M4)... UOC:AS1	1
4	PAOZZ	19200	9390014	LEVER,LOCK-RELEASE (M4)............ UOC:AS1	1
5	PAOZZ	19200	9390026	NUT,SELF-LOCKING, (M4)............. UOC:AS1	1
6	PAOZZ	96906	MS16562-202	PIN,SPRING (M4).................... UOC:AS1	1

END OF FIGURE

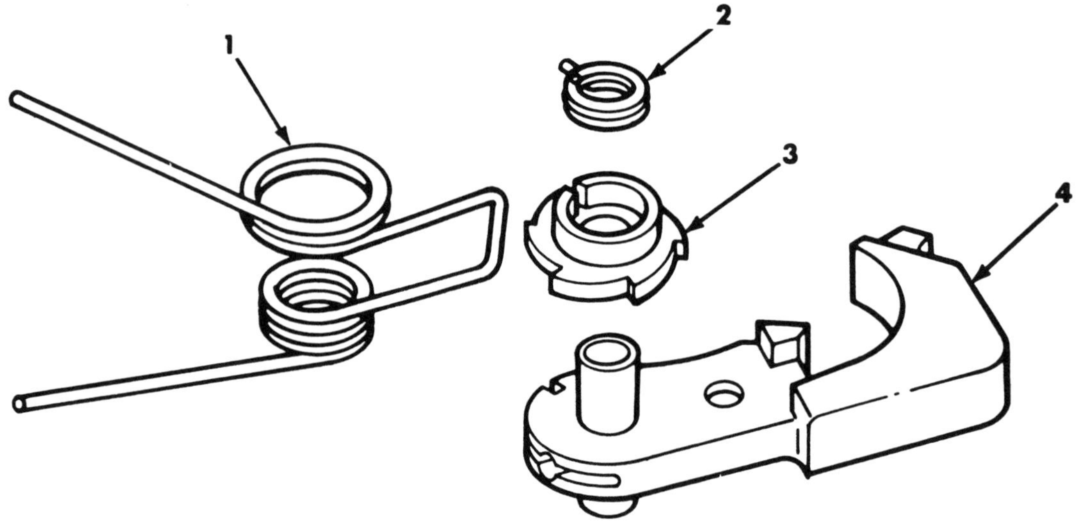

Figure C-13. Hammer Assembly M16A2 9349106 and M4 9390032.

(1) ITEM NO	(2) SMR CODE	(3) CAGEC	(4) PART NUMBER	(5) DESCRIPTION AND USABLE ON CODES(UOC)	(6) QTY
				GROUP 0402 HAMMER ASSEMBLY M16A2 9349106 AND M4 9390032 FIG. C-13. HAMMER ASSEMBLY	
1	PAFZZ	19200	9349107	SPRING,HELICAL, TORSION HAMMER.....	1
2	PAFZZ	19200	9349109	SPRING,HELICAL, TORSION, BURST CAM.	1
3	PAFZZ	19200	9349108	CAM,BURST (M16A2)................... UOC:AR8	1
3A	PAFZZ	19200	9390031	CAM,BURST (M4)..................... UOC:AS1	1
4	PAFZZ	19200	9349110	HAMMER AND HAMMER PIN RETAINER ASSEMBLY.............................	1

END OF FIGURE

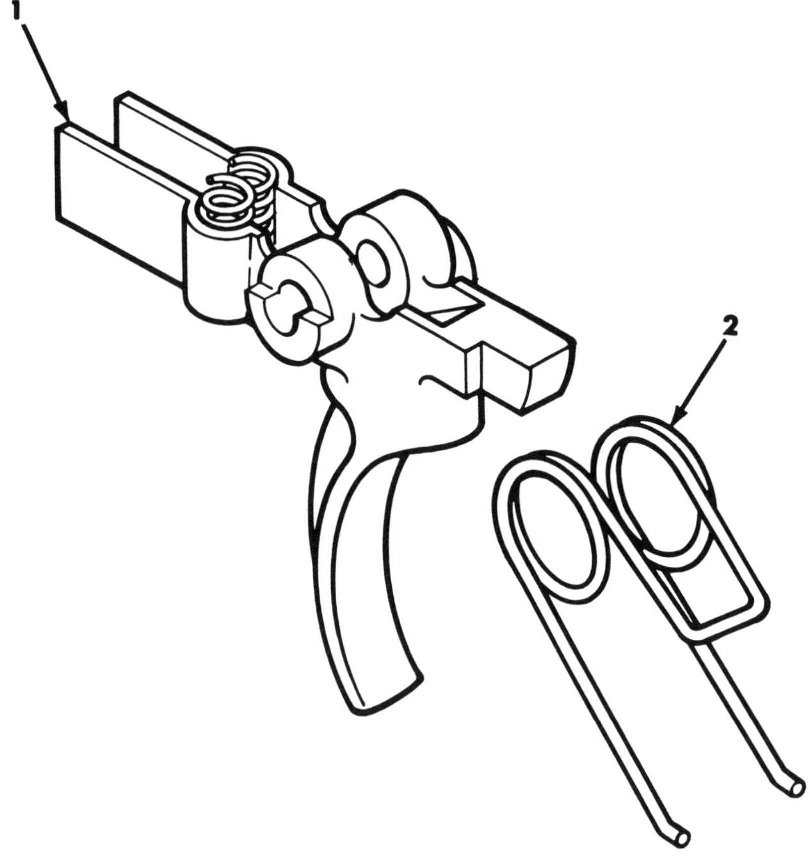

Figure C-14. Trigger Assembly 9349115.

SECTION II

(1) ITEM NO	(2) SMR CODE	(3) CAGEC	(4) PART NUMBER	(5) DESCRIPTION AND USABLE ON CODES (UOC)	(6) QTY
				GROUP 0403 TRIGGER ASSEMBLY 9349115 FIG. C-14. TRIGGER ASSEMBLY	
1	PAFFF	19200	9392518	TRIGGER SUB ASSEMBL	1
2	PAFZZ	19204	8448593	SPRING,HELICAL, TORSION, TRIGGER	1

END OF FIGURE

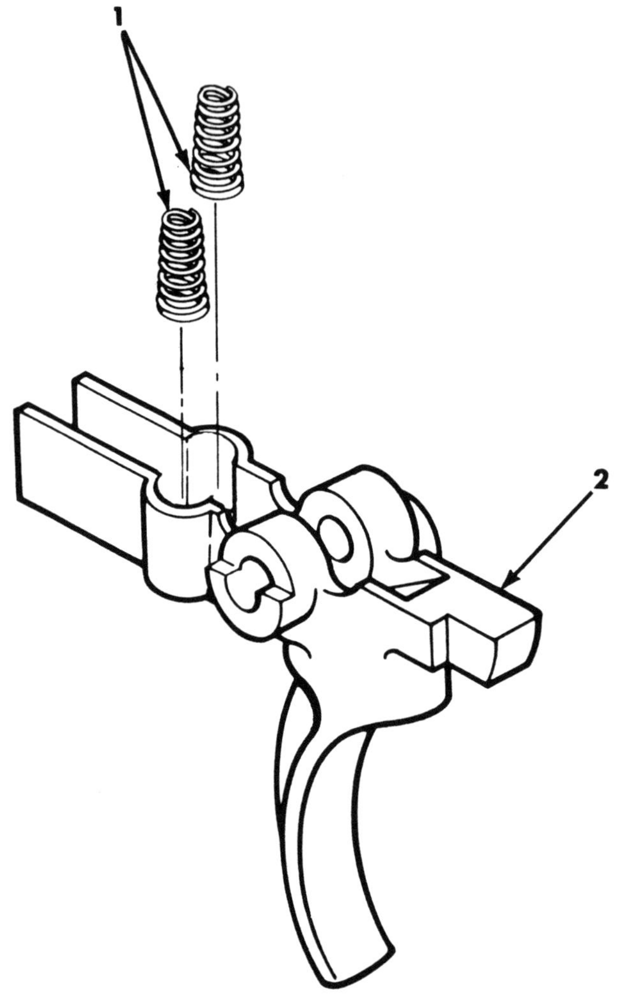

Figure C-15. Trigger Subassembly 9392518.

(1) ITEM NO	(2) SMR CODE	(3) CAGEC	(4) PART NUMBER	(5) DESCRIPTION AND USABLE ON CODES (UOC)	(6) QTY
				GROUP 040301 TRIGGER SUBASSEMBLY 9392518 FIG. C-15. TRIGGER SUBASSEMBLY	
1	PAFZZ	19200	9349116	SPRING,HELICAL ,DISCONNECT	2
2	XAFZZ	19200	9390736	TRIGGER	1

END OF FIGURE

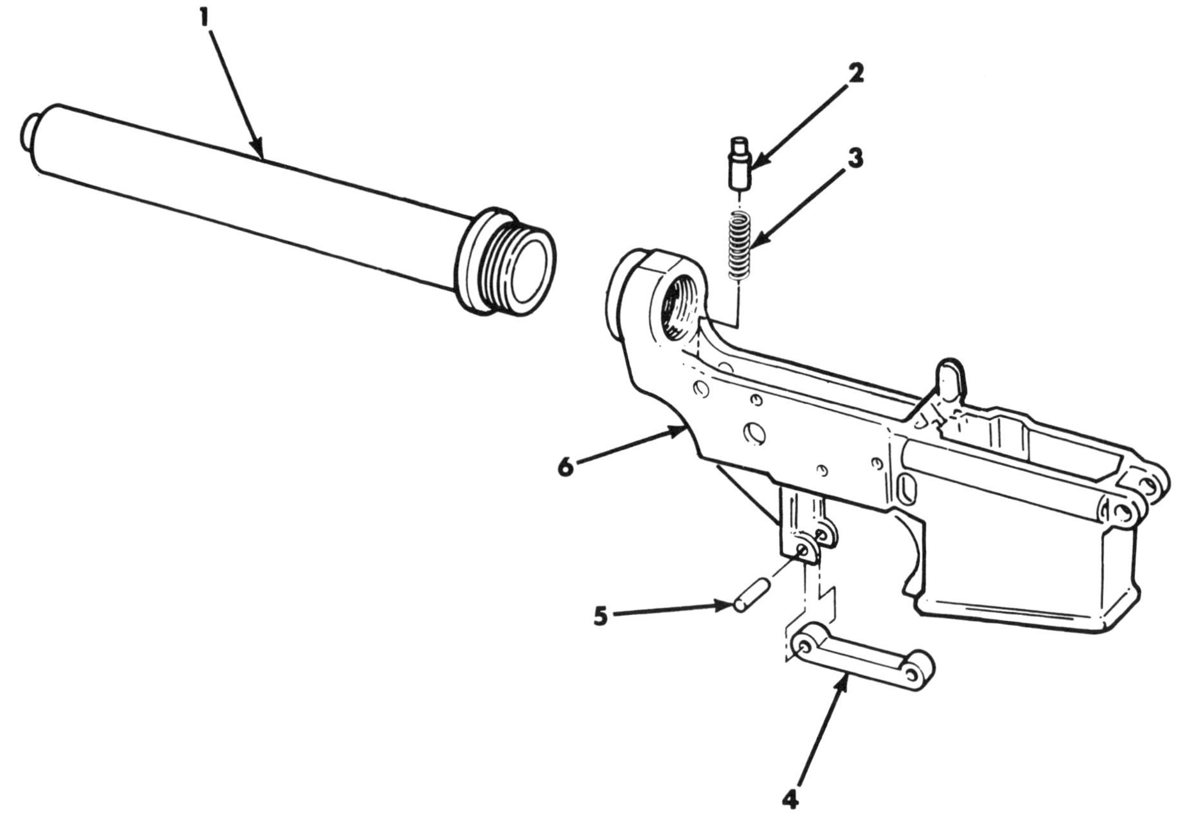

Figure C-16. M16A2 Lower Receiver and Receiver Extension
Assembly 9349101.

(1) ITEM NO	(2) SMR CODE	(3) CAGEC	(4) PART NUMBER	(5) DESCRIPTION AND USABLE ON CODES(UOC)	(6) QTY
				GROUP 0404 M16A2 LOWER RECEIVER AND RECEIVER EXTENSION ASSEMBLY 9349101 FIG. C-16. M16A2 LOWER RECEIVER AND RECEIVER EXTENSION ASSEMBLY	
1	PAFZZ	19200	8448581	EXTENSION,LOWER RECEIVER (LOWER RECEIVER EXTENSION)................. UOC:AR8	1
2	PAFZZ	19204	8448582	PIN,SHOULDER, HEADLESS, BUFFER RETAINER............................ UOC:AR8	1
3	PAFZZ	19200	8448583	SPRING,HELICAL, COMPRESSION, BUFFER RETAINER..................... UOC:AR8	1
4	PAFZZ	19204	8448587	GUARD,TRIGGER.......................	1
5	PAFZZ	96906	MS16562-129	PIN,SPRING,TRIGGER GUARD........... UOC:AR8	1
6	XAFDA	19200	9349102	RECEIVER,CARTRIDGE.................. UOC:AR8	1

END OF FIGURE

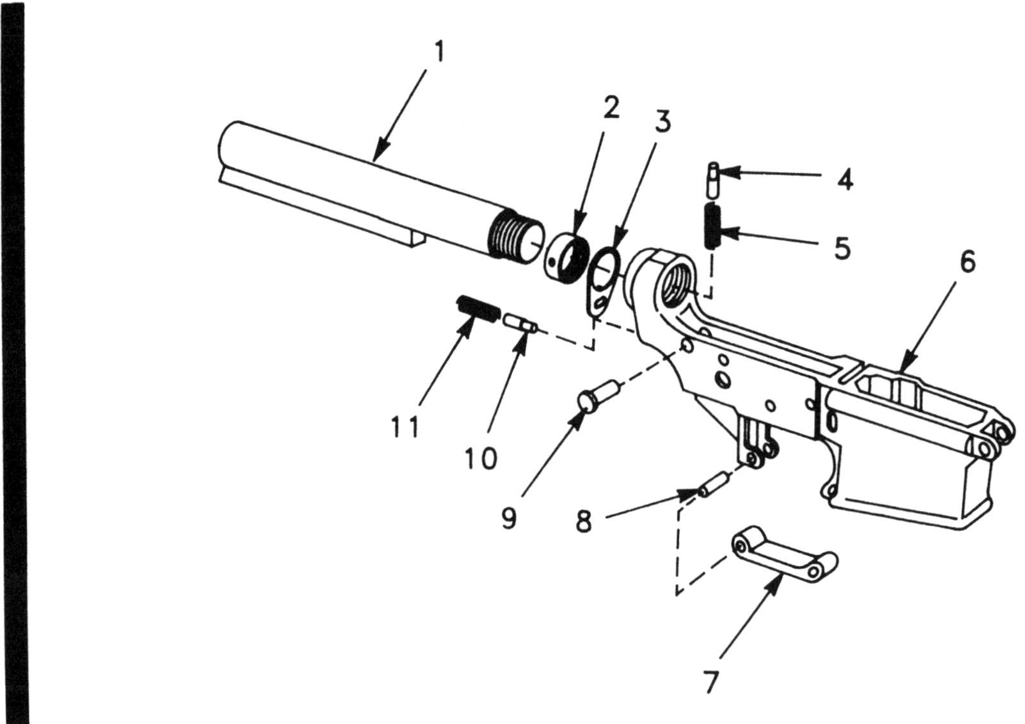

Figure C16A. M4 Lower Receiver and Receiver Extension
Assembly (NP).

(1) ITEM NO	(2) SMR CODE	(3) CAGEC	(4) PART NUMBER	(5) DESCRIPTION AND USABLE ON CODES(UOC)	(6) QTY
				GROUP 0404A M4 LOWER RECEIVER AND RECEIVER EXTENSION ASSEMBLY (NPN) FIG. C16A. M4 LOWER RECEIVER AND RECEIVER EXTENSION ASSEMBLY	
1	PAFZZ	19200	9390019	EXTENSION,LOWER RECEIVER (M4)...... UOC:AS1	1
2	PAFZZ	19200	9390020	NUT,PLAIN,ROUND (M4)............... UOC:AS1	1
3	PAFZZ	19200	9390021	PLATE,RECEIVER END (M4)............ UOC:AS1	1
4	PAFZZ	19204	8448582	PIN,SHOULDER, HEADLESS (M4)........ UOC:AS1	1
5	PAFZZ	19200	8448583	SPRING,HELICAL, COMPRESSION (M4)... UOC:AS1	1
6	XAFDA	19200	9390015	RECEIVER,CARTRIDGE (M4)............ UOC:AS1	1
7	PAFZZ	19204	8448587	GUARD,TRIGGER (M4).................	1
8	PAFZZ	96906	MS16562-129	PIN,SPRING.......................... UOC:AS1	1
9	PAOZZ	19204	8448584	PIN,GROOVED,HEADED (TAKE DOWN) (M4)	1
10	PAOZZ	19204	8448585	PIN,STRAIGHT, HEADLESS (M4)........	1
11	PAOZZ	19204	8448586	SPRING,HELICAL, COMPRESSION (M4)...	1

END OF FIGURE

Section III. SPECIAL TOOLS.

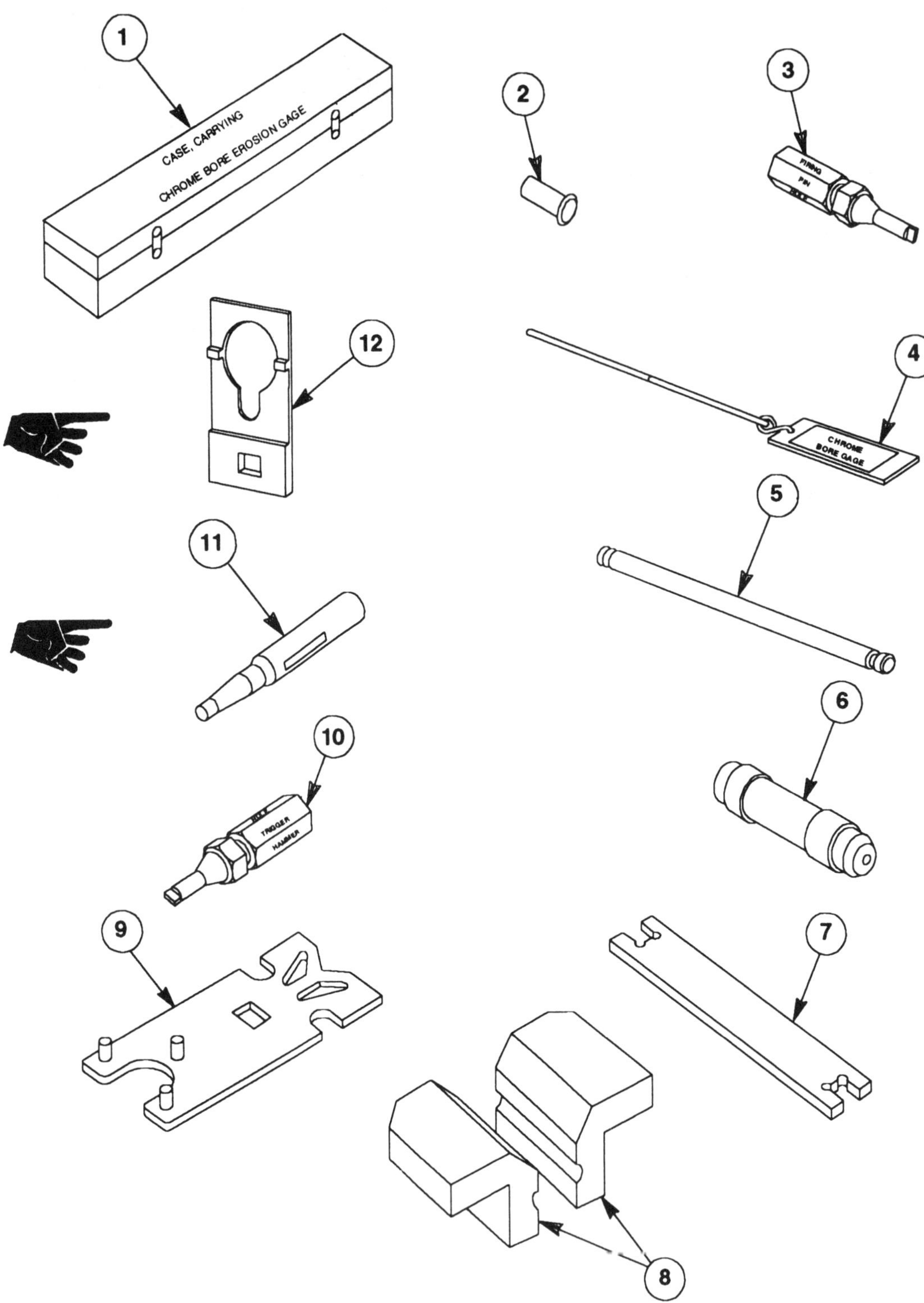

Figure C-17. Special Tools.

(1) ITEM NO	(2) SMR CODE	(3) CAGEC	(4) PART NUMBER	(5) DESCRIPTION AND USABLE ON CODES(UOC)	(6) QTY
				GROUP 9500 SPECIAL TOOLS FIG. C-17. SPECIAL TOOLS	
	PEFZZ	19204	8426685	MAINTENANCE KIT,GUN DS/GS SUPPORT MAINTENANCE FOR 5.56MM RIFLE, M16 RIFLE SERIES BOI: 2 PER SUPPORTING DSU/GSU..............................	
1	PAFZZ	19204	12006359	CASE,BORE GAGE PART OF KIT P/N 8426685................................	
2	PAFZZ	19204	8448201	..REFLECTOR TOOL, CHAMBER,PART OF KIT P/N 8426685.....................	
3	PAFZZ	19200	12620101	..GAGE,PLUG,PLAIN PART OF KIT P/N 8426685.............................	
4	PAFZZ	19204	8448496	..GAGE,BARREL,EROSION BARREL EROSION PART OF KIT P/N 8426685(CHROME BARREL)..............	
5	PAFZZ	19204	8448202	..GAGE,STRAIGHTNESS PART OF KIT P/ N 8426685...........................	
6	PAFZZ	19204	7799734	..GAGE,HEADSPACE PART OF KIT P/N 8426685.............................	
7	PAFZZ	19204	7799735	..GAGE,FIRING PIN PROTRUSION PART OF KIT P/N 8426685..................	
8	PAFZZ	19204	11010032	..FIXTURE,BARREL REMOVAL PART OF KIT P/N 8426685.....................	
9	PAFZZ	19204	11010033	..WRENCH,COMBINATION PART OF KIT P/ N 8426685...........................	
10	PAFZZ	19204	12006472	..GAGE,PLUG,TAPER CYLINDER PART O KIT P/N 8426685.....................	
11	PAOZZ	19200	12926769	KEY,MACHINE KEY, 2 PER ORGANIZATIONAL OR DIRECT SUPPORT SHOP..................................	
12	PAFZZ	19200	9390035	WRENCH,SPANNER 2 PER DS/GS SUPPORT SHOP A/R FOR M4 CARBINE............. UOC:AS1	

END OF FIGURE

CROSS-REFERENCE INDEXES

NATIONAL STOCK NUMBER INDEX

STOCK NUMBER	FIG.	ITEM	STOCK NUMBER	FIG.	ITEM
5315-00-017-9537	C-11	15	5360-00-992-6652	C16A	5
3040-00-017-9539	C-9	5	5315-00-992-6653	C-11	26
1005-00-017-9540	C-9	4		C16A	9
5360-00-017-9541	C-8	13	5315-00-992-6654	C-11	12
1005-00-017-9543	C-7	4		C-11	12A
1005-00-017-9546	C-1	1		C16A	10
1005-00-017-9547	C-2	1	5360-00-992-6655	C-11	11
1005-00-017-9548	C-11	8		C-11	11A
1005-00-056-2201	C-11	10		C16A	11
5360-00-056-2246	C-11	6	5360-00-992-6665	C-11	1
1005-00-056-2247	C-11	7	1005-00-992-6667	C-11	23
1005-00-056-7106	C-17		5315-00-992-7283	C-4	2
5315-00-058-6044	C-6	7	5305-00-992-7284	C-4	1
5315-00-058-6081	C-16	5	1005-00-992-7285	C-2	4
	C16A	8	1005-00-992-7287	C-3	1
5315-00-058-6678	C-10	1	1005-00-992-7288	C-3	7
5220-00-070-7814	C-17	6	1005-00-992-7290	C-3	3
6695-00-070-7815	C-17	7	1005-00-992-7291	C-3	6
4933-00-070-9151	C-17	8	5360-00-992-7292	C-3	5
5120-00-070-9152	C-17	9		C-11	22
1005-00-087-8998	C-6	8	5315-00-992-7294	C-2	3
3110-00-183-9175	C-8	4	5340-00-992-7297	C-16	1
	C-10	5	1005-00-992-7299	C-16	4
5220-00-221-9391	C-17	5		C16A	7
1005-00-403-0964	C-12	5	5360-00-992-7301	C-11	13
5340-00-463-3892	C-12	3	1005-00-992-7302	C-11	14
5315-00-463-3894	C-12	6	5360-00-992-7308	C-14	2
5360-00-523-8084	C-9	3	5315-00-992-7309	C-11	24
5310-00-527-3634	C-11	20	5360-00-999-0404	C-5	3
5315-00-582-3170	C-7	8	5340-00-999-0405	C-5	2
5315-00-597-5086	C-3	4	5365-00-999-0863	C-6	10
1005-00-738-6213	C-4	3	5365-00-999-0864	C-8	7
1005-00-760-3768	C-3	8	1005-00-999-1509	C-2	2
4933-00-800-7508	C-17	2	5220-01-014-8183	C-17	4
5315-00-812-3312	C-11	9	5315-01-027-4759	C-8	11
5315-00-843-9487	C12A	6		C-8	12
1005-00-921-5004	C-1	4	4933-01-035-5607	C-17	1
1005-00-937-3078	C-11	2	5220-01-043-9473	C-17	10
1005-00-978-1022	C-8	9	5315-01-048-9372	C-5	1
5315-00-978-1023	C-8	8		C-9	2
5360-00-978-1025	C-8	10	5320-01-063-7635	C-7	5
5360-00-978-1036	C-6	9	5220-01-075-5004	C-17	3
4710-00-978-1038	C-6	3	1005-01-134-3621	C-8	1
5315-00-979-3930	C-7	2	5305-01-134-3622	C-8	2
5360-00-979-3931	C-7	3	1005-01-134-3625	C-7	1
1005-00-992-6649	C-11	4	5355-01-134-3627	C-10	8
5315-00-992-6650	C-11	25	1005-01-134-3629	C-6	2
5315-00-992-6651	C-16	2	1005-01-134-3630	C-13	4
	C16A	4	1005-01-134-3631	C-10	2
5360-00-992-6652	C-16	3	1005-01-134-3633	C-6	6

CROSS-REFERENCE INDEXES

NATIONAL STOCK NUMBER INDEX

STOCK NUMBER	FIG.	ITEM	STOCK NUMBER	FIG.	ITEM
1005-01-134-3701	C-8	6			
5360-01-134-3710	C-8	15			
5360-01-135-0353	C-15	1			
1005-01-135-3697	C-10	6			
5355-01-135-4972	C-8	16			
1005-01-135-4973	C-11	28			
5360-01-136-5471	C-13	2			
1005-01-144-1468	C-9	1			
5305-01-144-1490	C-10	3			
5360-01-144-1492	C-13	1			
5305-01-144-1494	C-12	7			
5365-01-144-1496	C-6	5			
5340-01-144-1499	C-11	17			
5340-01-145-7910	C-11	16			
1005-01-146-7684	C-6	4			
1005-01-146-7685	C-12	2			
5305-01-147-8585	C-11	29			
1005-01-148-0172	C-13	3			
5360-01-148-1751	C-8	5			
	C-10	4			
1005-01-148-4805	C-11	19			
1005-01-216-4510	C-1	5			
1005-01-219-2402	C-14	1			
5340-01-225-8339	C-11	5			
1005-01-228-8504	C-12	1			
1005-01-231-3138	C-11	2A			
1005-01-233-8529	C-6	4A			
1005-01-233-8530	C16A	3			
1005-01-233-8531	C16A	1			
5315-01-233-8608	C12A	2			
5360-01-233-8616	C12A	3			
5360-01-233-8617	C-11	1A			
5310-01-233-8625	C16A	2			
5310-01-233-8626	C12A	5			
1005-01-233-8636	C12A	1			
4710-01-233-8637	C-6	3A			
5340-01-233-8638	C12A	4			
1005-01-234-2297	C-6	2A			
1005-01-247-7964	C-8	6A			
3040-01-247-7969	C-13	3A			
1010-01-264-6517	C-7	9			
5340-01-264-6530	C-7	7			
5365-01-267-2169	C-11	27			
5305-01-268-1191	C-11	21			
5315-01-310-0370	C-17	11			
5120-01-324-6631	C-17	12			
1005-01-368-9852	C-1	5A			

CROSS-REFERENCE INDEXES

PART NUMBER INDEX

CAGEC	PART NUMBER	STOCK NUMBER	FIG.	ITEM
88044	AN501D416-18	5305-01-268-1191	C-11	21
96906	MS16562-103	5315-00-058-6678	C-10	1
96906	MS16562-106	5315-00-058-6044	C-6	7
96906	MS16562-119	5315-00-812-3312	C-11	9
96906	MS16562-121	5315-01-027-4759	C-8	11
			C-8	12
96906	MS16562-129	5315-00-058-6081	C-16	5
			C16A	8
96906	MS16562-202	5315-00-843-9487	C12A	6
96906	MS16562-98	5315-00-597-5086	C-3	4
96906	MS16626-3137	5365-00-999-0863	C-6	10
96906	MS16632-3012	5365-00-999-0864	C-8	7
96906	MS19060-4808	3110-00-183-9175	C-8	4
			C-10	5
96906	MS35335-61	5310-00-527-3634	C-11	20
96906	MS39086-41	5315-00-582-3170	C-7	8
99999	NPN		C-11	30A
19204	11010032	4933-00-070-9151	C-17	8
19204	11010033	5120-00-070-9152	C-17	9
19204	12006359	4933-01-035-5607	C-17	1
19204	12006472	5220-01-043-9473	C-17	10
19200	12011987		C-10	7
19200	12011996	1005-01-368-9852	C-1	5A
19200	12597640	5365-01-267-2169	C-11	27
19200	12598617	1010-01-264-6517	C-7	9
19200	12598618	5340-01-264-6530	C-7	7
19200	12620101	5220-01-075-5004	C-17	3
19204	12624561	1005-01-216-4510	C-1	5
19200	12926769	5315-01-310-0370	C-17	11
19204	7799734	5220-00-070-7814	C-17	6
19204	7799735	6695-00-070-7815	C-17	7
19204	8426685	1005-00-056-7106	C-17	
19204	8448201	4933-00-800-7508	C-17	2
19204	8448202	5220-00-221-9391	C-17	5
19204	8448496	5220-01-014-8183	C-17	4
19204	8448501		C-1	2
19204	8448502	5315-00-992-7294	C-2	3
19204	8448503	1005-00-017-9547	C-2	1
19204	8448504	1005-00-999-1509	C-2	2
19204	8448505		C-2	5
19200	8448506	5315-00-992-7283	C-4	2
19200	8448507	1005-00-738-6213	C-4	3
19204	8448508	5305-00-992-7284	C-4	1
19200	8448509	1005-00-992-7285	C-2	4
19204	8448510		C-3	2
19204	8448511	1005-00-992-7287	C-3	1
19204	8448512	1005-00-992-7288	C-3	7
19204	8448513	1005-00-992-7290	C-3	3
19204	8448515	1005-00-992-7291	C-3	6
19204	8448516	5360-00-992-7292	C-3	5
			C-11	22

CROSS-REFERENCE INDEXES

PART NUMBER INDEX

CAGEC	PART NUMBER	STOCK NUMBER	FIG.	ITEM
19204	8448517	1005-00-017-9546	C-1	1
19204	8448518		C-5	4
19200	8448519	5340-00-999-0405	C-5	2
19204	8448520	5360-00-999-0404	C-5	3
19204	8448521-2	5315-01-048-9372	C-5	1
			C-9	2
19204	8448525	1005-00-978-1022	C-8	9
19204	8448532	5360-00-978-1025	C-8	10
19204	8448533	5315-00-978-1023	C-8	8
19200	8448540	5360-00-017-9541	C-8	13
19200	8448542	5360-00-523-8084	C-9	3
19204	8448543	3040-00-017-9539	C-9	5
19204	8448544	1005-00-017-9540	C-9	4
19204	8448555	5360-00-978-1036	C-6	9
19200	8448567	4710-00-978-1038	C-6	3
19204	8448571	1005-00-017-9543	C-7	4
19204	8448573	5315-00-979-3930	C-7	2
19204	8448574	5360-00-979-3931	C-7	3
19200	8448581	5340-00-992-7297	C-16	1
19204	8448582	5315-00-992-6651	C-16	2
			C16A	4
19200	8448583	5360-00-992-6652	C-16	3
			C16A	5
19204	8448584	5315-00-992-6653	C-11	26
			C16A	9
19204	8448585	5315-00-992-6654	C-11	12
			C-11	12A
			C16A	10
19204	8448586	5360-00-992-6655	C-11	11
			C-11	11A
			C16A	11
19204	8448587	1005-00-992-7299	C-16	4
			C16A	7
19204	8448593	5360-00-992-7308	C-14	2
19200	8448595	1005-00-992-6649	C-11	4
19204	8448599	5315-00-992-6650	C-11	25
19204	8448609	5315-00-992-7309	C-11	24
19200	8448615	1005-00-937-3078	C-11	2
19204	8448621	5315-00-017-9537	C-11	15
19200	8448628	1005-00-017-9548	C-11	8
19204	8448629	5360-00-992-6665	C-11	1
19204	8448631	1005-00-992-6667	C-11	23
19204	8448633	5360-00-056-2246	C-11	6
19204	8448634	1005-00-056-2247	C-11	7
19204	8448636	1005-00-992-7302	C-11	14
19204	8448637	5360-00-992-7301	C-11	13
19204	8448638	1005-00-056-2201	C-11	10
19204	8448652	1005-00-403-0964	C-12	5
19200	8448653	5340-00-463-3892	C-12	3
19204	8448655	5315-00-463-3894	C-12	6
19200	8448670	1005-00-921-5004	C-1	4

CROSS-REFERENCE INDEXES

PART NUMBER INDEX

CAGEC	PART NUMBER	STOCK NUMBER	FIG.	ITEM
19204	8448697	5320-01-063-7635	C-7	5
19204	8448712	1005-00-087-8998	C-6	8
19200	8448755	1005-00-760-3768	C-3	8
19200	9349050		C-1	3
19200	9349051	1005-01-134-3633	C-6	6
19200	9349052	5365-01-144-1496	C-6	5
19200	9349054		C-7	6
19200	9349056	1005-01-134-3625	C-7	1
19200	9349059	1005-01-134-3629	C-6	2
19200	9349062		C-6	1
19200	9349063	1005-01-134-3701	C-8	6
19200	9349065	5305-01-134-3622	C-8	2
19200	9349066	1005-01-134-3621	C-8	1
19200	9349067	5355-01-135-4972	C-8	16
19200	9349069	5360-01-148-1751	C-8	5
			C-10	4
19200	9349070	5360-01-134-3710	C-8	15
19200	9349072		C-8	3
19200	9349074	1005-01-134-3631	C-10	2
19200	9349075	1005-01-135-3697	C-10	6
19200	9349076	5305-01-144-1490	C-10	3
19200	9349077	5355-01-134-3627	C-10	8
19200	9349085	1005-01-144-1468	C-9	1
19204	9349086		C-8	14
19200	9349100		C-1	6
19200	9349101		C-11	30
19200	9349102		C-16	6
19200	9349106		C-11	3
19200	9349107	5360-01-144-1492	C-13	1
19200	9349108	1005-01-148-0172	C-13	3
19200	9349109	5360-01-136-5471	C-13	2
19200	9349110	1005-01-134-3630	C-13	4
19200	9349113	5340-01-144-1499	C-11	17
19200	9349114	5340-01-145-7910	C-11	16
19200	9349115		C-11	18
19200	9349116	5360-01-135-0353	C-15	1
19200	9349119	1005-01-135-4973	C-11	28
19200	9349120	5305-01-144-1494	C-12	7
19200	9349121		C-12	4
19200	9349124	1005-01-146-7684	C-6	4
19200	9349127	1005-01-148-4805	C-11	19
19200	9349128	5305-01-147-8585	C-11	29
19200	9349130	1005-01-146-7685	C-12	2
19200	9381367	5340-01-225-8339	C-11	5
19200	9381380	1005-01-228-8504	C-12	1
19200	9390001		C-1	3A
19200	9390003	1005-01-234-2297	C-6	2A
19200	9390007	1005-01-233-8529	C-6	4A
19200	9390009		C-7	6A
19200	9390011		C-1	6A
19200	9390012		C-11	28A

CROSS-REFERENCE INDEXES

PART NUMBER INDEX

CAGEC	PART NUMBER	STOCK NUMBER	FIG.	ITEM
19200	9390013	1005-01-233-8636	C12A	1
19200	9390014	5340-01-233-8638	C12A	4
19200	9390015		C16A	6
19200	9390016	4710-01-233-8637	C-6	3A
19200	9390019	1005-01-233-8531	C16A	1
19200	9390020	5310-01-233-8625	C16A	2
19200	9390021	1005-01-233-8530	C16A	3
19200	9390022	5360-01-233-8617	C-11	1A
19200	9390023	1005-01-231-3138	C-11	2A
19200	9390025	5315-01-233-8608	C12A	2
19200	9390026	5310-01-233-8626	C12A	5
19200	9390027	5360-01-233-8616	C12A	3
19200	9390028	1005-01-247-7964	C-8	6A
19200	9390029		C-6	1A
19200	9390031	3040-01-247-7969	C-13	3A
19200	9390032		C-11	3A
19200	9390035	5120-01-324-6631	C-17	12
19200	9390736		C-15	2
19200	9392518	1005-01-219-2402	C-14	1

CROSS-REFERENCE INDEXES

FIGURE AND ITEM NUMBER INDEX

FIG.	ITEM	STOCK NUMBER	CAGEC	PART NUMBER	
C-1	1	1005-00-017-9546	19204	8448517	
C-1	2		19204	8448501	
C-1	3		19200	9349050	
C-1	3A		19200	9390001	■
C-1	4	1005-00-921-5004	19200	8448670	
C-1	5	1005-01-216-4510	19204	12624561	
C-1	5A	1005-01-368-9852	19200	12011996	■
C-1	6		19200	9349100	
C-1	6A		19200	9390011	■
C-2	1	1005-00-017-9547	19204	8448503	
C-2	2	1005-00-999-1509	19204	8448504	
C-2	3	5315-00-992-7294	19204	8448502	
C-2	4	1005-00-992-7285	19200	8448509	
C-2	5		19204	8448505	
C-3	1	1005-00-992-7287	19204	8448511	
C-3	2		19204	8448510	
C-3	3	1005-00-992-7290	19204	8448513	
C-3	4	5315-00-597-5086	96906	MS16562-98	
C-3	5	5360-00-992-7292	19204	8448516	
C-3	6	1005-00-992-7291	19204	8448515	
C-3	7	1005-00-992-7288	19204	8448512	
C-3	8	1005-00-760-3768	19200	8448755	
C-4	1	5305-00-992-7284	19204	8448508	
C-4	2	5315-00-992-7283	19200	8448506	
C-4	3	1005-00-738-6213	19200	8448507	
C-5	1	5315-01-048-9372	19204	8448521-2	
C-5	2	5340-00-999-0405	19200	8448519	
C-5	3	5360-00-999-0404	19204	8448520	
C-5	4		19204	8448518	
C-6	1		19200	9349062	
C-6	1A		19200	9390029	■
C-6	2	1005-01-134-3629	19200	9349059	
C-6	2A	1005-01-234-2297	19200	9390003	■
C-6	3	4710-00-978-1038	19200	8448567	
C-6	3A	4710-01-233-8637	19200	9390016	■
C-6	4	1005-01-146-7684	19200	9349124	
C-6	4A	1005-01-233-8529	19200	9390007	■
C-6	5	5365-01-144-1496	19200	9349052	
C-6	6	1005-01-134-3633	19200	9349051	
C-6	7	5315-00-058-6044	96906	MS16562-106	
C-6	8	1005-00-087-8998	19204	8448712	
C-6	9	5360-00-978-1036	19204	8448555	
C-6	10	5365-00-999-0863	96906	MS16626-3137	■
C-7	1	1005-01-134-3625	19200	9349056	
C-7	2	5315-00-979-3930	19204	8448573	
C-7	3	5360-00-979-3931	19204	8448574	
C-7	4	1005-00-017-9543	19204	8448571	
C-7	5	5320-01-063-7635	19204	8448697	
C-7	6		19200	9349054	
C-7	6A		19200	9390009	■
C-7	7	5340-01-264-6530	19200	12598618	

CROSS-REFERENCE INDEXES

FIGURE AND ITEM NUMBER INDEX

FIG.	ITEM	STOCK NUMBER	CAGEC	PART NUMBER
C-7	8	5315-00-582-3170	96906	MS39086-41
C-7	9	1010-01-264-6517	19200	12598617
C-8	1	1005-01-134-3621	19200	9349066
C-8	2	5305-01-134-3622	19200	9349065
C-8	3		19200	9349072
C-8	4	3110-00-183-9175	96906	MS19060-4808
C-8	5	5360-01-148-1751	19200	9349069
C-8	6	1005-01-134-3701	19200	9349063
C-8	6A	1005-01-247-7964	19200	9390028
C-8	7	5365-00-999-0864	96906	MS16632-3012
C-8	8	5315-00-978-1023	19204	8448533
C-8	9	1005-00-978-1022	19204	8448525
C-8	10	5360-00-978-1025	19204	8448532
C-8	11	5315-01-027-4759	96906	MS16562-121
C-8	12	5315-01-027-4759	96906	MS16562-121
C-8	13	5360-00-017-9541	19200	8448540
C-8	14		19204	9349086
C-8	15	5360-01-134-3710	19200	9349070
C-8	16	5355-01-135-4972	19200	9349067
C-9	1	1005-01-144-1468	19200	9349085
C-9	2	5315-01-048-9372	19204	8448521-2
C-9	3	5360-00-523-8084	19200	8448542
C-9	4	1005-00-017-9540	19204	8448544
C-9	5	3040-00-017-9539	19204	8448543
C12A	1	1005-01-233-8636	19200	9390013
C12A	2	5315-01-233-8608	19200	9390025
C12A	3	5360-01-233-8616	19200	9390027
C12A	4	5340-01-233-8638	19200	9390014
C12A	5	5310-01-233-8626	19200	9390026
C12A	6	5315-00-843-9487	96906	MS16562-202
C16A	1	1005-01-233-8531	19200	9390019
C16A	2	5310-01-233-8625	19200	9390020
C16A	3	1005-01-233-8530	19200	9390021
C16A	4	5315-00-992-6651	19204	8448582
C16A	5	5360-00-992-6652	19200	8448583
C16A	6		19200	9390015
C16A	7	1005-00-992-7299	19204	8448587
C16A	8	5315-00-058-6081	96906	MS16562-129
C16A	9	5315-00-992-6653	19204	8448584
C16A	10	5315-00-992-6654	19204	8448585
C16A	11	5360-00-992-6655	19204	8448586
C-10	1	5315-00-058-6678	96906	MS16562-103
C-10	2	1005-01-134-3631	19200	9349074
C-10	3	5305-01-144-1490	19200	9349076
C-10	4	5360-01-148-1751	19200	9349069
C-10	5	3110-00-183-9175	96906	MS19060-4808
C-10	6	1005-01-135-3697	19200	9349075
C-10	7		19200	12011987
C-10	8	5355-01-134-3627	19200	9349077
C-11	1	5360-00-992-6665	19204	8448629
C-11	1A	5360-01-233-8617	19200	9390022

Change 3

CROSS-REFERENCE INDEXES

FIGURE AND ITEM NUMBER INDEX

FIG.	ITEM	STOCK NUMBER	CAGEC	PART NUMBER
C-11	2	1005-00-937-3078	19200	8448615
C-11	2A	1005-01-231-3138	19200	9390023
C-11	3		19200	9349106
C-11	3A		19200	9390032
C-11	4	1005-00-992-6649	19200	8448595
C-11	5	5340-01-225-8339	19200	9381367
C-11	6	5360-00-056-2246	19204	8448633
C-11	7	1005-00-056-2247	19204	8448634
C-11	8	1005-00-017-9548	19200	8448628
C-11	9	5315-00-812-3312	96906	MS16562-119
C-11	10	1005-00-056-2201	19204	8448638
C-11	11	5360-00-992-6655	19204	8448586
C-11	11A	5360-00-992-6655	19204	8448586
C-11	12	5315-00-992-6654	19204	8448585
C-11	12A	5315-00-992-6654	19204	8448585
C-11	13	5360-00-992-7301	19204	8448637
C-11	14	1005-00-992-7302	19204	8448636
C-11	15	5315-00-017-9537	19204	8448621
C-11	16	5340-01-145-7910	19200	9349114
C-11	17	5340-01-144-1499	19200	9349113
C-11	18		19200	9349115
C-11	19	1005-01-148-4805	19200	9349127
C-11	20	5310-00-527-3634	96906	MS35335-61
C-11	21	5305-01-268-1191	88044	AN501D416-18
C-11	22	5360-00-992-7292	19204	8448516
C-11	23	1005-00-992-6667	19204	8448631
C-11	24	5315-00-992-7309	19204	8448609
C-11	25	5315-00-992-6650	19204	8448599
C-11	26	5315-00-992-6653	19204	8448584
C-11	27	5365-01-267-2169	19200	12597640
C-11	28	1005-01-135-4973	19200	9349119
C-11	28A		19200	9390012
C-11	29	5305-01-147-8585	19200	9349128
C-11	30		19200	9349101
C-11	30A		99999	NPN
C-12	1	1005-01-228-8504	19200	9381380
C-12	2	1005-01-146-7685	19200	9349130
C-12	3	5340-00-463-3892	19200	8448653
C-12	4		19200	9349121
C-12	5	1005-00-403-0964	19204	8448652
C-12	6	5315-00-463-3894	19204	8448655
C-12	7	5305-01-144-1494	19200	9349120
C-13	1	5360-01-144-1492	19200	9349107
C-13	2	5360-01-136-5471	19200	9349109
C-13	3	1005-01-148-0172	19200	9349108
C-13	3A	3040-01-247-7969	19200	9390031
C-13	4	1005-01-134-3630	19200	9349110
C-14	1	1005-01-219-2402	19200	9392518
C-14	2	5360-00-992-7308	19204	8448593
C-15	1	5360-01-135-0353	19200	9349116
C-15	2		19200	9390736

Change 3

CROSS-REFERENCE INDEXES

FIGURE AND ITEM NUMBER INDEX

FIG.	ITEM	STOCK NUMBER	CAGEC	PART NUMBER
C-16	1	5340-00-992-7297	19200	8448581
C-16	2	5315-00-992-6651	19204	8448582
C-16	3	5360-00-992-6652	19200	8448583
C-16	4	1005-00-992-7299	19204	8448587
C-16	5	5315-00-058-6081	96906	MS16562-129
C-16	6		19200	9349102
C-17		1005-00-056-7106	19204	8426685
C-17	1	4933-01-035-5607	19204	12006359
C-17	2	4933-00-800-7508	19204	8448201
C-17	3	5220-01-075-5004	19200	12620101
C-17	4	5220-01-014-8183	19204	8448496
C-17	5	5220-00-221-9391	19204	8448202
C-17	6	5220-00-070-7814	19204	7799734
C-17	7	6695-00-070-7815	19204	7799735
C-17	8	4933-00-070-9151	19204	11010032
C-17	9	5120-00-070-9152	19204	11010033
C-17	10	5220-01-043-9473	19204	12006472
C-17	11	5315-01-310-0370	19200	12926769
C-17	12	5120-01-324-6631	19200	9390035

APPENDIX D
EXPENDABLE/DURABLE SUPPLIES AND MATERIALS LIST

Section I.   INTRODUCTION

D-1.   SCOPE. This appendix lists expendable/durable supplies and materials you will need to operate and maintain the M16A2 Rifle. This listing is for informational purposes only and is not authority to requisition the listed items. These items are authorized to you by CTA 50-970, Expendable/Durable Items (Except Medical, Class V, Repair Parts, and Heraldic Items), or CTA 8-100, Army Medical Department Expendable/Durable Items.

D-2.   EXPLANATION OF COLUMNS.

   a.   **Column (1)—Item Number.** This number is assigned to the entry in the listing and is referenced in the narrative instructions to identify the material (e.g., "Use cloth, abrasive, crocus, item 12, app D").

   b.   **Column (2)—Level.** This column identifies the lowest level of maintenance that requires the listed item.

   C—Operator/Crew Maintenance
   O—Unit Maintenance
   F--Direct Support Maintenance

   c.   **Column (3)—National Stock Number.** This is the National stock number assigned to the item; use it to request or requisition the item.

   d.   **Column (4)—Description.** Indicates the Federal item name and, if required, a description to identify the item. The last line for each item indicates the Commercial and Government Entity Code (CAGEC) in parentheses followed by the part number.

   e.   **Column (5)—Unit of Measure (U/M).** Indicates the measure used in performing the actual maintenance function. This measure is expressed by a two-character alphabetical abbreviation (e.g., ea, in., pr). If the unit of measure differs from the unit of issue, requisition the lowest unit of issue that will satisfy your requirements.

Section II.   EXPENDABLE/DURABLE SUPPLIES AND MATERIALS LIST

(1) Item Number	(2) Level	(3) National Stock Number	(4) Description	(5) U/M
1	F	8040-00-944-7292	ADHESIVE, KIT: (81348) MMM-A-1754	KT
2	O	8020-00-244-0153	BRUSH, ARTIST'S: metal ferrule, flat chisel edge 7/16 w, 1 1/8, ex- posed bristle (81348) H-B-241	EA
3	F	1005-00-716-2702	BRUSH, CLEANING, SMALL ARMS: (19205) 7162702	EA
4	C	1005-00-903-1296	BRUSH, CLEANING, SMALL ARMS: bore (19204) 11686340	EA
5	C	1005-00-999-1435	BRUSH, CLEANING, SMALL ARMS: chamber (19204) 8432358	EA
6	C	1005-00-444-6602	BRUSH, CLEANING, SMALL ARMS: tooth (19204) 8448462	EA
7	O	7920-00-205-2401	BRUSH, CLEANING, TOOLS AND PARTS: (81349) MILS43871	EA
8	O	6850-00-965-2332	CARBON REMOVING COMPOUND: (81348) P-C-111	GL
9			CLEANER, LUBRICANT AND PRESERV- ATIVE: (27412)	
	O	9150-01-079-6124	CLP- 4 oz (118.30 ml) bottle	EA
	O	9150-01-054-6453	CLP- 1 pt (0.47 l) bottle	EA
	O	9150-01-053-6688	CLP- 7 gal. (26.50 l) bottle	EA
10	C	9150-01-102-1473	CLEANER, LUBRICANT AND PRESERV- ATIVE: (81349) MIL-L-63460 1/2 oz (14.79 ml) bottle	EA
11	C	9920-00-292-9946	CLEANER, TOBACCO PIPE: cotton turf, wire core (89855) DILLSPIPE cleaner (36 per pkg)	EA
12			CLEANING COMPOUND, RIFLE BORE: small arms bore cleaning solution (RBC) (81349) MIL-C-372	
	C	6850-00-224-6656	2 oz (59.15 ml) bottle	OZ
	O	6850-00-224-6657	8 oz (236.59 ml) can	CN
	O	6850-00-224-6663	1 gal. (3.79 l) can	CN

EXPENDABLE/DURABLE SUPPLIES AND MATERIALS LIST (CONT)

(1) Item Number	(2) Level	(3) National Stock Number	(4) Description	(5) U/M
13	O	5350-00-221-0872	CLOTH, ABRASIVE: (58536) A-A-1206	SH
14	O	8010-00-181-7859	COATING COMPOUND, FLUORESCENT: paint for blank firing attachment (81349) MIL-P-21563 1 pt (0.47 l) can	EA
15	F	 6810-00-244-0290 6810-00-616-9188	DICHLOROMETHANE, TECHNICAL: (81349) MIL-D-6998 5 gal. (18.93 l) pail 600 lb (272.16 kg) drum	 CN DR
16	O	6850-00-281-1985	DRY CLEANING SOLVENT: (58536) A-A-711 1 gal. (3.79 l) can	GL
17	O	8010-00-297-0560	ENAMEL: olive drab no. 3407 (81348) TT-E-527 1 gal. (3.79 l) can	GL
18	O	 8415-00-823-7455 8415-00-823-7456 8415-00-823-7457	GLOVES, CHEMICAL AND OIL PROTECTIVE: (81348) ZZ-G-381 Size 9 Size 10 Size 11	 PR PR PR
19	F	9150-00-754-2595	GREASE, MOLYBDENUM DISULFIDE: (81349) MIL-G-21164	LB
20	C	1005-01-113-0321	HANDLE SECTION, CLEANING ROD, SMALL ARMS: (19204) 8436776	EA
21	O	9150 01 260 2534	LUBRICANT, SOLID FILM: (81349) MIL-L-23398 16 oz (473.18 ml) spray can	OZ
22	C	9150-00-292-9689	LUBRICATING OIL, WEAPONS: (LAW) (81349) MIL-L-14107 1 qt (0.95 l) can	QT

EXPENDABLE/DURABLE SUPPLIES AND MATERIALS LIST (CONT)

(1) Item Number	(2) Level	(3) National Stock Number	(4) Description	(5) U/M
23			LUBRICATING OIL, WEAPONS: (LSA), semifluid (81349) MIL-L-46000	
	C	9150-00-935-6597	2 oz (59.15 ml) plastic btl	OZ
	C	9150-00-889-3522	4 oz (118.30 ml) plastic btl	OZ
	O	9150-00-687-4241	1 qt (0.95 l) can	CN
	O	9150-00-753-4686	1 gal. (3.79 l) can	CN
24	O	4940-00-795-3595	PAN, WASH: (94453) 1211	EA
25	F	6850-00-826-0981	PENETRANT KIT: (81349) MIL-I-25135	KT
26	F	8135-01-019-1691	POLYETHYLENE SHEET: (84744) PE88-80-2	EA
27	C	7920-00-205-1711	RAG, WIPING: (58536) A-A-531 50 lb (22.68 kg) bdl	LB
28	C	1005-00-050-6357	ROD SECTION, CLEANING, SMALL ARMS: (19204) 8436775 (3 REQUIRED)	EA
29	F	8030-00-670-8553	SEALING COMPOUND: (16059) DEVCONF	KT
30	C	1005-00-937-2250	SWAB HOLDER SECTION, CLEANING ROD, SMALL ARMS: (19204) 11686327	EA
31	C	1005-00-912-4248	SWAB, SMALL ARMS: (19204) 11686408	EA
32	C	6920-01-253-4005	TARGET, 25 METER ZEROING, M16A2: (19200) 12597650	EA

APPENDIX E
ILLUSTRATED LIST OF MANUFACTURED ITEMS

E-1. INTRODUCTION.

a. This appendix includes complete instructions for making items authorized to be manufactured or fabricated at unit or direct support maintenance.

b. A part number in alphanumeric order is provided for cross-referencing the part number of the item to be manufactured to the figure which covers fabrication criteria.

c. All bulk materials needed for manufacture of an item are listed by part number or specification number in a tabular list on the illustration.

E-2. MANUFACTURED ITEMS PART NUMBER INDEX.

INDEX

Item	Figure Number
Front sight detent depressor	E-1
Front sight post removal and installation tool	E-2
Pivot pin removal tool	E-3
Pivot pin installation tool	E-4
Slave pin	E-5

E-3. MANUFACTURED ITEMS ILLUSTRATIONS—FORMAT AND CONTENT.

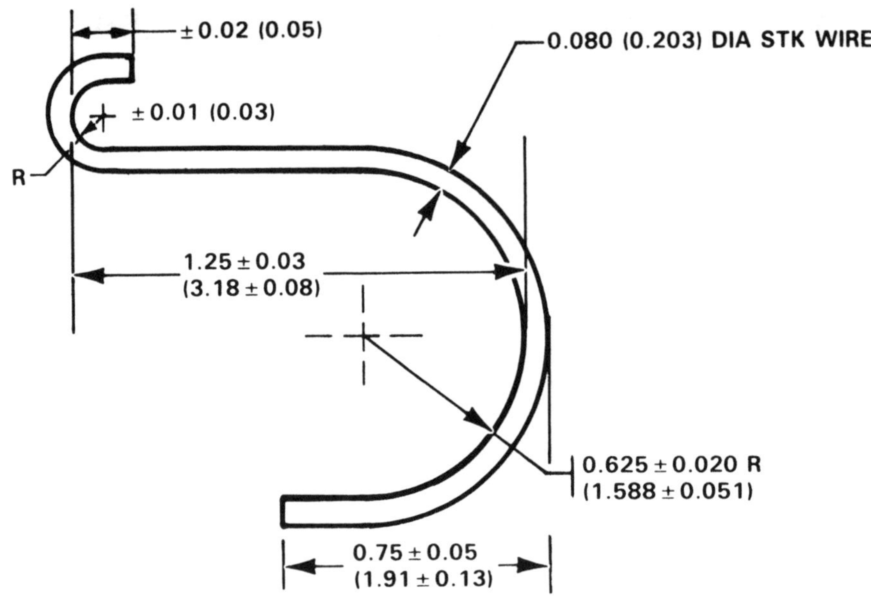

FABRICATE FROM 0.08 IN. MUSIC WIRE OR EQUIVALENT.
FINISH: NO. 5.3.1.2 OR 5.3.2.2 OF MIL-STD-17.

NOTE: ALL DIMENSIONS SHOWN ARE IN INCHES WITH THE METRIC CONVERSION TO CEN-
TIMETERS IN PARENTHESES.

Figure E-1. Front sight detent depressor.

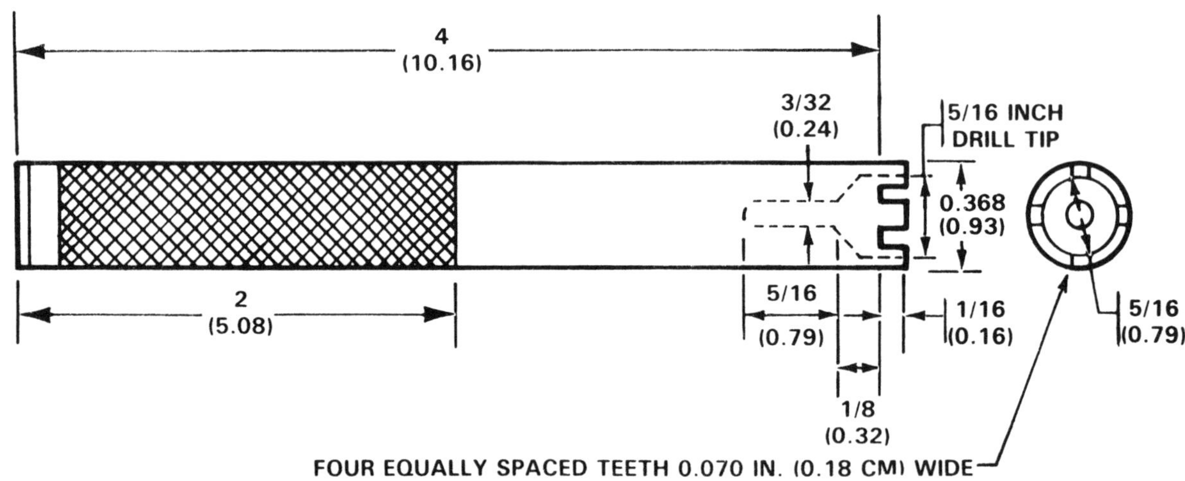

FABRICATE FROM 0.375 INCH
ROUND METAL BAR, ASTM A686,
FSCM 81346, GRADE C, CLASS W2-09,
NSN 9510-00-640-4407 OR EQUIVALENT.

NOTES: 1. ALL DIMENSIONS SHOWN ARE IN INCHES WITH THE METRIC CONVERSION TO
CENTIMETERS IN PARENTHESES.
2. TEETH MUST BE HAND FILED TO FIT FRONT SIGHT POST.

Figure E-2. Front sight post removal and installation tool.

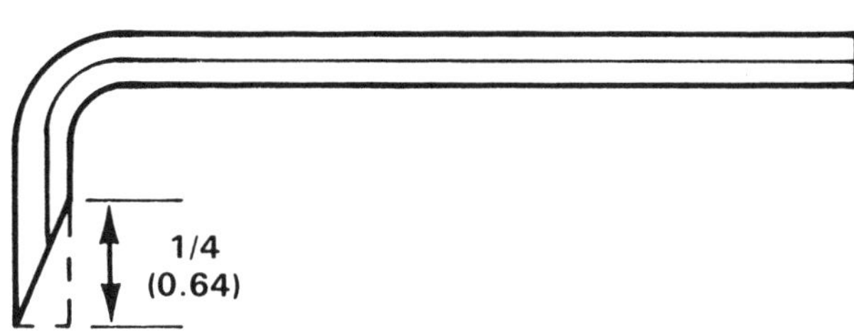

FABRICATE FROM 1/16 IN. SOCKET HEAD SCREW KEY
NSN 5120-00-198-5398 OR EQUIVALENT.

NOTE: ALL DIMENSIONS SHOWN ARE IN INCHES WITH THE METRIC CONVERSION TO CEN-
TIMETERS IN PARENTHESES.

Figure E-3. Pivot pin removal tool.

E-3. MANUFACTURED ITEMS ILLUSTRATIONS—FORMAT AND CONTENT (CONT).

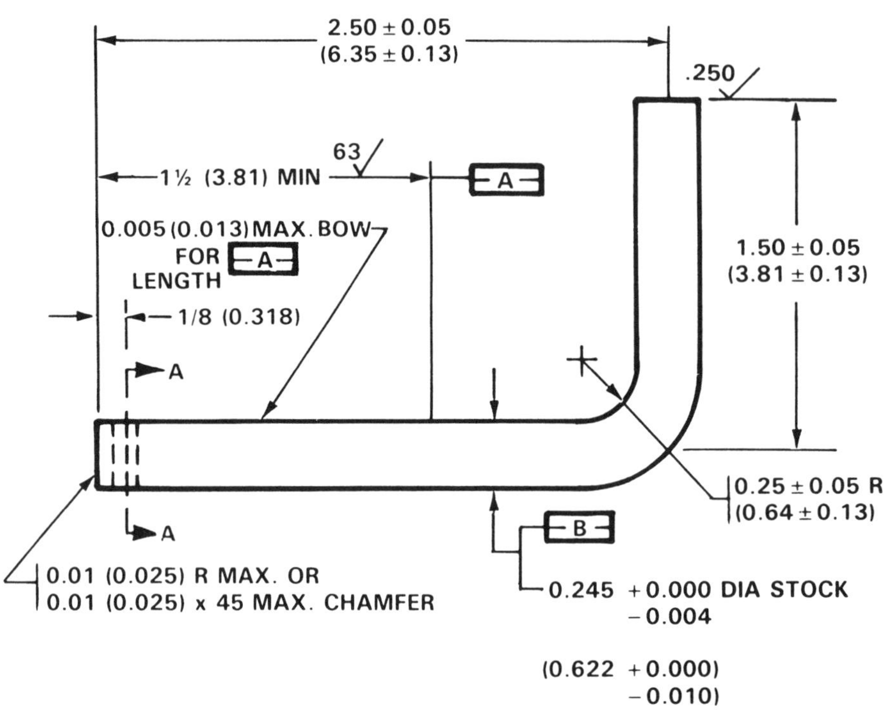

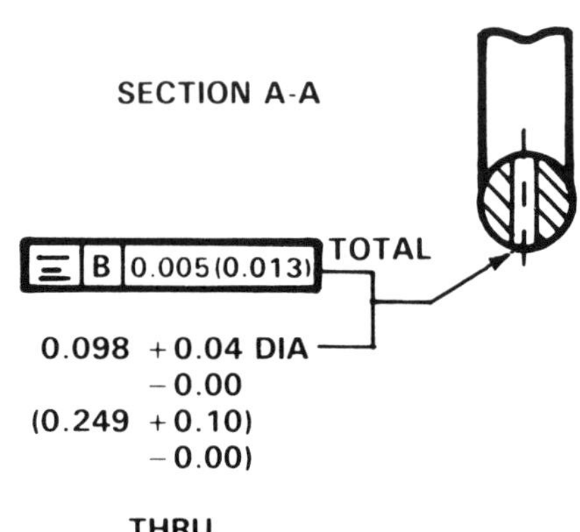

FABRICATE FROM 0.245 IN. STEEL AISI 1095 OR EQUIVALENT.
HARDEN AND TEMPER TO RC-57-61 FOR LENGTH -A-.
FINISH 5.3.1.2 OR 5.3.2.2 OF MIL-STD-171.

NOTE: ALL DIMENSIONS SHOWN ARE IN INCHES WITH THE METRIC CONVERSION TO CEN-
TIMETERS IN PARENTHESES.

Figure E-4. Pivot pin installation tool.

VIEW A

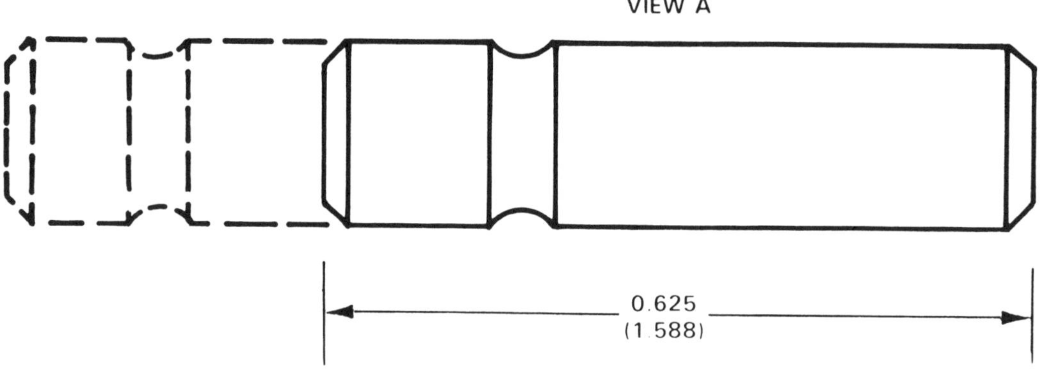

VIEW B

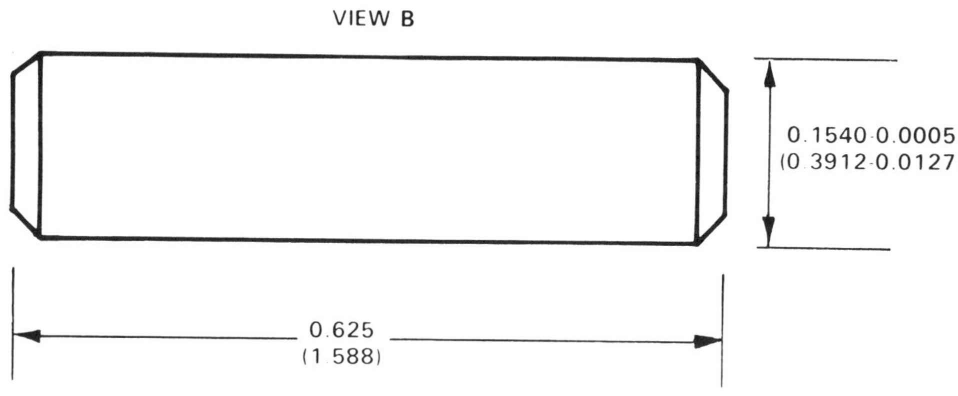

FABRICATE FROM OLD TRIGGER PIN (VIEW A) P/N
8448609 OR FABRICATE SLAVE PIN (VIEW B) FROM
MATERIAL BLOCK, WIRE, STEEL ALLOY, GRADE 4140,
ASTM-A547 OR EQUIVALENT.

NOTE: ALL DIMENSIONS SHOWN ARE IN INCHES WITH THE METRIC CONVERSION TO CEN-
TIMETERS IN PARENTHESES.

Figure E-5. Slave pin.

ALPHABETICAL INDEX

Subject   **Page**

A

B

C

ALPHABETICAL INDEX (CONT)

Subject	Page

L

M

O

P

ALPHABETICAL INDEX (CONT)

Subject   **Page**

R

S

T

U

Subject   **Page**

W

By Order of the Secretary of the Army:

CARL E. VUONO
General, United States Army
Chief of Staff

Official:

PATRICIA P. HICKERSON
Brigadier General, United States Army
The Adjutant General

By Order of the Secretary of the Air Force:

MERRILL A. McPEAK
General, United States Air Force
Chief of Staff

CHARLES C. McDONALD
General, United States Air Force
Commander, Air Force Logistics Command

DISTRIBUTION:

To be distributed in accordance with DA Form 12-40E, (0020), Unit and
Direct and General Support Maintenance Requirements for TM 9-1005-319-23&P.

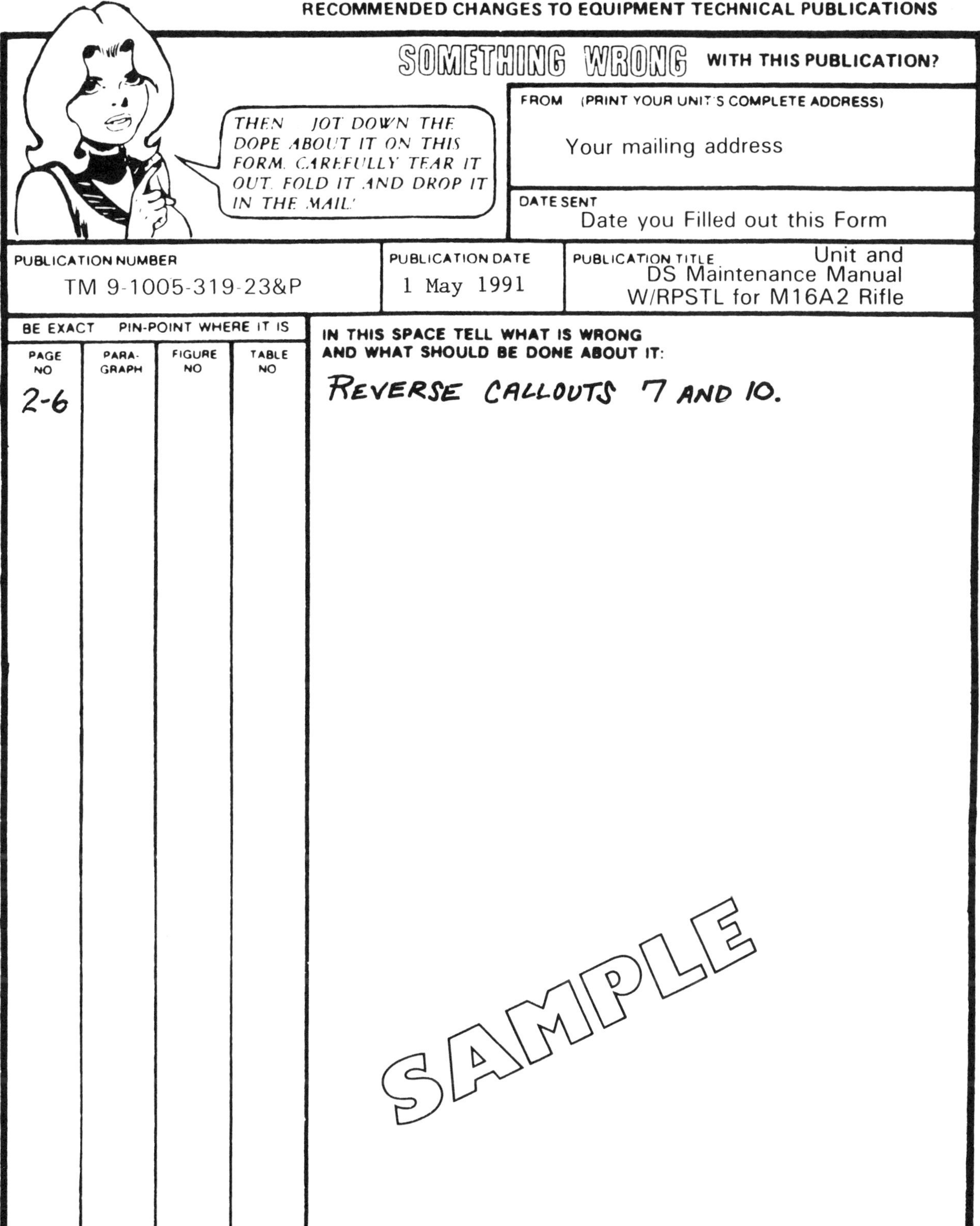
SOMETHING WRONG WITH THIS PUBLICATION?
THEN JOT DOWN THE DOPE ABOUT IT ON THIS FORM. CAREFULLY TEAR IT OUT. FOLD IT AND DROP IT IN THE MAIL!
FROM (PRINT YOUR UNIT'S COMPLETE ADDRESS)
Your mailing address
DATE SENT
Date you Filled out this Form
PUBLICATION NUMBER
TM 9-1005-319-23&P
PUBLICATION DATE
1 May 1991
PUBLICATION TITLE
Unit and DS Maintenance Manual W/RPSTL for M16A2 Rifle
BE EXACT   PIN-POINT WHERE IT IS
PAGE NO
PARA-GRAPH
FIGURE NO
TABLE NO
2-6
IN THIS SPACE TELL WHAT IS WRONG AND WHAT SHOULD BE DONE ABOUT IT:
REVERSE CALLOUTS 7 AND 10.
SAMPLE
PRINTED NAME, GRADE OR TITLE, AND TELEPHONE NUMBER
SIGN HERE
Your name
DA FORM 2028-2
1 JUL 79
PREVIOUS EDITIONS ARE OBSOLETE.
P S --IF YOUR OUTFIT WANTS TO KNOW ABOUT YOUR RECOMMENDATION MAKE A CARBON COPY OF THIS AND GIVE IT TO YOUR HEADQUARTERS

SOMETHING WRONG WITH THIS PUBLICATION?

FROM (PRINT YOUR UNIT'S COMPLETE ADDRESS)

DATE SENT

PUBLICATION NUMBER	PUBLICATION DATE	PUBLICATION TITLE
TM 9-1005-319-23&P	1 May 1991	Unit and DS Maintenance Manual W/RPSTL for M16A2 Rifle

BE EXACT   PIN-POINT WHERE IT IS

PAGE NO	PARA-GRAPH	FIGURE NO	TABLE NO

IN THIS SPACE TELL WHAT IS WRONG AND WHAT SHOULD BE DONE ABOUT IT:

PRINTED NAME, GRADE OR TITLE, AND TELEPHONE NUMBER

SIGN HERE

DA FORM 2028-2
1 JUL 79

PREVIOUS EDITIONS ARE OBSOLETE

P.S.--IF YOUR OUTFIT WANTS TO KNOW ABOUT YOUR RECOMMENDATION MAKE A CARBON COPY OF THIS AND GIVE IT TO YOUR HEADQUARTERS

TEAR ALONG PERFORATED LINE

FILL IN YOUR
UNIT'S ADDRESS

FOLD BACK

DEPARTMENT OF THE ARMY

OFFICIAL BUSINESS

**COMMANDER
US ARMY ARMAMENT MUNITIONS
AND CHEMICAL COMMAND
ATTN AMSMC-MAS
ROCK ISLAND IL 61201-9948**

Other Books Available From Desert Publications